Dietary Restriction:
Implications for the Design
and Interpretation of Toxicity
and Carcinogenicity Studies

Dietary Restriction: Implications for the Design and Interpretation of Toxicity and Carcinogenicity Studies

Edited by Ronald W. Hart, David A. Neumann, and Richard T. Robertson

ILSI
International Life Sciences INSTITUTE

ILSI Press
Washington, D.C.

ILSI PRESS
International Life Sciences Institute/ILSI Risk Science Institute
1126 Sixteenth Street, N.W.
Washington, D.C. 20036-4804 USA

Printed in the United States of America

Library of Congress Catalog Number 95-77852

ISBN 0-944398-34-0

About ILSI

The International Life Sciences Institute is a nonprofit, worldwide foundation established in 1978 to advance the understanding of scientific issues relating to nutrition, food safety, toxicology, risk assessment, and the environment.

By bringing together scientists from academia, government, industry, and the public sector, ILSI seeks a balanced approach to solving problems of common concern for the well-being of the general public.

Headquartered in Washington, D.C., ILSI is affiliated with the World Health Organization as a nongovernmental organization and has specialized consultative status with the Food and Agriculture Organization of the United Nations.

ILSI accomplishes its work through its branches and institutes. ILSI's branches currently include Argentina, Australasia, Brazil, Europe, Japan, Mexico, North America, Southeast Asia, and Thailand, and a focal point in China. The ILSI Health and Environmental Sciences Institute focuses on global environmental issues. The ILSI Research Foundation institutes include:

ILSI Allergy and Immunology Institute
ILSI Human Nutrition Institute
ILSI Pathology and Toxicology Institute
ILSI Risk Science Institute

Contents

Contributors

ORGANIZING COMMITTEE

William T. Allaben
National Center for Toxicological
    Research
3900 NCTR Road (HFT-30)
Jefferson, AR 72079

Joseph F. Contrera
Center for Drug Evaluation and Research
U.S. Food and Drug Administration
Room 13B16
5600 Fishers Lane (HFD-400)
Rockville, MD 20857

Ronald W. Hart
National Center for Toxicological
    Research
3900 NCTR Road
Jefferson, AR 72079

David G. Hattan
Center for Food Safety and Applied
    Nutrition
U.S. Food and Drug Administration
200 C Street, SW (HFS-225)
Washington, DC 20204

Donald H. Hughes
The Procter & Gamble Company
Miami Valley Laboratories
P.O. Box 398707
Cincinnati, OH 45239-8707

David A. Neumann
ILSI Risk Science Institute
1126 Sixteenth Street, NW
Washington, DC 20036

Stephen S. Olin
ILSI Risk Science Institute
1126 Sixteenth Street, NW
Washington, DC 20036

Richard T. Robertson
Department of Safety Assessment
Merck Research Laboratories
Building 45-206
West Point, PA 19486

Bernard A. Schwetz
National Center for Toxicological
    Research
3900 NCTR Road (HFT-1)
Jefferson, AR 72079

Richard L. Sprott
National Institute on Aging
Gateway Building 2C231
7201 Wisconsin Avenue
Bethesda, MD 20892

EDITORIAL BOARD

William T. Allaben
National Center for Toxicological
    Research
3900 NCTR Road (HFT-30)
Jefferson, AR 72079

Ronald W. Estabrook
Biochemistry Department
University of Texas Southwestern
    Medical Center
5323 Harry Hines Blvd.
Dallas, TX 75235-9038

Ronald W. Hart
National Center for Toxicological
    Research
3900 NCTR Road
Jefferson, AR 72079

Donald H. Hughes
The Procter & Gamble Company
Miami Valley Laboratories
P.O. Box 398707
Cincinnati, OH 45239-8707

David A. Neumann
ILSI Risk Science Institute
1126 Sixteenth Street, NW
Washington, DC 20036

Richard T. Robertson
Department of Safety Assessment
Merck Research Laboratories
Building 45-206
West Point, PA 19486

Bernard A. Schwetz
National Center for Toxicological
    Research
3900 NCTR Road (HFT-1)
Jefferson, AR 72079

Richard B. Setlow
Brookhaven National Laboratory
Biology Department, Building 463
P.O. Box 5000
Upton, NY 11973-5000

Richard L. Sprott
National Institute on Aging
Gateway Building 2C231
7201 Wisconsin Avenue
Bethesda, MD 20892

AUTHORS

Kamal M. Abdo
National Institute of Environmental
    Health Sciences
P.O. Box 12233
Research Triangle Park, NC 27709

William T. Allaben
National Center for Toxicological
    Research
3900 NCTR Road (HFT-30)
Jefferson, AR 72079

Thomas F. Berg
National Center for Toxicological
    Research
3900 NCTR Road (HFTDE2)
Jefferson, AR 72079

Kenneth J. Blank
Pathology and Laboratory Medicine,
    MS 435
Hahnemann School of Medicine
Broad & Vine Streets
Philadelphia, PA 19102-1192

Ernice B. Blann
Division of Nutritional Toxicology
National Center for Toxicological
    Research
3900 NCTR Road (HFT-140)
Jefferson, AR 72079

David L. Busbee
Division of Cell Biology
Department of Anatomy
College of Veterinary Medicine
Texas A&M University
College Station, TX 77843-4485

F. Chen
Department of Anatomy
University of Arkansas for Medical
    Sciences
Little Rock, AR 72079

Shu Chen
National Center for Toxicological
    Research
3900 NCTR Road
Jefferson, AR 72079

X. W. Chen
National Center for Toxicological
    Research
3900 NCTR Road
Jefferson, AR 72079

Joseph F. Contrera
Center for Drug Evaluation and Research
U.S. Food and Drug Administration
Room 13B16
5600 Fishers Lane (HFD-400)
Rockville, MD 20857

V. Desai
Division of Biometry and Risk
    Assessment
National Center for Toxicological
    Research
3900 NCTR Road
Jefferson, AR 72079

Zora Djuric
Department of Hematology/Oncology
Wayne State University
P.O. Box 02143
Detroit, MI 48201

Peter H. Duffy
National Center for Toxicological
    Research
3900 NCTR Road (HFTDE2)
Jefferson, AR 72079

William H. Farland
Office of Health and Environmental
    Assessment
U.S. Environmental Protection Agency
401 M Street, SW (RD-689)
Washington, DC 20460

Ritchie J. Feuers
National Center for Toxicological
    Research
3900 NCTR Road
Jefferson, AR 72079

Lynn T. Frame
National Center for Toxicological
    Research
3900 NCTR Road
Jefferson, AR 72079

B. Guntapalli
Department of Microbiology
    and Immunology
Texas A&M University Medical Center
College Station, TX 77843

Astrid Gutsmann
Geriatric Research, Education
    and Clinical Center
Audie L. Murphy Memorial VA Hospital
Division of Geriatrics and Gerontology
Department of Medicine
University of Texas Health Science
    Center
San Antonio, TX 78284

Ronald W. Hart
National Center for Toxicological
    Research
3900 NCTR Road
Jefferson, AR 72079

Joseph K. Haseman
Statistics and Biomathematics Branch
National Institute of Environmental
    Health Sciences
P.O. Box 12233 (MD A3-03)
Research Triangle Park, NC 27709

Bruce S. Hass
National Center for Toxicological Research
3900 NCTR Road (HFT-140)
Jefferson, AR 72079

David G. Hattan
Center for Food Safety and Applied
    Nutrition
U.S. Food and Drug Administration
200 C Street, SW (HFS-225)
Washington, DC 20204

Yuzo Hayashi
Biological Safety Research Center
National Institute of Health Sciences
1-18-1, Kamiyoga, Setagaya-ku
Tokyo 158, Japan

L. K. Heilbrun
Division of Hematology and Oncology
Department of Internal Medicine
Wayne State University
Detroit, MI 48201

Ahmad R. Heydari
Department of Medicine
University of Texas Health Science Center
Audie Murphy VA Hospital, GRECC-182
7400 Merton Minter Boulevard
San Antonio, TX 78284

S. Holt
Department of Microbiology and
    Immunology
Texas A&M University Medical Center
College Station, TX 77843

S. Jill James
National Center for Toxicological Research
3900 NCTR Road
Jefferson, AR 72079

Frank W. Kari
National Institute of Environmental Health
    Sciences
P.O. Box 12233
Research Triangle Park, NC 27709

Kevin P. Keenan
Department of Safety Assessment
Merck Research Laboratories
Sumneytown Pike
West Point, PA 19486

Joseph W. Kemnitz
Department of Medicine
University of Wisconsin
Madison, WI 53706

Nina Kozlovskaya
National Center for Toxicological
    Research
3900 NCTR Road
Jefferson, AR 72079

David Kritchevsky
The Wistar Institute
3601 Spruce Street
Philadelphia, PA 19104

Julian E.A. Leakey
National Center for Toxicological
    Research
3900 NCTR Road (HFTDE2)
Jefferson, AR 72079

Min-Young Lee
National Center for Toxicological
    Research
3900 NCTR Road
Jefferson, AR 72079

S. M. Lewis
National Center for Toxicological
    Research
3900 NCTR Road
Jefferson, AR 72079

M. H. Lu
National Center for Toxicological
    Research
3900 NCTR Road
Jefferson, AR 72079

D. A. Luongo
Division of Hematology and Oncology
Department of Internal Medicine
Wayne State University
Detroit, MI 48201

Beverly D. Lyn-Cook
National Center for Toxicological
    Research
3900 NCTR Road
Jefferson, AR 72079

Takashi Makinodan
Geriatric Research Education and Clinical
    Center (11G)
West Los Angeles VA Medical Center
Wilshire and Sawtelle Boulevards
Los Angeles, CA 90073

Mikhail Manjgaladze
National Center for Toxicological
    Research
3900 NCTR Road
Jefferson, AR 72079

Edward J. Masoro
Department of Physiology
The University of Texas Health Science
    Center
7703 Floyd Curl Drive
San Antonio, TX 78284-7756

E. Merriam
Department of Anatomy and Public
    Health
College of Veterinary Medicine
Texas A&M University
College Station, TX 77843

Susan Miller
Division of Cell Biology
Department of Anatomy & Public Health
College of Veterinary Medicine
Texas A&M University
College Station, TX 77848

Donna M. Murasko
Department of Microbiology
    and Immunology
Medical College of Pennsylvania
2900 Queen Lane
Philadelphia, PA 19129

David A. Neumann
ILSI Risk Science Institute
1126 Sixteenth Street, NW
Washington, DC 20036

E. Oriaku
Florida A&M University
P.O. Box 5761
Tallahassee, FL 32314

William R. Pendergrass
Department of Pathology SM-30
University of Washington
Seattle, WA 98195

J. W. Pipkin
Division of Genetic Toxicology
National Center for Toxicological
    Research
3900 NCTR Road
Jefferson, AR 72079

Erika Randerath
Department of Pharmacology
Baylor College of Medicine
One Baylor Plaza
Houston, TX 77030

K. Randerath
Division of Toxicology
Department of Pharmacology
Baylor College of Medicine
Houston, TX 77030

Ghanta N. Rao
National Toxicology Program
National Institute of Environmental
    Health Sciences
P.O. Box 12233 (MD A0-01)
Research Triangle Park, NC 27709

B. A. Reading
Division of Hematology and Oncology
Department of Internal Medicine
Wayne State University
Detroit, MI 48201

Crissy L. Rhodes
National Center for Toxicological
    Research
3900 NCTR Road
Jefferson, AR 72079

Arlan Richardson
Geriatric Research, Education
    and Clinical Center
Audie L. Murphy Memorial VA Hospital
Division of Geriatrics and Gerontology
Department of Medicine
University of Texas Health Science
    Center
San Antonio, TX 78284

M. Schroeder
Department of Anatomy and Public
    Health
College of Veterinary Medicine
Texas A&M University
College Station, TX 77843

Bernard A. Schwetz
National Center for Toxicological
    Research
3900 NCTR Road (HFT-1)
Jefferson, AR 72079

John Seng
National Center for Toxicological
    Research
3900 NCTR Road
Jefferson, AR 72079

J. G. Shaddock
Division of Genetic Toxicology
National Center for Toxicological
    Research
3900 NCTR Road
Jefferson, AR 72079

William E. Sonntag
Bowman Gray School of Medicine
Wake Forest University
Medical Center Blvd.
Winston-Salem, NC 27157-1083

Keith A. Soper
Merck Research Laboratories
Sumneytown Pike
West Point, PA 19486

V. Srivastava
Department of Anatomy and Public
    Health
College of Veterinary Medicine
Texas A&M University
College Station, TX 77843

Deneen R. Stewart
Department of Microbiology
    and Immunology
Medical College of Pennsylvania
Philadelphia, PA

Ryoya Takahashi
Geriatric Research, Education
    and Clinical Center
Audie L. Murphy Memorial VA Hospital
Division of Geriatrics and Gerontology
Department of Medicine
University of Texas Health Science
    Center
San Antonio, TX 78284

Angelo Turturro
National Center for Toxicological
    Research
3900 NCTR Road (HFTDE2)
Jefferson, AR 72079

Hideo Uno
Wisconsin Regional Primate Research
    Center
University of Wisconsin
Madison, WI 53715

Richard Weindruch
Department of Medicine
University of Wisconsin
2638 Medical Sciences Center
Madison, WI 53706

V. Wilson
Department of Microbiology
  and Immunology
Texas A&M University Medical Center
College Station, TX 77843

Norman S. Wolf
Department of Pathology, SM-30
University of Washington
Seattle, WA 98195

Shijun Xia
National Center for Toxicological
  Research
3900 NCTR Road
Jefferson, AR 72079

Xiaowei Xu
Department of Physiology
  and Pharmacology
Bowman Gray School of Medicine
Wake Forest University
Winston-Salem, NC 27157-1083

Shenghong You
Geriatric Research, Education
  and Clinical Center
Audie L. Murphy Memorial VA Hospital
Division of Geriatrics and Gerontology
Department of Medicine
University of Texas Health Science
  Center
San Antonio, TX 78284

G.-D. Zhou
Division of Toxicology
Department of Pharmacology
Baylor College of Medicine
Houston, TX 77030

OTHERS

John R. Bucher
National Toxicology Program
National Institute of Environmental
  Health Sciences
P.O. Box 12233
Research Triangle Park, NC 27709

Robert A. Good
University of South Florida
Allergy & Clinical Immunology Program
All Children's Hospital
801 Sixth Street South
St. Petersburg, FL 33701-4899

Alex Malaspina
International Life Sciences Institute
1126 Sixteenth Street, NW
Washington, DC 20036

Margaret A. Miller
Center for Veterinary Medicine
U.S. Food and Drug Administration
Metro Park 2 (HFV-100)
7500 Standish Place
Rockville, MD 20855

Gerald Rhodes
SmithKline Beecham Pharmaceuticals
P.O. Box 1539
King of Prussia, PA 19406

M. David C. Scales
Glaxo Group Research Limited
Park Road
Ware, Hertfordshire SG12 0DP
United Kingdom

Foreword

While Paracelsus is renowned for noting that dose distinguishes a poison from a remedy, other early scholars contributed substantively to the study of poisons and their antidotes. The 12th-century physician and philosopher Moses Maimonides described how dietary components such as dairy products could modulate the effects of certain poisons, presumably by altering absorption. Observations that diet can modulate acute toxicity have been supplemented by more recent demonstrations that diet can also affect chronic toxicity. In the context of safety assessment, where toxicity and carcinogenicity are routinely determined in rodents given unrestricted access to food, restricted or controlled food intake can dramatically alter the outcome of a study, as is amply demonstrated in these proceedings.

Animals placed on restricted or controlled levels of dietary intake that provide adequate numbers of calories and appropriate amounts of vitamins and minerals appear to be healthier and live longer than their ad libitum–fed counterparts. Such observations have profound implications for the safety assessment process. Specifically, if dietary control is used during routine bioassays, how would controlled food intake affect metabolism and the responses to potential toxins and carcinogens, and how should the results of studies employing dietary control be interpreted with respect to identifying human health hazards? These and related questions formed the basis for the conference "Dietary Restriction: Implications for the Design and Interpretation of Toxicity and Carcinogenicity Studies." These proceedings present the most recent findings of leading investigators from a variety of scientific disciplines who use dietary modulation to study an array of biological processes and responses. It is focused, multidisciplinary approaches such as this that will advance our understanding of the relationships between diet, health, and safety.

These proceedings represent the third quadrennial conference on this topic organized by ILSI, the International Life Sciences Institute. Based in Washington, D.C., ILSI is a nonprofit, worldwide foundation established in 1978 to advance the understanding of scientific issues relating to nutrition, food safety, toxicology, risk assessment, and the environment by bringing together scientists from academia, government, industry, and the public sector to solve problems with broad implications for the well-being of the general public. ILSI is committed to understanding the role of diet in health and disease in both animals and humans and will continue to facilitate the presentation and discussion of the latest scientific findings that address these concerns.

Alex Malaspina, President
International Life Sciences Institute

Preface

During the 1960s standard protocols were established for performing long-term toxicity and carcinogenicity studies in rodents as part of the routine safety assessment of chemicals. The protocols addressed issues of animal husbandry, test agent dosage and delivery, and data collection and interpretation. Bioassays typically were of 2 years duration and utilized B6C3F1 mice or Sprague-Dawley or Fischer 344 rats, although other strains were also used. This standardization was intended to promote intra- and interlaboratory reproducibility by reducing experimental variability.

In the interest of standardization, rats and mice, the species of choice for these studies, were provided with unrestricted access to food. It was assumed that ad libitum feeding not only would satisfy the nutritional needs of the test animals, but also would reduce husbandry costs. Examination of the data that have accrued under these standardized conditions reveals that untreated or control animals of the commonly used strains currently exhibit more rapid growth and have increased body weights relative to their strain-, sex-, and age-matched counterparts used during the 1960s and 1970s. Obesity, increased incidence of spontaneous tumors and intercurrent disease, and diminished life span are now common and likely are associated with this accelerated growth. These observations raise concerns about the reliability and utility of data obtained from long-term bioassays and about the impact of such confounding factors on the sensitivity of the bioassay.

Although the increased size of rodents used in standard bioassays may be attributable, in part, to breeding practices that select for rapidly growing animals that reproduce early, there is substantial evidence that ad libitum feeding results in obesity and a variety of concomitant adverse health effects. Indeed, rodents are the only mammalian species that are not fed prescribed amounts of food under laboratory conditions. The amount of food consumed by test animals remains one of the few uncontrolled variables under the standard bioassay protocol. Food consumption, measured in mass or calories, by test animals can vary considerably, even under conditions of ad libitum feeding. This often reflects differences in access to food with respect to feeding device operation, housing conditions, duration of feed presentation, and other laboratory practices.

Restricting or controlling the amount of food consumed by test rodents results in slower growth and reduced body weight. Values comparable to those of historical controls can be achieved by moderate control of food intake. Evidence suggests that relative to rodents fed ad libitum, those on calorically controlled feeding regimens are healthier, have a lower incidence of spontaneous tumors and intercurrent disease, live longer, and exhibit physiologic and metabolic attributes associated with younger animals. An increase in time until tumor detection has been observed in calorically restricted rodents exposed to carcinogens.

Long-term rodent bioassays provide an experimental approach to identifying potential human health hazards. Detection of toxic or carcinogenic effects of a test compound administered to the rodents is considered evidence for similar effects in humans exposed to the same compound. However, the variability in food uptake and the confounding consequences of ad libitum feeding raise substantive concerns about the interpretation of bioassay results. Similarly, although the sensitivity of the bioassay may be compromised by the relatively high incidence of spontaneous neoplasms among ad libitum–fed control animals, the increase in time until tumor detection associated with feed-restricted rodents may also affect assay sensitivity. These issues constituted the basis for the ILSI Risk Science Institute's conference "Dietary Restriction: Implications for the Design and Interpretation of Toxicity and Carcinogenicity Studies," which took place in Washington, D.C., February 28–March 2, 1994.

This two-and-one-half-day conference focused on the feasibility and practical implications of the routine use of dietary restriction in safety assessment. To facilitate this approach, a series of questions posed by scientists from the U.S. Food and Drug Administration and the U.S. Environmental Protection Agency was circulated to the speakers in advance of the conference. Speakers were invited to consider the 10 questions listed below when preparing their presentations and the companion manuscripts that appear in this volume.

1. Why is there concern about the use of ad libitum feeding, and is adoption of dietary restriction protocols an appropriate response to those concerns?
2. Are validation studies needed to compare the results of studies conducted under conditions of ad libitum feeding with those obtained under dietary restriction, and if so, how should such studies be designed?

3. Is there a need to establish a standardized protocol for dietary restriction studies? Should diet formulation be standardized, and if so, what formulation is most appropriate?
4. Because dietary restriction often results in increased longevity and in the delayed onset of tumorigenesis relative to what is observed in animals fed ad libitum, should the duration of the study be changed?
5. Because dietary restriction can reduce the incidence of both spontaneous and chemically induced tumors, is there a threshold level of dietary restriction that will increase longevity while having little or no impact on the sensitivity of tumor detection?
6. Should dietary restriction be used during supporting toxicity, metabolic, pharmacokinetic, and dose ranging studies necessary for dose selection and the interpretation of carcinogenicity studies?
7. Does dietary restriction affect the minimum toxic dose, the no-observed-adverse-effect level, the maximum tolerated dose, or any other toxicologically relevant parameter?
8. What effect does modest (10–30%) dietary restriction have on physiologic functions, e.g., endocrine activity, metabolism of endogenous and exogenous substances, reproductive function, etc.? For a given level of restriction, are there species- or strain-specific effects?
9. Is there an appropriate role for dietary restriction in developmental and reproductive toxicological studies, and when during development should dietary restriction be imposed on animals in long-term bioassays?
10. Does the use of dietary restriction increase the relevance of animal studies for human health risk assessment?

Because of the diversity of research objectives of the various investigators, no single report addresses each of the 10 questions listed above, yet a careful reading of all of the reports suggests that answers to at least some of these questions may be forthcoming. Moreover, information presented during the conference and in this volume addresses many fundamental biological questions about the influence of dietary intake on metabolic and pharmacokinetic activity, genetic stability, molecular and physiologic processes, compound toxicity, and pathogenesis. In addition to their impact on the design and interpretation of long-term safety studies, these findings will likely prove important to scientists interested in cancer biology, aging, and the overall health of laboratory animals.

The conference and these proceedings were made possible by the support of the International Life Sciences Institute (ILSI), the ILSI Risk Science Institute, the seven centers of the U.S. Food and Drug Administration, the U.S. Environmental Protection Agency's Office of Health and Environmental Assessment, the National Cancer Institute, the National Institute of Environmental Health Sciences, the National Institute on Aging, and the American Industrial Health Council. Additional support was provided by The Coca-Cola Company, Hoffmann-La Roche, Inc., Merck, Sharp and Dohme Research Laboratories, PepsiCo, Inc., The Procter & Gamble Company, and Schering-Plough Research Institute.

The organizing committee members were generous in contributing their time and expertise to developing and planning the conference. The session chairs were instrumental in keeping the speakers and audience focused on the issues, and performed double duty by serving on the editorial board for the proceedings. Whatever impact the conference and proceedings have will reflect the enthusiasm, interest, and commitment of the presenters-authors and their collaborators and staffs.

The infinite details associated with a two-and-one-half-day conference involving nearly 200 scientists were admirably handled by Diane Dalisera and the late Deborah Wilson of the ILSI Risk Science Institute. Roberta Gutman of ILSI Press provided the guidance that transformed nearly 30 disparate manuscripts into this volume. Finally, our thanks to everyone else who contributed to the conference and its proceedings.

Ronald W. Hart
National Center for Toxicological Research

David A. Neumann
ILSI Risk Science Institute

Richard T. Robertson
DuPont Merck Pharmaceutical Company

Dietary Restriction: An Update

Ronald W. Hart and Angelo Turturro
National Center for Toxicological Research

Introduction

As early as the studies of Kennaway and colleagues in the 1920s (reviewed in Weisburger and Weisburger 1967), rodents were being used to delineate the carcinogenic potential of chemical agents. From these efforts onward, animal strains have been developed and evaluated as to their resistance or sensitivity to a number of model carcinogens and the ability of these agents to induce various forms of cancer. The focus of this research was to identify and better understand the mechanisms of chemical carcinogenesis. In the early 1960s a call arose from the American public to identify and eliminate cancer-causing chemicals from the environment (Council on Environmental Quality 1986). As a consequence of this public pressure, new laws were passed, agencies were created, and a large-scale systematic testing program using a relatively standard protocol to identify such agents was launched by the National Cancer Institute (Hart et al. 1985). This testing program was the predecessor of our modern-day chronic animal bioassay program.

Despite the fact that the protocol initially used, and in many ways currently maintained, was never designed to screen for chemical carcinogens, it has nevertheless proven useful. For example, the known human carcinogens that have been tested in this system are to one degree or another positive, i.e., associated with tumor formation in animals under test (Hart et al. 1985). Due to interspecies differences in metabolism and other factors, however, this does not mean that all chemicals found to be carcinogenic in rodents will be carcinogenic in humans. Even if they are carcinogenic in humans, the site of action may not be the same as in rodents. Despite this, from a regulatory standpoint, finding of carcinogenicity in rodents is considered proof that the chemical of concern does have a carcinogenic potential in humans. As noted in Hart et al. (1985), "in the absence of adequate data on humans, it is reasonable, for practical purposes, to regard chemicals for which there is sufficient evidence of carcinogenicity in animals as if they presented a carcinogenic risk to humans." As a consequence, without a suitable alternative to the chronic animal assay, our best course of action might be to recognize the limitations of the current model and to work to overcome these limitations. The experience of the last two decades has helped us to focus on a number of issues and has provided us with a basis for developing some common procedures for conducting a more consistent long-term bioassay.

Based on the premise that the animal bioassay is only a surrogate for humans and is complicated by both theoretical problems in interpretation and practical problems in execution, we can aim to increase the utility of this model by 1) improving the consistency of the model across time, location, and procedure; and 2) sufficiently understanding the differences between the model and the human so that more reasonable extrapolations can be made.

In doing so, however, it is important to appreciate that species and strain were originally selected for the bioassay in order to achieve greater consistency among studies. In selecting animals, a number of major issues received careful consideration (Hart et al. 1985, chapter 3). Rodents, for example, were selected because they are readily available, have mammalian metabolism, and are cost-effective and time-efficient. The requirements for the test species include 1) a life span of reasonable length, i.e., short enough to be economically feasible yet long enough for tumors to develop; 2) a capacity to breed well in captivity and be relatively small in size; 3) similarity, as much as possible, to humans in metabolism and pathology; and 4) good health, yet susceptible to chemically induced cancer. Reducing variability and enhancing reproducibility is paramount to regulatory decision making; indeed, it is in part for this reason that inbred strains are generally used, in the belief that they exhibit less variability than outbred strains and, hence, presumably exhibit more stability of response. Either the Sprague-Dawley (SD) rat (an outbred albino chosen mainly for historical reasons and a lack of testicular tumor pathology), the Fischer 344 (F344) rat (an inbred albino with a low mammary tumor incidence in females), or the B6C3F1 hybrid mouse (more vigorous than its C57BL/6 parent and not as prone to murine mammary tumor virus infection as its C3H parent) are often used in toxicity testing. Despite the acknowledged limitations of these genotypes, they have acquired a significant advantage over all other strains because massive databases have been accumulated on their toxicological responses to numerous agents under various conditions. Indeed, thanks to these databases we can characterize problems with these strains and rapidly identify shifts in their biological behavior (Rao et al. 1990).

In order to increase reproducibility of the chronic bioassay, not only was careful attention given to strain and species selection, but stringent controls were developed over environmental conditions, animal care procedures, dietary composition/purity, etc. Animal bedding and feed are often analyzed and monitored. The rooms in which the animals are bred, housed, and tested are well-ventilated and controlled for noise, light, temperature, and humidity, with access limited to authorized personnel. Rack and cage configurations are often rotated to avoid any source of variability such as retinal and testicular degeneration due to differential light exposure. Animals are usually acclimated for 1 to 2 weeks before the onset of a study, and if the animals are not raised on site they are generally purchased from the same supplier in order to avoid disease and intrastrain differences. If the animals are found to be diseased, the study is often terminated to avoid the confounding effects that any therapeutic treatment might have on the bioassay interpretation. All cages, racks, and other equipment are regularly cleaned, and animal care personnel undergo continuous training. A number of on-site and off-site groups have been established to ensure compliance with all appropriate rules, regulations, and husbandry practices (for example, see National Toxicology Program 1992, Material and Methods).

Finally, when it comes to diet, the same degree of attention is given to all details of food composition and analysis for trace contaminants, even to the point that the cost of analysis in many cases equals or exceeds the cost of the diet. Until recently, however, dietary intake was not controlled, and its effects on chemical toxicity were not fully appreciated. This lack of control over dietary intake was surprising for two reasons: 1) the well-documented impact of reduced dietary intake on the occurrence rate of both spontaneous and chemically induced cancers (Turturro and Hart 1992, Weindruch and Walford 1988); and 2) the zealous efforts to reduce or eliminate sources of variability in the test.

With such a wide array of controls and oversight in place, it was assumed that individual variability had been minimized, while the reproducibility of results across laboratories and time had been optimized. This does not appear to be the case, however. Food consumption is often not measured, or not measured well, although body weight (BW) is almost always measured. The relationship of food consumption and BW allows the latter to be used as a surrogate for consumption. As seen in Figure 1, the average peak BWs (usually those observed near the end of the experiment) of cohorts of ad libitum SD control rats taken from only nine studies can vary by nearly twofold, e.g., from less than 500 to almost 1000 grams per animal (Turturro et al. 1994). Furthermore, the higher peak weights in the control animals are correlated with lower levels of animal survival at 24 months on test. This is consistent with the observation of Keenan et al. (1992), who showed, in SD males used as controls in studies of drug toxicity, that a trend toward heavier animals started a number of years ago and is associated with an increased age-specific incidence of age-associated diseases and with decreased survival. Such changes in the physiology of SD rats have caused major problems in using this rat strain in 2-year chronic animal bioassays due to their poor health and low survival at 24 months of age. Increasingly, similar problems have also been identified with the other major rat strains used in the chronic animal bioassays. Rao et al. (1990) observed that both BW and diseases such as pituitary tumors are increasing in the F344 rat, and there is a concomitant decrease in survival at 24 months on test.

The observation that survival is inversely proportional to BW and directly proportional to poor health is not surprising since Ross (1976) demonstrated some time ago that early BW in SD rats correlated well with pituitary tumor incidence and inversely with survival. Among F344 rats, we found that, unlike SD rats, BW at 12 months on test correlated better with survival than BW at an earlier age (Turturro et al., this volume). Additionally, it has recently been demonstrated that in humans late BW is inversely correlated with mortality when smoking and illness at the start of the study are taken into account (Lee et al. 1993).

For regulatory purposes, there is the additional problem, based on the observations of Turturro et al. (this volume), that significant variability in BWs occurs across chronic bioassays. They also show that even relatively minor changes in BW can significantly alter both survival at 24 months on test and the incidence of specific pathologic conditions, such as liver tumors. For instance, Turturro et al. (this volume, Figure 2) report that weight differences as small as five grams between 35 and 40 grams BW can double the incidence of liver tumors from a rate of 10% to 20%, in male mice. Corroboration of these results has recently been published (Seilkop, 1995).

Retrospectively, it appears that this relationship may account for much of the variability observed in liver tumor occurrence among the National Toxicology Program (NTP) historical control animals. Failure to control for these factors can complicate data interpretation.

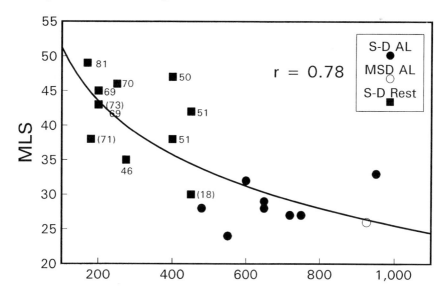

Figure 1. Peak body weight (BW), maximal life span (MLS), and percentage of calorie or BW restriction in male Sprague-Dawley (S-D) rats. BW, MLS (average life span of cohort's longest lived decile) for studies using ad libitum (AL) (filled circle) and diet-restricted DR (filled squares) S-D male rats in caloric restriction experiments. Open circle (MSD AL) is from a recent study in S-D male rats, with the MLS estimated from the survival curve (Keenan et al, 1992). The numbers in parentheses denote DR estimated by the percentage of BW of the AL controls, while the other numbers near the squares denote DR performed by feeding the specified percent of calories consumed by the AL controls. The point with two numbers is actually two points from different experiments with the same BW and MLS. Note that there is significant variability in the AL-fed controls and that, as a result, DR studies reported to have restricted calories to 50% of AL controls resulted in animals with similar BW as AL controls from other studies. Curve is logarithmic regression, which fits the data better than a linear model. r = regression coefficient. Data from Table 2.4 in Weindruch and Walford (1988) supplemented by information from Berg and Simms (1961).

Dietary Control and Dietary Restriction

One model that provides information about the impact of different levels of dietary intake involves studies using dietary restriction (DR). Dietary restriction is simply feeding the restricted animals less than the amount consumed by ad libitum (AL) feeding—whatever that is—of control animals, while insuring nutritional adequacy. Dietary control (DC), in contrast, uses a population dietary allotment to achieve certain fixed growth or consumption targets for a group, while insuring nutritional adequacy. DR often involves DC as a practical matter; however, a number of DR paradigms, especially in primates, use a percentage of an individual animal's initial food consumption as the determinant of "restricted" consumption and so do not involve DC. DC is a practical and cost-effective method for reducing variability in BW, food consumption, and physiological parameters, resulting in a more uniform test animal (Turturro et al., this volume).

With the exception of rodents and rabbits, limiting feed to a certain fixed amount is common practice for maintenance in mammalian species. The reasons for these exceptions appear to have arisen from a concern that use of individually housed animals would inordinately raise the cost of doing a study due to additional labor costs. For example, methods such as pair feeding were considered to be too labor-intensive and thus too costly for the relatively large number of rodents used in a long-term bioassay. With the development of automated technology, however, such concerns have been largely resolved, and it is now feasible to raise a large number of individually housed rodents, using DC and DR protocols, at a relatively reasonable cost.

Studies of the dietary modulation of toxicity have a long history (McCay 1947). In the 1940s, experiments by Tannenbaum and Silverstone (1949) demonstrated the impact of DR on chemically induced toxicity and defined a number of important physiological factors (e.g., BW and body composition) that might be altered by restriction. Keys (1949) demonstrated that many of the DR's positive effects could be extrapolated to humans. Later work by Ross (1976), Weindruch and Walford (1988), Kritschevsky and Klurfield (1987), Himeno et al. (1992), and others refined many of these earlier observations and extended them to other strains and species of rodents. These studies also demonstrated that DR can modulate the onset and severity of a number of diseases, lengthen life span, and inhibit or greatly reduce the toxicity induced by numerous physical, chemical, and viral agents.

The impact of DR on these endpoints will depend on the extent and duration of the restriction, the disease or induced toxicity of interest, the genotype and species tested, and the agent being evaluated. For instance, Weindruch and Walford (1988) have shown that the effect of DR on life span diminishes as a function of age of onset. A number of studies (reviewed in Weindruch and Walford 1988) and more recent ones performed at the National Center for Toxicological Research (Turturro et al. 1994, this volume) have shown that different pathological endpoints in the F344 rat exhibit a differential response to DR. For example, as demonstrated by Turturro et al. (1994), with a 40% caloric restriction the onset of pituitary tumors is delayed much longer than the onset of male-specific leukemia in

F344 rats. Similarly, the effect of equivalent restriction in different genotypes and species differs (Turturro and Hart 1991, Sheldon et al. 1994),
because similar levels of restriction can increase median life span by either 30% or less than 10%, depending on genotype.

Effect of Variability on Toxicity

From a mechanistic standpoint, these data are interesting because they
illuminate the influence of BW, which is related to organ growth and cellular replication, on spontaneous tumor occurrence and possibly survival.
From a regulatory standpoint they also suggest that 1) the control animal
data from a carcinogenicity study may be inappropriate if, at certain times
in the study, the BWs of dosed and control animals are different; 2) lack of
control over diet may result in significant variability in the background
tumor response between laboratories or between studies at the same laboratory; and 3) interpretation of results is not straightforward because these
factors can vary as a function of sex, strain, and pathology evaluated.

As a consequence of these observations, it is not only important that we
increase the survival and improve the health of our test animals (which
may be achieved by a number of means), but that we consider the desir-
ability of decreasing the variability of our test results by reducing the variability in the BW of animals on test. These considerations should guide us
as we consider the following recommended approaches for addressing the
problems of diminished survival in the rodent bioassay: 1) earlier termination of the bioassay, 2) use of a new strain, 3) breeding back to the earlier
weight in the strain used, 4) gang housing, 5) mathematical adjustment of
data, and 6) dietary control (DC).

Potential Solutions

Each potential solution has its own set of advantages and disadvantages. It is important to carefully consider each course because of the impacts it would have.

Earlier Termination of Chronic Toxicity Testing

The present protocol of exposing mice and rats for 2 years is a compromise that attempts to balance lifelong exposure to the test agent (thereby
increasing test sensitivity) with the problems of both concurrent disease
and differentiation of age- versus agent-induced pathologies. (An expo-
nential increase of disease occurs near the end of the life span in these
animals, which increases background disease significantly.) Terminating

the chronic assays at 18 months would not only increase the number of animals surviving to the end of the study but would also decrease animal care costs of a chronic bioassay. This solution could be thought of as simply altering what is already a compromise. The problems with shortening the bioassay are significant, however. One would lose sensitivity as a result of lowering the total dose (by approximately 25%) and restricting the time during which potential tumors could develop to noticeable size, which would complicate the pathological analysis. This effect on time-to-tumor may be more serious than the dose reduction per se, since the incidences of some tumors (e.g., lung tumors from smoking) have been shown to rise as a sixth or greater power of age in humans (Armitage and Doll 1954), suggesting that shortening the time for tumor development may have a devastating effect on the bioassay's capacity to detect the tumor potential of an agent. In addition, if the process that results in the increasing rate of mortality in rodents remains unchecked, it may be only a matter of time before significant mortality starts to occur earlier.

Use of a New Rodent Genotype

The genotypes presently used in chronic tests are selected for a variety of reasons, e.g., to facilitate comparison of potency of different agents and to avoid genotype-specific pathologies that could interfere with evaluation of target organ(s) of interest. Switching to a long-lived genotype would allow continued use of standard procedures; only the animal used would be changed. The two major problems with this solution are the loss of comparability with the many chronic assays conducted in the past and with numerous existing studies on the metabolism, pharmacokinetics, oncogene expression, etc. of the genotypes in current use. In addition, the background incidences of tumors would have to be defined under standard conditions for this genotype, and the relevance to humans of toxicity information derived in this genotype would have to be established. Further, unless a change is made in the process(es) that result in acceleration of mortality in rodents, it again may be only a matter of time before significant mortality starts to occur at 24 months on test.

Breeding of Rodents with Body Weights Similar to Historical Control Animals

Attempts to recreate the rodent with BWs comparable to those used in earlier studies by selective breeding of animals within a restricted weight range has already begun (Rao, this volume). This approach is based on the belief that, since the problem arose as a result of inadvertent selection for faster-growing animals, back-selection for smaller animals will reverse

the process. This approach has the advantage of using the same procedures and nominally the same genotypes as is currently done. The major
problem with altering the genetics of the test animal by selective breeding
is that there is no a priori reason to believe that selection for a complex
physiological parameter, such as BW, will reverse the specific changes
that led to the observed increase in growth rate. For instance, growth rate
could be increased by selecting for increased food efficiency, while a slower
growth rate could be achieved by selecting for decreased food absorption
in the gut. At the end of the selection process, the resulting animal will be,
in essence, a new strain, and will have to be characterized anew. The mechanism of slow growth will have to be evaluated and, unlike the plethora of
information available on DR, there will be almost nothing known about
this new animal. Furthermore, selection for slower growth may be associated with undesirable traits seen in slow-growing or small mouse mutants
(e.g., runting, malabsorption, disease, elimination of anterior pituitary function, etc.) that would further complicate interpretation of any chronic animal bioassay performed with these animals. The difficulty therefore is one
of analogy versus homology.

Gang Housing

The use of gang or group housing has been proposed as a means of
reducing the BWs associated with rodent bioassays while at the same time
reducing the costs associated with such studies. Indeed, gang housing is
traditionally used in bioassays for rats and, until a few years ago, for female mice. (The NTP recently decided to gang house female mice again.)
This housing paradigm results in a smaller animal, presumably because of
behavioral interactions among the animals. Expanded use of gang housing
would reduce the cost of the bioassay, but there are problems associated
with gang housing: 1) it is already done with rats and therefore would be
useless to modulate animal weight in tests using that species; 2) male mice
are difficult to gang house without fights, which result in trauma and
changes in disease incidence; and 3) in feeding studies there is a variability in food consumption (and thus resultant BW and dose) in gang-housed
animals (Turturro et al., this volume). For example, the dominant male
will eat more and receive more agent. Both of these factors (e.g., dose and
nutrition, Hart et al. 1985) are important factors in tumor variability.

Mathematical Adjustment

One method to accommodate the changes induced by variability in food
consumption is to adjust tumor incidences in the rodents on test by factors derived from previous studies. Although such adjustments may be

necessary for comparisons among chronic studies already performed, the major problem with this approach is that the effects of any given agent on food consumption/BW may be mediated through different toxic processes that are not modeled well by procedures used to model the spontaneous tumor incidences, which would be the basis of any adjustments. Basing mathematical adjustments on toxic or chemical class may be effective but will require an extensive research effort.

Dietary Control and Restriction

Controlling diet is an effective way to modulate BW, toxicity, and survival. DC is a practical and cost-effective method for reducing variability in BW, food consumption, and physiological parameters, resulting in a more uniform test animal (Turturro et al., this volume). Using DR, in contrast, the AL level is not reproduced between studies, which results in a DR animal that varies from study to study. DR can be used to reduce BW, to decrease occurrence of various spontaneous tumors, and to increase survival, while permitting the investigator to control the dose. The major problem associated with use of DR is the potential to reduce the toxicity of certain chemicals to varying degrees, thereby altering the sensitivity of a test. Using DC is, in a sense, like developing a new strain of test animal; as a result, such an approach requires characterization. This concern is offset by the realization that, in many but not all cases, the BW drift already occurring in the various test strains may necessitate recharacterization of these essentially new strains. An advantage of DC, however, is that this approach stabilizes the strain, permanently reduces weight drift, and enhances the reproducibility that was the basis for using inbred strains (or their hybrids) in the first place.

Conference Focus

This conference focuses on the potential use of DC and DR in the animal bioassay because:

1) lack of DC appears to have induced major variability in spontaneous tumor incidence in the historical control and, perhaps, treated animals;

2) dietary restriction appears to lengthen the life span and improve the health of mammals, thus overcoming the progressive increase in mortality observed in AL animal models presently being used for chronic toxicity testing;

3) diet is a primary factor in most degenerative disease processes in humans (Surgeon General 1988); and

4) dietary restriction alters the incidence level of both spontaneous and chemically induced tumors (Hart et al. 1992).

Conference Issues

Sufficient data are now available to consider incorporating the control of dietary intake into the design and conduct of animal studies. How to use this powerful tool to improve the value of chronic and short-term tests for safety assessment is complicated, however, by 1) the differential effects of DR on toxicity, 2) the inability to extrapolate DR results to humans due to lack of DR data in primates, and 3) the need to evaluate these studies in the context of other studies using AL feeding. The purpose of this conference is to address these difficult issues in a thoughtful and scientifically appropriate manner to derive practical insights into their regulatory consequences. It will provide us with the information needed not only to enhance our understanding of the relationships between diet, health, and toxicity, but to help us resolve the various dilemmas we face with the present chronic bioassay.

References

Armitage P, Doll R (1954) The age distribution of cancer and a multistage theory of carcinogenesis. Brit J Cancer 8:1–15

Berg B, Simms H (1961) Nutrition and longevity in the rat. III. Food restriction beyond 800 days. J Nutr 74:23–32

Council on Environmental Quality (1986) Environmental quality. 15th annual report of the Council on Environmental Quality, Supt. of Document, Washington, DC

Hart R, Chou M, Feuers R, et al. (1992) Caloric restriction and chemical toxicity/carcinogenesis. Qual Assur Good Prac Regulation Law 1:120–131

Hart R, Weisburger E, Turturro A (eds) (1985) Chemical carcinogens: a review of the science and associated principles. Fed Regist 50:10,371–10,442

Himeno Y, Engleman R, Good RA (1992) Influence of calorie restriction on oncogene expression and DNA synthesis during liver regeneration. Proc Nat'l Acad Sci U S A 89:5497–5501

Keenan K, Smith P, Ballam G, et al. (1992) The effect of diet and dietary optimisation on survival in carcinogenicity studies—an industry viewpoint. In McAuslane J, Lumley C, Walker S (eds), The carcinogenicity debate. Quay Publishing, London, pp 77–102

Keys A (1949) Nutrition Ann Rev Biochem 18:487–534

Kritschevsky D, Klurfield D (1987) Caloric effects in experimental mammary tumorigenesis. Am J Clin Nutr 45:236–242

Lee I, Manson JE, Hennekens CH, Paffenbarger R (1993) Body weight and mortality: a 27-year follow-up of middle-aged men. JAMA 270(23):2863–2828

McCay C (1947) Effects of restricted feeding upon aging and chronic diseases in rats and dogs. Am J Public Health 37:521–528

National Toxicology Program (1992) NTP technical report on the toxicology and carcinogenesis studies of HC Yellow 4 (CAS No. 59820-43-8) in F-344/N rats and B6C3F1 mice (feed studies). NTP Technical Report No. 419, NIH Publication No. 92-3150, National Institutes of Health, Bethesda, MD

Rao GN, Haseman JK, Grumbein S, et al. (1990) Growth, body weight, survival and tumor trends in F344/N rats during an eleven-year period. Toxicol Pathol 18:61–70

Ross M (1976) Nutrition and longevity in experimental animals. In Winick M (ed), Nutrition and aging. J Wiley and Sons, New York, pp 23–41

Seilkop S (1995) The effect of body weight on tumor incidence and carcinogenicity testing in B6C3F1 mice and F-344 rats. Fundam Appl Toxicol 24:247–259

Sheldon W, Blackwell B, Bucci T, Turturro A (1994) Effect of ad libitum feeding and forty percent food restriction on body weight, longevity, and neoplasia in B6C3F1, C57Bl6 and B6D2F1 mice [abstract]. Conference on Dietary restriction: implications for the design and interpretation of toxicity and carcinogenicity studies, Washington, DC, February 28–March 2.

Surgeon General (1988) The surgeon general's report on nutrition and health. Publication No. 88-50210, Department of Health and Human Services, Washington, DC

Tannenbaum A, Silverstone H (1949) The influence of the degree of caloric restriction on the formation of skin tumors and hepatomas in mice. Cancer Res 9:724–727

Turturro A, Blank K, Murasko D, Hart R (1994) Mechanisms of caloric restriction affecting aging and disease. Ann N Y Acad Sci 719:159–170

Turturro A, Hart R (1991) Longevity-assurance mechanisms and caloric restriction. Ann N Y Acad Sci 621:363–372

Turturro A, Hart R (1992) Dietary alteration in the rate of cancer and aging. Exp Gerontol 27:583–592

Weindruch R, Walford R (1988) Retardation of aging and disease by dietary restriction. C.C. Thomas, Springfield, IL

Weisburger J, Weisburger E (1967) Test for chemical carcinogens. In Busch H (ed), Methods in cancer research. Academic Press, New York, pp 307, 398

Control of Dietary Intake: Implications for Drug Evaluation

Joseph F. Contrera
FDA Center for Drug Evaluation and Research

The requirement for carcinogenicity studies in two rodent species for human drug products is the most costly and time-consuming nonclinical regulatory requirement for marketing. The results of rodent carcinogenicity studies can have a profound effect on the approval of a product. Even when benefit-risk considerations warrant marketing with appropriate labeling, adverse findings can seriously affect the competitive status of a given drug compared with others for the same indication.

The shortcomings of the rodent carcinogenicity study have been debated for decades, but no suitable alternative has been developed. Problems arising from the decreasing life span of rodents commonly employed in carcinogenicity studies have led to the application of dietary restriction (DR) as a means of extending the life span of assay animals. The FDA Center for Drug Evaluation and Research (CDER) is aware of a steadily increasing number of carcinogenicity studies in progress employing some form of DR or optimization. The CDER Carcinogenicity Assessment Committee (CAC) is also receiving an increasing number of carcinogenicity study proposals employing dietary optimization for consideration. Before initiating DR studies, drug registrants are encouraged to consult with the CDER CAC.

At present the decision to carry out a DR or optimized carcinogenicity study rests with the sponsor and is generally motivated by concern about adequate 2-year survival in the rodent strains available to the sponsor. The issue is no longer whether DR or optimized studies will be submitted to meet regulatory requirements, but how best to realistically proceed in an environment where both ad libitum (AL) and DR studies are employed by the pharmaceutical industry. Some of the specific regulatory concerns of the CDER are discussed below.

Study Duration

Dietary restriction increases the life span of rodents and appears to delay the onset of spontaneous tumors. Evidence presented in this volume demonstrates that the time to tumor formation for spontaneous tumors, e.g., mammary tumors in female rats, is increased in animals with restricted dietary intake. The incidence rate of such spontaneous tumors and their rate of progression do not appear to be altered by DR. Mammary tumors are ideally suited because they are palpable and therefore their appearance and rate of growth can be easily followed during the course of a study. What is learned for mammary tumors may apply to other tumor types.

If the appearance of spontaneous tumors is delayed under conditions of DR and the rate of tumor progression is unaltered, at the time of sacrifice at 2 years tumors in DR animals may be smaller than those in AL animals. This may potentially reduce the sensitivity of the bioassay. Because step (serial) sectioning is not routinely done in bioassays, there is a concern that smaller tumors associated with DR are more likely to be missed in the course of routine histopathological examination of tissues.

One approach to resolving this problem is to reconsider the duration of the carcinogenicity study. Is 2 years the appropriate duration for rodent carcinogenicity studies under conditions of DR? Should the time of termination of such bioassays be based on some accepted survival rate in dosed groups, e.g., when 50% survival is reached? A critical issue to consider in this context is whether it is reasonable to assume that the time to tumor formation for chemically induced tumors will also be delayed, as is the case for spontaneous tumors in studies employing DR protocols.

Study Design

There are significant variations in the design of DR protocols, with some studies carried out by restricting feed to 60–80% of the average AL level and others by limiting access to food to 5.5–6.5 hours per day. A more uniform study design should be developed and adopted for implementing DR studies.

The hour of the day when a drug is administered and whether it is administered before or after feeding is a significant variable that can influence systemic exposure and the maximum tolerated dose (MTD), especially in bioassays employing DR protocols. An optimized dosing schedule should be developed for each compound tested.

There is concern that the results of carcinogenicity studies performed under conditions of DR will be difficult to compare with the results of AL studies, thus diminishing the usefulness of future carcinogenicity databases. Some would argue that this is also an unrecognized reality and source of assay variability for AL studies. The utility of current rodent carcinogenicity databases has been questioned because the outcome of AL rodent carcinogenicity studies carried out in different laboratories at different times may vary significantly owing to the lack of control of dietary intake, variations in housing (single or group housing), steadily increasing body weight, and decreasing life span, possibly owing to genetic drift.

The Two-Rodent Species Requirement

The standard carcinogenicity testing requirement for drug products calls for 2-year studies of rats and mice of both sexes, a total of four study cells. The work of Gold et al. (1993), Tennant (1993), Ashby and Tennant (1991), Huff et al. (1991), and others supports the view that carcinogenicity potency and possible relevance of tumor findings to humans may be related to the number of study cells that are positive in rat and mouse bioassays. Ideally, both rat and mouse carcinogenicity studies should employ either AL or DR designs. Can the results of rat and mouse studies be validly compared in this manner if one study is carried out with AL feeding and the other under conditions of DR?

Dose Ranging and Other Toxicity Studies

Dietary restriction can alter the bioavailability and clearance of a compound and therefore can influence toxicity. It is essential that carcinogenicity dose-ranging studies employ the same feeding protocol that will be used in the carcinogenicity study.

The toxicity profile for a compound derived from standard 6- or 12-month AL rat toxicity studies may not be applicable to studies employing DR protocols. Rodent toxicity studies should be considered not only necessary to anticipate clinical toxicity but also as a resource to assist in interpreting the results of carcinogenicity studies. Because of the possible effect of DR on the toxicity profile of a compound, information derived from AL toxicity studies may be of little use in interpreting the results of a carcinogenicity study performed under conditions of DR. The decision to employ DR in carcinogenicity studies has implications for the entire toxicology program. Ideally, rodent chronic toxicity studies should be carried out under conditions of DR when it is anticipated that DR will be employed in carcinogenicity studies.

Assay Sensitivity and Interpretation

A further concern is whether DR results in an assay animal that is less prone to xenobiotically induced tumors than assay animals fed AL. From what we have learned from the published literature and at this meeting, the answer will vary with the compound tested. If specific metabolic or kinetic processes linked to the carcinogenicity of a particular compound are reduced by DR, fewer tumors should be produced, and the sensitivity of the bioassay may be viewed as being decreased compared with the AL condition. Conversely, if metabolic or kinetic processes linked to carcinogenicity are enhanced by DR, more tumors may be produced, and the sensitivity of the bioassay may be viewed as increased.

This brings us to the question of interpreting divergent results from carcinogenicity studies of the same compound using both AL and DR protocols. How do we evaluate such findings from a regulatory perspective given that the ultimate objective of animal toxicity studies is to identify potential hazards to human health? From a regulatory perspective, the most important consideration is which study condition (AL or DR) best models the human metabolic and pharmacokinetic profile for the test compound. Sufficient comparative human and rodent metabolic and pharmacokinetic information would be required to make such a regulatory decision.

Many of the preceding considerations and concerns also apply to other measures under consideration by the scientific community for increasing the life span of rodents used in carcinogenicity studies. These measures include the manipulation of dietary constituents in AL studies, the development of alternative rodent strains, and back breeding of existing rodent strains.

In conclusion, the following observations warrant scientific evaluation:
1) The life spans of rodent strains commonly used for the bioassay have steadily decreased to a level in some colonies where adequate survival in 2-year carcinogenicity studies can no longer be practically achieved.
2) The declining life span of rodents employed in the bioassay is associated with a significant increase in food consumption and body weight.
3) Genetic drift caused by selective breeding for rapid growth may have contributed to the steadily decreasing life span and increasing body weight of rodents used in carcinogenicity studies.

4) Genetic drift has reduced the usefulness of rodent historical control carcinogenicity data.

5) Ad libitum feeding is a significant source of variation in the bioassay because of the large individual variation in body weight and food consumption of assay animals within a study.

6) One option for improving the survival of rodents in the bioassay and reducing variation is the application of moderate caloric or dietary control.

7) There is a need to develop an alternative to the percentage of control food intake currently used as a criterion for the degree of DR. Consideration should be given to developing ideal growth curves and food consumption curves for commonly used rodent strains that can be applied to restore body weight and longevity to an agreed-upon level. An effort should also be made to arrive at a consensus on an optimal body weight range for rat and mouse strains commonly employed in carcinogenicity studies that will extend life span sufficiently without compromising assay sensitivity.

8) Carcinogenicity dose-ranging studies must be carried out at the same degree of dietary control and under the same housing conditions that are employed in the bioassay.

9) The feeding and dosing schedule in studies where diet is controlled should be selected to achieve optimal systemic drug exposure.

10) Rodent toxicity studies should be carried out under the same dietary control and housing conditions that are employed in the carcinogenicity study.

References

Ashby J, Tennant RW (1991) Definitive relationships among chemical structure, carcinogenicity and mutagenicity for 301 chemicals tested by the U.S. NTP. Mutat Res 257:229–306

Gold LS, Slone TH, Stern BR, Bernstein L (1993) Comparison of target organs of carcinogenicity for mutagenic and non-mutagenic chemicals. Mutat Res 286:75–100

Huff J, Cirvello J, Haseman J, Bucher J (1991) Chemicals associated with site-specific neoplasia in 1394 long-term carcinogenesis experiments in laboratory rodents. Environ Health Perspect 93:247–270

Tennant RW (1993) Stratification of rodent carcinogenicity bioassay results to reflect relative human hazard. Mutat Res 286:111–118

Control of Dietary Intake: Implications for Food Safety Evaluation

David G. Hattan

FDA Center for Food Safety and Applied Nutrition

The Center for Food Safety and Applied Nutrition (CFSAN) has a number of important reasons for interest in the area of dietary restriction (DR) as applied to toxicity and carcinogenicity testing. The first reason is to provide industry with a potentially useful interim measure to address the difficult problem of reduced survivorship currently being demonstrated in chronic feeding studies. The second reason is to provide "baseline" data on effects to be expected from this particular alteration in animal diets. This later type of information is needed to allow CFSAN to more clearly distinguish compound-related from dietary-mediated effects in animal studies. Finally, study of the diet relationships, including DR, may allow CFSAN important insights relating to interpreting growth and toxicologic responses potentially expressed when feeding designer foods to experimental animals.

Both industry and government have noted that in the past several years the extent of rat survivorship at the end of 2-year toxicity and carcinogenicity studies has been decreasing (Clayson et al. 1991). Where at one time it was not uncommon to observe 2-year survival of 60% or more, it is not uncommon now to find that the level of survival has plummeted to 20–30%, and in a few cases even lower. As the degree of survivorship has been reduced, the size of the animals in these studies has increased (Clayson et al. 1991). That is, both body weight and overall skeletal size have increased. Indeed, there are certain strains of rats (e.g., Sprague-Dawley derived) that appear to continue to grow virtually throughout the whole 2-year time period of the chronic study. As the animals have grown, they have also begun to suffer additional types and incidences of spontaneous pathology. For example, there appears to be a correlation between elevated body weights and greater incidences of pituitary and mammary tumors as well as enhanced levels of cardiomy-opathy and nephropathy (Turturro et al. 1994).

One might forecast that increases in the size of the test animal would imply a more robust subject; however, this has not been the case. These large animals actually appear to be obese by the end of the chronic study and suffer elevated rates of spontaneous non-neoplastic and neoplastic lesions compared to the smaller animals from earlier studies. For example, the Fischer 344 (F344) rats now experience an earlier onset of mammary and pituitary tumors as their overall body weight has increased (Turturro et al. 1994). Controlling food intake by F344 rats results in amelioration of a number of these spontaneously manifested lesions (Hart et al. 1993). Dietary control or restriction reduced the incidences of leukemia, testicular and pituitary adenoma formation, and the severity of nephropathy and cardiomyopathy (Rao, this volume).

It has also been noted that the incidence of chemically induced tumors can be attenuated by reducing the food intake of the rat. The reduction in occurrence of spontaneous tumors might have the salutary effect of reducing background animal pathology; however, the fact that dietary restriction may concomitantly reduce the probability of chemically induced tumors raises a potentially disturbing question for government regulators (Clayson et al. 1991). What is the balance of effect on the animal to decrease the likelihood of tumor expression? If it is found for any kind of tumor that DR comparatively lowers chemically induced tumors more than spontaneously occurring ones at a particular site, then the sensitivity for assessing the carcinogenicity of that agent has been lowered. Thus far, the degree of DR used in these studies has been relatively large, on the order of a 40% decrease from ad libitum (AL) feeding. It may be that there are lesser degrees of DR that will increase longevity of the test subjects without mediating an intolerable decrease in the sensitivity of the rats to chemically induced tumors.

In addition to changing the amount of the diet that the animal consumes, there can be unintended modifications to the dietary formulation that can affect the occurrence of various kinds of spontaneous and compound-induced adverse effects in the rat. For example, in the process of restricting the amount of diet consumed, certain modified diet regimens attempt to compensate for the attendant decrease in availability of micronutrients in the food consumed by elevating the concentration of vitamins in the ration. Thus, while the level of dietary fats (including polyunsaturated fatty acids) may be decreased, the amounts of vitamin E consumed are increased. This particular dietary manipulation results in an increase in the antioxidant-to-fat ratio and may lower the probability of tumor expression by certain types of chemicals.

Following the decision to use DR in toxicological testing of a chemical, regulators and those generating the study data may face other questions and issues that may surface as an outgrowth of this decision. Will a differential degree of change in physiological, biochemical, and toxicological parameters be induced by a specified level of DR from one rat strain to another? Will differing degrees of change to the sensitivity of expression of toxic effects be mediated in different rat strains? If there is a requirement for use of other species in the testing of a compound (a highly likely occurrence), how does one extrapolate the effects observed in the DR species to other species that have not been diet-restricted? Or is it required that all species have at least some level or type of DR testing in order to be able to make comparative assessments? In current practice, nonrodent species used for testing have restricted access to food.

The CFSAN may also be able to use the information derived from these DR studies to provide baseline data for interpreting studies performed to test the relative toxicological responses of macroadditive-type food additives. These compounds are foodlike and are usually intended to replace or substitute for normal macronutrients, such as fat, protein, and/or carbohydrate. These materials can contribute the same properties (such as smooth texture and bulk) to the food matrix as the usual macronutrients. Because these new macroadditives are used in such relatively large quantities, it will not be possible to feed large multiples of the amounts expected to be ingested by humans. Thus, there will be at most a dose exaggeration of only a few fold compared to the level of ingestion of the macronutrient for which it substitutes. Even considering these limitations to the potential level of testing, these poorly absorbed materials may mediate unwanted dilution of the diet, leading to nutritional effects that will make the safety assessment of these materials more difficult than that of traditional food additives. There is a real possibility that adverse effects noted in a study may be due to nutritionally induced effects rather than "true" toxic responses.

For the future, it may be that a number of areas of research will need additional development so that a clearer impression of DR's utility may emerge. For example, more information is needed on the comparative biochemical, metabolic, and physiological effects of DR on different strains of rats and using different (and most likely less restrictive) levels of caloric control. Additional studies are warranted to assess the validity, reproducibility, and interpretability of the responses elicited thus far.

For the CFSAN, the potential influence of administering differing concentrations of micronutrients in combination with varying amounts of macronutrients needs to be assessed. With respect to macroadditives, what treatment protocol would allow adequate control for the different types of macroadditives fed and potentially differential effects on the degree of absorption of the various fats and water soluble vitamins? If there is an effect on the absorbability of micronutrients and some form of reconstitution of vitamin availability must be assured, how does one develop the appropriate protocol design and determine the most relevant model species? High-exposure food additives ordinarily must undergo reproduction and teratology testing as well as in utero administration of the test material as an included phase of some chronic studies. What design factors should be considered before the study is started?

Earlier in this paper a question was raised about how data from one test species could be compared and/or assessed in light of the experimental results from another test species, but there is another question as well. Which human state or condition are we modeling with DR? Or is it related to the relative degree of caloric restriction imposed on the respective rat strain? Perhaps a 20% decrease in the AL feeding of an F344 rat models a different human condition than the same level of DR imposed on a Sprague-Dawley–derived rat. It may be that certain qualitative changes seen in either humans or a model species cannot be duplicated in the other. It would be very important to find such qualitative differences in responses to DR because they could exert important influences on interpreting effects observed during treatment. This in turn might allow effects to be distinguished as adverse or merely adaptive and within the possible range of normal physiological response. It seems certain determining what the clear and certain role for caloric or dietary restriction is awaits further investigation and insight.

References

Clayson DB, Rao GN, Roe FJC, et al. (1991) Impact of dietary restriction on bioassays and recommendations for future research: panel discussion. In Fishbein L (ed), Biological effects of dietary restriction. Springer-Verlag, Berlin

Hart RW, Leakey JEA, Allaben WT, et al. (1993) Role of nutrition and diet in degenerative processes. Int J Toxicol Environ Health 1:26–32

Turturro A, Blank K, Murasko D, Hart R (1994) Mechanisms of caloric restriction affecting aging and disease. Ann N Y Acad Sci 719:159–170

Dietary Restriction and Long-Term Rodent Bioassays: Implications for Evaluating Environmental Agents

William Farland
U.S. Environmental Protection Agency

The types of information considered by the Environmental Protection Agency (EPA) during regulatory review differ somewhat from those associated with drug and food safety evaluations. The agency deals with multimedia exposures by multiple routes and very often evaluates a broad distribution of exposures within the population. Thus, issues such as high- to low-dose extrapolation become particularly important as the EPA begins to look at questions arising from the evaluation of environmental agents.

Before turning our attention to the implications of the modulation of bioassay animal dietary intake on human health effects assessment, I should point out that the EPA has a much broader mandate. We also are concerned about ecological effects assessment. Thus, questions about the relationships between diet and nutrition and the potentially toxic effects of chemicals on animals and humans is of considerable interest to us. Nevertheless, this discussion will address human health effects assessment.

When we evaluate potential environmental carcinogens at the EPA, we look at the full range of information that may be available to us (Table 1). Guidelines for developing such data have been published for both our pesticides and toxics programs. The implication, of course, is that as we begin to think about changing guidance for conducting long-term bioassays, it will be extremely important to evaluate the implications of such changes for our programs. This will require measures that lead to consistency in developing data so that they can be used in our regulatory reviews.

In our overall assessment of a compound's health effects, we first consider the available epidemiologic data. Frequently, we do not find epidemiologic studies on the environmental agents that are of concern to us. In cases where epidemiologic studies are available, it is important to

Table 1. Sources of information used by the Environmental Protection Agency in human health effects assessments

Epidemiologic studies
Long-term bioassays
Short-term bioassays
In vitro assays
Structure-activity relationships
Pharmacokinetics/pharmacodynamics
Exposure route, rate, and duration

understand that one of the potential confounding issues that often compromises use of that information is the nutritional status of the particular epidemiologic group(s) under study. For example, when we consider reports that skin cancer develops in a Taiwanese population widely exposed to arsenic, the question arises whether the skin cancer was related to the nutritional status of that population and to the arsenic exposure or simply to the arsenic exposure alone. Until this question is addressed, it is not clear how to extrapolate these findings to similarly exposed human populations in the United States that have very different nutritional status.

In the absence of epidemiologic data, the results of long-term bioassays become a very important part of our database. As pointed out by Dr. Hart (Hart and Turturro, this volume), we think it is important that long-term rodent bioassays be performed using a consistent approach regarding the strain of animals used and the specific test protocol, recognizing that various factors enter into decisions about the appropriate experimental design. Nevertheless, it is critical to maintain the ability to compare studies over a long period of time and to facilitate multiple evaluations of individual chemicals. The EPA has been working to improve its ability to interpret the results of long-term rodent bioassays both in the historical database and in these recently completed studies.

In addition, we have been focusing our attention on short-term in vivo and in vitro tests that often provide mechanistic information that helps in interpreting both long-term bioassays and epidemiologic studies. This type of information is particularly important as we begin to ask questions about dose and response, and particularly as we begin to extrapolate across a population and consider both high- and low-dose exposure scenarios.

If we begin to collect data from long-term assays that reflect a change in caloric intake, i.e., the use of dietary restriction (DR) or control, then in fact most of the mechanistic information that we consider would also need to be collected from animals subjected to similar types of modulation in order for it to be relevant to the long-term bioassay data. This issue warrants careful consideration and is the focus of several reports in this volume (e.g., Duffy et al., Keenan and Soper).

Structure-activity relationships also play a large role in the EPA's health assessment evaluations. This reflects the idea that one builds a database, draws inferences from data that have been collected in the past, and makes decisions based on that data. This judgment process becomes particularly difficult when one changes protocols to collect additional information. Several questions must be asked: for example, When is there sufficient information in the new database to support the evaluation process? What is the relationship between the new database and its predecessor? This is not to say that we should not change protocols or should not move forward; I am just pointing out that our understanding of structure-activity relationships will suffer as we begin to change our testing protocols.

It is also important that we consider the issues of exposure, dose, pharmacokinetics, and pharmacodynamics. Clearly, the way that rodents handle chemicals to which they are exposed will have an impact on the overall outcome of the bioassays and/or on the mechanistic information that is being developed. Similarly, it is quite clear that as we begin to change assay conditions, such as controlling dietary intake during short-term in vivo studies, both the pharmacokinetics and pharmacodynamics of the test material are likely to change. Thus as we begin to evaluate chemicals, we need to look closely at issues related to exposure route, rate, and duration. If one begins to change the diet, the test animals, or other aspects of the test protocol as has been proposed (summarized by Hart and Turturro, this volume), the changes will likely have a profound impact on evaluating and interpreting the data relative to health assessment (Contrera, this volume; Hattan, this volume).

While some laws under which the EPA operates require specific data development, the majority of our evaluations have to be done with the information that is available to us. Thus, we frequently do not have access to a complete set of data such as that associated with drug and in some cases food safety evaluation. Regardless of quality and quantity of available data, we are expected to make regulatory decisions. Only our pesticides program has data submission requirements that are quite similar to those of certain FDA regulatory offices.

With regard to how data are evaluated, there has been a real evolution of hazard characterization within the EPA since the 1980s. We have begun to focus on a weight-of-the-evidence (WOE) approach, which allows us to move simply from evaluation of traditional toxicologic test results, e.g., scoring individual groups in a bioassay, to a position of trying to look at all available information, including mechanistic and exposure data. The WOE approach enables decisions about the hazard (carcinogenic) potential as well as the dose-response characteristics of a particular chemical.

This emphasis on WOE has given us an opportunity to really focus on hazard characterization as opposed to hazard identification. Without putting too much emphasis on the semantic differences, it is important to understand that we are not simply trying to use the most sensitive assay to detect an effect and using that as the basis for calling a specific material a hazard. Rather, we are attempting to characterize hazards with regard to their relevance to humans.

There are three issues that have considerable impact on our increasing use of the WOE approach. The first is the sensitivity of the particular tests we are using to detect a response, which we have to understand in order to interpret the responses. Second, information about the mechanistic basis of the response helps us to understand why chemicals are affecting biological systems as they are and to evaluate the relevance of these mechanisms to the much more complex multiple, biological processes that characterize chronic neoplastic disease. Finally, knowledge about the relevance to humans of all of the various test results and other data is critical to performing health hazard evaluations.

As I have mentioned, it is important that we have a sensitive screening process or test to begin to evaluate hazard potential. In the context of DR studies, we may find less sensitivity because of the extended life span and the concomitant changes associated with age-related pathology in animals undergoing DR. Conversely, the argument has been made that reducing spontaneously arising tumors through DR might increase the sensitivity of the bioassay (e.g., Hart and Turturro, this volume).

The literature is rife with discussion of how DR may affect responses such as general toxicity and carcinogenicity. From the perspective of developing an understanding of the mechanistic basis for the observed responses, however, the specific mechanisms responsible for the observed changes in carcinogenesis in animals subjected to DR remain to be elucidated. At this time, the observed modulating effect of DR is subject to multiple explanations, and it is premature to attribute specific effects to particular mechanisms. Clearly, many of the reports in this volume contribute significantly to improved understanding of the underlying mechanisms.

With respect to the relevance to human hazard, I think it is quite likely that use of "controlled diets," to borrow Dr. Hart's term (Hart and Turturro, this volume), for long-term bioassays may be most relevant to average, or perhaps optimum, human nutrition. A tremendous amount of data will be required for us to determine the relationships between various levels of DR in rodent bioassays and their relevance to humans or specific human subpopulations.

In addition to qualitative changes that one might see, I believe it is important that we think about some of the quantitative implications of DR studies. Dr. Hart noted that DR will likely result not only in alteration of various biological functions and processes but also in changes in the rates of these functions and processes. Both of these types of changes will have an impact on bioassay sensitivity, mechanistic understanding, and relevance to humans, and they need to be considered as we begin to evaluate both toxicity and carcinogenicity data from studies in which animals are subjected to DR.

The use of DR also has implications for defining maximum tolerated dose (MTD). On the one hand, we may find ourselves increasing the estimated MTDs simply because of the types of data that we get. For instance, if the sensitivity of the bioassay is diminished, greater doses may be required to elicit demonstrable effects. On the other hand, there may be additional, more subtle effects that are representative of toxicity occurring at lower levels, but these would not necessarily be detected if the MTD is defined in nontraditional terms, e.g., life shortening, reduced weight gain, etc. Consequently, one might perceive the need to test at higher doses in order to reach an MTD using the more traditional definitions. If we continue to abide by those definitions, the problems associated with high-dose to low-dose extrapolation of carcinogenic events in rodents to human populations will likely be exacerbated.

Such considerations also apply to determining the no-observed-adverse-effect level (NOAEL). Again, the questions relate to changes in physiologic processes and to defining adverse effects with regard to diet modulation. These issues certainly will need clarification and further examination before we can establish and interpret NOAELs based on studies using DR. Similarly, calculation of cancer potency values or slope factors based on rodent bioassays employing DR measures is a particularly difficult issue because we may, as noted earlier, need to consider the possibility of increasing doses to attain an MTD, and this will affect the degree of extrapolation required for human health effects assessment. Moreover, if DR results in changes in both background and induced tumors, it is likely that the slopes also will be changing. The question, of course, is whether one can generate a slope factor of greater relevance to humans through use of DR studies.

I would like to close by mentioning a few of the implications that emerge from this discussion. Risk assessment and guidance procedures will be emphasizing full characterization and will be looking for additional data to carry this out. The opportunity to do full characterization in many cases will require traditional toxicological approaches as well as some new approaches to evaluating hazards and risks. We are certainly expanding the role of mechanistic information in the evaluation process. Where dietary information can be folded into an understanding of mechanisms, we will do so. As we begin to look at the dose-response approach, the questions surrounding both life expectancy and longevity of bioassay animals become an important issue in terms of understanding their relationship to MTD and other key parameters. These issues will need to receive careful consideration during the evaluation process.

Recently, in our revisions of the cancer risk assessment guidelines, we have suggested that we will take a two-step dose-response approach to risk assessment. We will be looking at the range of observations in animals as a surrogate for highly exposed individuals within human populations and will examine the uncertainties within that range. This will allow us to move independently into the range of inference or low-dose extrapolations. It will be important for us to understand the implications of DR for both of those ranges with respect to the dose-response curve for widely exposed human populations.

Although few studies using DR have been submitted, it is likely that they will come to the attention of the EPA. The fact that the EPA probably has reviewed fewer of these types of studies than has the FDA reflects the nature of programs administered by the EPA. We frequently deal with studies that are done according to our specific guidance for industrial chemicals. Thus, we will gradually gain experience in evaluating and understanding these types of studies.

Regardless, I think it will be helpful, in order to fully characterize hazard, to evaluate traditional studies and, where data are available, studies using DR. Both types of information should be considered in the hazard characterization process. It is quite clear that a large database will need to be assembled before studies employing DR will replace the traditional long-term studies that use ad libitum feeding.

The utility of studies employing DR is highly dependent on establishing their relevance to humans. My personal feeling is that DR studies may be more important in understanding the range of risks for human populations than in identifying hazards. Studies employing DR may provide an opportunity to understand the sensitivity of populations across a range rather than simply determining whether, under some conditions, a chemical is likely to produce a toxic or carcinogenic response.

Finally, DR studies may have an impact on risk reduction strategies for certain environmental agents. For example, consider the idea of the ubiquitous pollutant that may be very difficult to control. If we have appropriate data suggesting that the toxic insults associated with exposure to that chemical may be reduced by certain types of nutritional or dietary intakes, it may actually be useful for us to take such an approach to controlling the toxicity of that pollutant rather than embarking on the daunting task of trying to reduce environmental levels of that pollutant.

The Merits and Demerits of Dietary Restriction in Animal Carcinogenicity Tests: A Regulatory Overview

Yuzo Hayashi
National Institute of Health Sciences, Japan

In 1968, a World Health Organization scientific working group consisting of world experts in the field of chemical carcinogenesis was convened to discuss scientific principles for the evaluation of the carcinogenicity of drugs. It prepared a technical report entitled *Principles for the Testing and Evaluation of Drugs for Carcinogenicity,* which covers scientific issues concerning testing methods as well as interpretation of test results (World Health Organization 1969). This report has provided the basis for carcinogenicity testing guidelines for pharmaceuticals and other chemicals issued in various countries, including the United States, United Kingdom, Canada, and Japan. All of these guidelines require performance of life span studies in two rodent species (usually rats and mice) at the maximum tolerated dose. Regarding diets, however, the WHO report states requirements only for the control of impurities, such as naturally occurring carcinogens or pesticide residues. It recommends future studies for the development of nutritionally adequate, defined diets but gives no description of optimum amounts of caloric intake in long-term rodent studies, despite the awareness at that time of an inverse correlation between animal longevity and food consumption (Berg and Simms 1961). Consequently, carcinogenicity testing guidelines have adopted, in common, free access to nutritionally adequate diets as the standard feeding procedure.

After 20 years of experience with long-term carcinogenicity tests in rodents and concomitant advances in knowledge of the mechanisms of carcinogenesis, it is now recognized that the present methods and strategies for assessing the carcinogenicity of chemicals are in need of urgent revision, particularly with respect to dose levels, dose periods, and animal species (Monro 1993). It also has been recommended that novel methods be designed and applied to replace the currently used free access to nutritionally adequate diets, i.e., ad libitum (AL) feeding, so that overnutrition can be avoided (Keenan et al. 1992).

A conference on the "Biological Effects of Dietary Restriction" was convened March 5–7, 1989, in Washington D.C., and it was evident then that there had been tremendous progress in the area of dietary restriction (DR) research in the preceding two decades. However, it was also apparent that further scientific knowledge and technological development were necessary for determining optimum calorie intake in long-term rodent studies and for standardizing a DR model (Henry et al. 1991).

In consideration of this historical background, this paper aims at addressing 1) the scientific evidence and knowledge of the influence of DR on the outcome of carcinogenicity tests in rodents and 2) regulatory considerations concerning the development of strategies for incorporating a DR regimen into carcinogenicity testing guidelines.

Scientific Evidence

Influence of Dietary Restriction on the Outcome of Carcinogenicity Tests

Influence on the occurrence of spontaneous tumors. Since the pioneering work of McCay et al. (1935), the influence of dietary factors on longevity and spontaneously occurring diseases in rodents has been extensively studied. The findings commonly indicate that DR can prolong longevity and retard the occurrence of age-associated nonneoplastic and neoplastic lesions. Selected data from these studies are summarized in Table 1 and indicate that DR significantly retards the appearance of mammary tumors, pheochromocytomas of the adrenal medulla, C-cell tumors of the thyroid, and bronchioalveolar tumors of the lung in Fischer 344 (F344) rats, as well as hepatocellular tumors, lymphomas, and pulmonary adenomas in B6C3F1 mice. Recently, Imai et al. (1991) also showed that DR could retard the occurrence of putative preneoplastic enzyme-altered liver cell foci in F344 rats and Sprague-Dawley rats.

Influence on the occurrence of induced tumors. Dietary and caloric restriction is also known to reduce the occurrence of certain types of carcinogen-induced tumors in short-term animal bioassays. After a critical appraisal of animal studies, Boutwell (1992) concluded that in the case of carcinogenesis of the mammary glands in rats and skin in mice, the incidence of tumors might be determined by the absolute caloric intake rather than by the percentage of fat in the diet. This was argued on the basis of data from four experiments where a comparison was made of the effects

Table 1. Comparison of spontaneous tumor occurrence in F344 rats and B6C3F1 mice fed ad libitum (AL) or restricted (R) diets

Tumor	F344 rats (24 months)[a]				F344 rats (30 months)[b]				B6C3F1 mice (30 months)[b]			
	Female		Male		Female		Male		Female		Male	
	AL	R(30%)	AL	R(30%)	AL	R(25%)	AL	R(25%)	AL	R(25%)	AL	R(25%)
Breast tumor	6	0	1	2	6	1	2	0	—	—	—	—
Adrenal pheo-chromocytoma	3	1	13	4	3	0	4	2	—	—	—	—
Thyroid C-cell tumor	2	4	10	3	1	0	2	0	—	—	—	—
Pulmonary adenoma	1	0	7	1	—	—	—	—	1	0	6	0
Liver cell tumor	0	0	1	0	—	—	—	—	2	0	5	3
Lymphoma	—	—	—	—	—	—	—	—	4	0	3	0

[a]Imai et al. 1991.
[b]Witt et al. 1989.

on carcinogenesis of feeding a low-fat (AL) diet with a high-fat diet restricted in total calories. As shown in Table 2, the incidence of both skin tumors and mammary tumors was found to be dependent on the caloric intake of the animal but not on the percentage of the fat in the diet or on the amount of fat consumed per day.

At the 1989 DR conference Ip (1991) reviewed the influence of caloric restriction (CR) on mammary cancer development in a dimethylbenzanthracene (DMBA) rat model and concluded that CR, even in the presence of a high fat intake, exerts a more striking influence than does a decrease in dietary fat in terms of suppressed cancer.

Beth et al. (1987) reported that delayed onset and reduced incidence and multiplicity of mammary cancers are observed with DR in rats examined 6 months after N-methyl-N-nitrosourea (MNU) administration. Recently, Chevalier et al. (1993) further demonstrated that graded levels of DR significantly reduced mammary tumor incidence and multiplicity in rats examined 30 weeks after treatment with MNU in a dose-response manner. In this experiment, DR was associated with significant changes in the organ/tissue distribution of retinoids. The authors concluded, however, that these modifications were not necessarily related to the cancer preventive activity of DR.

Pollard et al. (1984) studied the inhibitory effects of DR on intestinal tumorigenesis in methylazoxymethanol (MAM)-treated rats. Male Sprague-

Table 2. Effects of caloric restriction on occurrence of carcinogen-induced tumors

Carcinogen	Animal	Calorie intake (% AL)	Fat level (% diet)	Fat intake (g/animal/day)	Tumor occurrence Site	Tumor occurrence Incidence (%)	Ref.
Methylchol-anthrene	Mouse	100 66	5 15	0.06 0.24	Skin	54 28	Lavik and Baumann (1943)
Benzo(a)-pyrene	Mouse	100 83	2 27	0.07 0.54	Skin	82 72	Boutwell et al. (1949)
7,12-Dimethyl benz(a)-anthracene	Rat	100 81	5 30	0.6 2.2	Mammary	43 7	Boissonneault et al. (1986)

Note: AL = ad libitum.

Dawley rats were inoculated once with MAM at a dose of 30 mg/kg and groups were fed restricted diets daily for the subsequent 140 days. Groups A and B were fed a 12-g diet daily (25% restricted) from day 10 and day 63, respectively, after MAM exposure. Control MAM-treated rats were fed the same diet AL. When examined at day 140, there was a significant reduction of intestinal tumors in group A rats that had been placed on the restricted diet on day 10. However, there were no significant differences in tumor incidences between the control group and group B rats that had been maintained on the restricted diet beginning on day 63.

Lagopoulos et al. (1991) investigated the effects of alternating AL feeding and 30% DR on the development of diethylnitrosamine (DEN)-induced hepatic neoplasia in mice 24 weeks after DEN exposure. Dietary restriction retarded the growth of glucose-6-phosphatase–deficient (G6Pd) preneoplastic foci and subsequently that of hepatocellular adenomas and adenocarcinomas. Dietary restriction started early in life was particularly efficient at inhibiting the development of G6Pd foci. Conversely, the growth of foci was stimulated when the mice first had restricted access to food and thereafter were fed AL. The plasma insulin concentrations were much higher in AL-fed mice compared with DR mice, and the authors suggested that this may have contributed to the observed promotion of DEN-induced liver tumors in the mice. It was also found that DR could inhibit the development of azaserine-induced pancreatic acinar cell tumors in rats (Roebuck et al. 1981).

Effects and Mechanisms of Dietary Restriction

The mechanisms by which DR exerts its anti-aging and cancer-preventive actions still remain to be elucidated, but it has been suggested that its effects are mediated by altering endocrine functions, the immune system, the neuroregulatory system, and intermediary metabolism of exogenous and endogenous substances. This possibility was thoroughly described and discussed at the 1989 DR conference (Fishbein 1991); the following covers some selected pertinent details.

There is general consensus that carcinogenesis involves genetic alterations in somatic cells with activation of oncogenes and/or inactivation of cancer suppressor genes. Thus, carcinogens are defined as chemicals that can cause, either directly or indirectly, such genetic alterations in target cells. The term "genotoxic carcinogen" indicates a chemical capable of producing cancer by directly altering the genetic material of target cells, whereas "nongenotoxic carcinogen" represents a chemical capable of producing cancer by some secondary mechanism not related to direct gene damage (Hayashi 1992).

Genotoxic carcinogens usually require bioactivation to reactive intermediates before eventually producing DNA adducts in target cells. Leakey et al. (1991) indicated that CR appears to evoke changes in hepatic drug-metabolizing isozyme expression resulting in altered rates of carcinogen activation and detoxification. In the case of aflatoxin B_1, for example, Chou et al. (1991) demonstrated that CR can reduce formation of DNA adducts and the subsequent occurrence of DNA strand breaks, strongly suggesting inhibition of carcinogenic initiation.

General criteria for nongenotoxic carcinogens have not yet been established, primarily because many different kinds of examples, acting through a variety of mechanisms, have been reported. After a critical review of the literature, however, it was deduced that long-lasting epigenetic effects causing cell proliferation or sustained cellular hyperfunction or dysfunction may predispose to neoplastic transformation. Therefore, chemicals capable of producing such epigenetic effects may act in their target sites as nongenotoxic carcinogens (Hayashi 1992). Thus, it is presumed that the effects of DR on cellular proliferation might represent a mechanism of cancer prevention. Recently, Clayson et al. (1991) examined the effects of 25% DR on ^{3}H-thymidine incorporation into mammary ductal cells, esophageal epithelial cells, urinary bladder mucosal cells, and mucosal epithelial cells of the jejunum, duodenum, and colon, and found significantly lower labeling indices in the restricted group than in the control group.

Active oxygen-mediated carcinogenesis also deserves mention in terms of a possible mechanism of chemical-induced carcinogenesis (Hayashi 1992). That DR can reduce the formation of active oxygen species in the body, eventually decreasing the occurrence of oxidative DNA damage (Simic and Bergtold 1991), is interesting in this context.

Regulatory Considerations

General Principles for Designing Toxicity Test Guidelines

Ideally, toxicity test guidelines should be designed so that they minimize both false-positive and false-negative outcomes. Therefore, it is important for regulatory scientists to consider whether the incorporation of DR into the test protocol can improve the accuracy and/or sensitivity of carcinogenicity tests.

It is well documented that CR in long-term animal tests is associated with prolongation of survival and reduced occurrence of age-related neoplastic and nonneoplastic disease; consequently, the occurrence of false-positive outcomes would be expected to be decreased. On the other hand, various papers (Chou et al. 1991, Leakey et al. 1991, Clayson et al. 1991) indicate that DR can reduce sensitivity of animals to carcinogens through various mechanisms such as interfering with the metabolism of active intermediates and reducing mitotic rates in target cells. The use of animals with lowered sensitivity might therefore lead to higher rates of false-negative outcomes.

It should be stressed that it is usually possible or at least feasible to distinguish false-positive from true-positive outcomes, but it is sometimes impossible or extremely difficult to make the same distinction in the case of negative data. Therefore, the acceptance of DR as a routine procedure in carcinogenicity tests depends upon future elucidation of the optimum levels of caloric intake with respect to balancing between the advantages and disadvantages of DR against the desire to minimize false-negative outcomes.

Points for Future Investigation

In general, carcinogenicity test results are evaluated on the basis of intergroup comparisons of tumor occurrence in individual organs. This may cause, in some cases, a false-negative evaluation when weak carcinogens are studied in a testing system that uses low-sensitivity animals such as those maintained on a DR regimen. This means that adoption of DR in carcinogenicity tests should be considered together with incorporation of new parameters or strategies applicable to routine carcinogenicity tests to elevate their sensitivity.

It is generally agreed that cancers develop through multiple steps. Therefore, pathologic examination or evaluations integrating the occurrence of preneoplastic lesions, antecedent benign tumors, and fully developed cancers can increase the sensitivity and accuracy of the final carcinogenicity judgment. In the case of nongenotoxic carcinogens, mitotic rates in target cells, endocrinologic disturbances, and possible immunosuppression can also be useful parameters for the carcinogenicity evaluation.

Development of a practical model must be undertaken as a necessary technical requirement for incorporating DR into carcinogenicity tests. Two kinds of models can be considered: 1) feeding of restricted amounts of nutritionally adequate diets, including those currently used as standard, and 2) free access to calorically lowered diets. In the case of the former, the use of individual cages equipped with an automatic feeding facility device is necessary to avoid cannibalism and consumption of food as a bolus (Hayashi et al. 1979). In the case of the latter, the development of an appropriate macroadditive is required.

Summary and Conclusions

Dietary restriction in carcinogenicity tests is associated with longer survival and retarded occurrence of age-related diseases leading to an improvement of test results and eventually to a decrease in the occurrence of false-positive outcomes. It is also known that DR can reduce the susceptibility of animals to carcinogens through various mechanisms such as interfering with the metabolism of active intermediates or reducing mitotic rates in target cells, which can increase the possibility of false-negative outcomes. Therefore, adoption of DR as a routine procedure in carcinogenicity tests should be considered only together with incorporation of new parameters and strategies for increasing the sensitivity of test systems.

Acknowledgments

The author is grateful to Dr. Malcolm A. Moore, Inter-Mal Nagoya, and Dr. Fumio Furukawa and Miss Emiko Hattori, National Institute of Health Sciences, Tokyo, for comments on the manuscript. This work was supported in part by a grant-in-aid for cancer research from the Ministry of Health and Welfare and a grant-in-aid for a comprehensive 10-year strategy for cancer control from the Ministry of Health and Welfare, Japan.

References

Berg BN, Simms S (1961) Nutrition and longevity in the rat. J Nutr 74:23–32
Beth M, Berger MR, Aksoy M, Schämhl D (1987) Comparison between the effects of dietary fat level and calorie intake on methylnitrosourea-induced mammary carcinogenesis in female SD rats. Int J Cancer 39:737–744

Boissonneault GA, Elson C, Pariza MW (1986) New energy effects of dietary fat on chemically induced mammary carcinogenesis in F344 rats. J Natl Cancer Inst 76:335–338

Boutwell RK (1992) Caloric intake, dietary fat level and experimental carcinogenesis. In Jacobs MM (ed), Exercise, calories, fat and cancer. Plenum Press, New York, pp 95–101

Boutwell RK, Brush MK, Rusch HP (1949) The stimulating effects of dietary fat on carcinogenesis. Cancer Res 9:741–746

Chevalier S, Tuchweber B, Bhat PV, Lacroix A (1993) Dietary restriction reduces the incidence of MNU-induced mammary tumors and alters retinoid tissue concentrations in rats. Nutr Cancer 20:187-196

Chou NW, Pegram RA, Gao P, Allaben WT (1991) Effects of caloric restriction on aflatoxin B_1 metabolism and DNA modification in Fischer 344 rats. In Fishbein L (ed), Biological effects of dietary restriction. ILSI Monographs. Springer-Verlag, New York, pp 42–54

Clayson DB, Scott FW, Mongeau R, et al. (1991) Dietary and caloric restriction: its effect on cellular proliferation in selected mouse tissues. In Fishbein L (ed), Biological effects of dietary restriction. ILSI Monographs. Springer-Verlag, New York, pp 55–64

Fishbein L, ed (1991) Biological effects of dietary restriction. ILSI Monographs. Springer-Verlag, New York

Hayashi Y (1992) Overview of genotoxic carcinogens and nongenotoxic carcinogens. Exp Toxicol Pathol 44:465–472

Hayashi Y, Kato M, Otsuka H (1979) Inhibitory effects of diet restriction on monocrotaline intoxication in rats. Toxicol Lett 3:151–155

Henry CJ, Clayson DB, Rao GN, et al. (1991) Impact of dietary restriction on bioassays and recommendations for future research: panel discussion. In Fishbein L (ed), Biological effects of dietary restriction. ILSI Monographs. Springer-Verlag, New York, pp 321–336

Imai K, Yoshimura S, Hashimoto K, Boorman GA (1991) Effects of dietary restriction on age-associated pathological changes in Fischer 344 rats. In Fishbein L (ed), Biological effects of dietary restriction. ILSI Monographs. Springer-Verlag, New York, pp 87–98

Ip C (1991) The impact of caloric restriction on mammary cancer development in an experimental model. In Fishbein L (ed), Biological effects of dietary restriction. ILSI Monographs. Springer-Verlag, New York, pp 65–72

Keenan K, Smith P, Ballam G, et al. (1992) The effects of diet and dietary optimization (caloric restriction) on survival in carcinogenicity studies—an industry viewpoint. In McAuslane JAN, Lumley CE, Walker SR (eds), The carcinogenicity debate. CMR Workshop Series. Quay Publishing, Lancaster, UK, pp 77–102

Lagopoulos L, Sunahara GI, Würzner H, et al. (1991) The effects of altering dietary restriction and ad libitum feeding of mice on the development of diethylnitrosamine-induced liver tumors and its correlation to insulinaemia. Carcinogenesis 12:311–315

Lavik PS, Baumann CA (1943) Further studies on the tumor promoting action of fat. Cancer Res 3:749–756

Leakey LEA, Barare JJ, Harman JR, et al. (1991) Effects of long-term caloric restriction on hepatic drug-metabolizing enzyme activities in the Fischer 344 rat. In Fishbein L (ed), Biological effects of dietary restriction. ILSI Monographs. Springer-Verlag, New York, pp 207–216

McCay CM, Crowell MF, Maynard LA (1935) The effect of retarded growth upon the length of life span and upon the ultimate body size. J Nutr 10:63–79

Monro A (1993) How useful are chronic (lifespan) toxicology studies in rodent in identifying pharmaceuticals that pose a carcinogenic risk to humans? Adverse Drug React Toxicol Rev 12:5–34

Pollard M, Luckert PH, Pan GY (1984) Inhibition of intestinal tumorigenesis in methylazoxymethanol-treated rats by dietary restriction. Cancer Treat Rep 68:405–408

Roebuck JD, Yager JD, Longnecker DS, Wilpone SA (1981) Promotion by unsaturated fat on azaserine-induced pancreatic carcinogenesis in the rat. Cancer Res 41:3961–3966

Simic MG, Bergtold DS (1991) Urinary biomarkers of oxidative DNA-base damage and human caloric intake. In Fishbein L (ed), Biological effects of dietary restriction. ILSI Monographs. Springer-Verlag, New York, pp 217–225

Witt WM, Brand CD, Attwood VG (1989) A nationally supported study on caloric restriction of rodents. Lab Anim 18:37–43

World Health Organization (1969) Principles for the testing and evaluation of drugs for carcinogenicity. Technical Report Series No. 426. WHO, Geneva

Design Issues in the Use of the Diet-Restricted Rodent Model

Edward J. Masoro
University of Texas Health Science Center

Introduction

Restricting the food intake of rodents below that of ad libitum (AL) rodents is called dietary restriction (DR). DR of about 40% has yielded an animal model that has become a major tool for aging research (Masoro 1988a). This dietary manipulation markedly slows the rate of aging (Weindruch and Walford 1988). The survival curves in Figure 1 show that the increase in longevity by DR is robust and reproducible. The two studies on male Fischer 344 (F344) rats by Yu et al. (1982, 1985) depicted in Figure 1 were carried out 4 years apart. Assessment of four studies carried out by four different laboratories using male rats of different strains that were fed either AL or restricted to about 60% of the AL intake revealed a mortality rate doubling time of 102 days (with a range of 99 to 104 days) for the AL rats and 197 days (with a range of 187 to 210 days) for the DR rats (Masoro 1992). Findings such as these provide the evidence underlying the claim that DR slows aging processes.

Choice of Diet

Reduction of energy (caloric) intake is the major factor underlying the anti-aging action of DR. Much of the evidence for this claim comes from studies carried out in our laboratory in which use of a standard semisynthetic diet enabled this issue to be readily explored.

One approach to examining the anti-aging effects of DR was to reduce individual components of the diet so that intake of a component by AL rats was the same as that occurring during DR. Survival curves for rats in which dietary intake of fat or minerals were reduced are presented in Figures 2

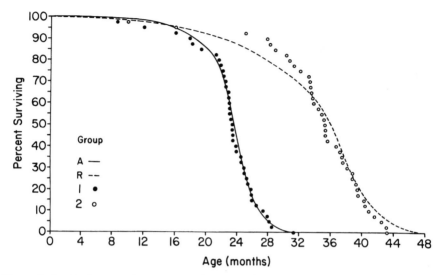

Figure 1. Survival curves from two studies carried out 4 years apart on ad libitum (AL) and diet-restricted (DR, 60% of AL intake) male F344 rats. First study: Group A (*n* = 115), AL; Group R (*n* = 115), DR. Second study: Group 1 (*n* = 40), AL; Group 2 (*n* = 40), DR. From Yu et al. (1985).

and 3, respectively (Iwasaki et al. 1988). Reducing the intake of fat or minerals did not influence longevity; of course, for interpretation of these studies to be unambiguous, energy intake of the rats on the diets restricted in fat or minerals must be the same as that of rats on the standard diet. Fortunately, that was the case.

Another approach was to restrict all dietary components by 40% except for one. The effects of restricting all components except for protein on longevity characteristics are presented in Table 1 (Masoro et al. 1989). The ability of DR to increase longevity was not affected in the absence of protein restriction.

In the design of DR studies, therefore, the composition of the diet used is not important except for one caveat: Malnutrition due to deficiency of a nutrient must be avoided.

Age of Initiation

In most studies initiation of DR has occurred at or soon after weaning. Indeed, for many years it was believed that DR slowed the aging process by retarding growth and development as first proposed by McCay et al. (1935). Recent studies have shown this not to be true (Weindruch and Walford 1982, Yu et al. 1985). The study of Yu et al. (1985) with male

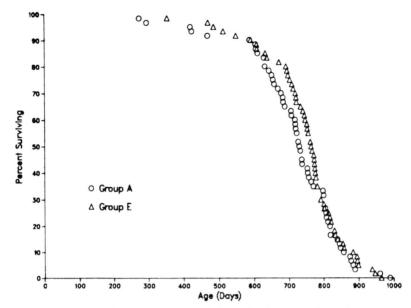

Figure 2. Survival curves for male F344 rats fed standard semisynthetic diet (Group A) and those fed a diet that reduced fat intake by 40% (Group E). From Iwasaki et al. (1988).

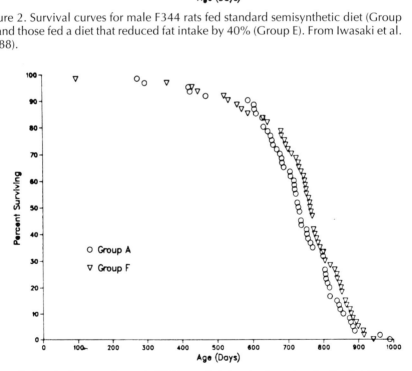

Figure 3. Survival curves for male F344 rats fed standard semisynthetic diet (Group A) and those fed a diet that reduced mineral intake by 40% (Group F). From Iwasaki et al. (1988).

Table 1. Influence of restriction of energy intake by 40% with or without restriction of protein intake on longevity of male Fischer 344 rats.

Dietary regimen	Number of rats	Median length of life (days)	Age of 10th-percentile survivors (days)	Maximum length of life (days)
Ad libitum (AL)	60	730	857	989
Dietary restriction (DR, 40% reduction in energy and protein intake)	60	936	1121	1275
DR (40% reduction in energy but not protein intake)	60	956	1158	1295

From Masoro et al. (1989).

F344 rats is summarized in Figure 4. DR started at 6 months of age (when skeletal development is almost complete in this rat strain) is as effective in extending life span (age of 10th-percentile survivors and maximum life span) as when started at 6 weeks of age. DR imposed when the animals were 6 weeks to 6 months of age (i.e., during the rapid growth period) did not markedly increase life span. In a mouse study, Weindruch and Walford (1982) found that DR (a 44% reduction in caloric intake) initiated in midlife has most of the antiaging actions observed when DR was started at weaning, although the magnitude of these effects is somewhat reduced.

For most toxicological studies, DR should probably be initiated after the developmental period but before occurrence of serious age-associated pathologic lesions. Based on these criteria, 6 months of age would probably be a good age to initiate DR in male F344 rats. The choice of the appropriate age to start DR will differ among species and strains, however, with sex and with the goals of the study. Therefore, the age at which DR is started must be empirically determined for each application.

Severity of Restriction

Most DR studies have used a 40% or so reduction in food intake because it results in marked increases in life span without causing an appreciable number of deaths when the animals are young. However, it may be that such a marked slowing of aging processes is not a desired outcome when carrying out toxicologic studies. Indeed, less severe levels of DR may be the model of choice for many toxicologic studies. In establishing the age of onset and the level of restriction, however, criteria must be established for evaluating the effects of DR.

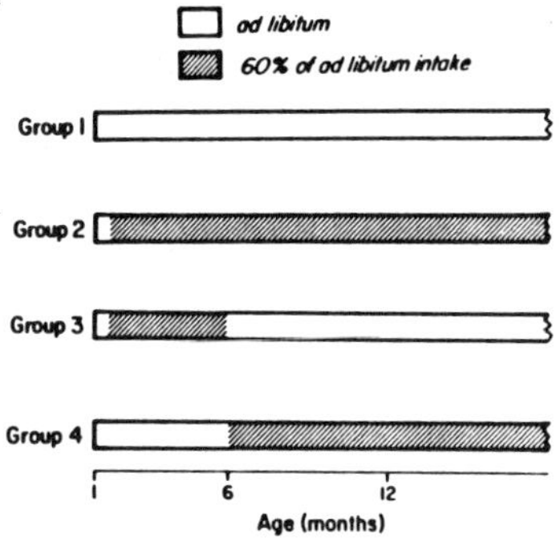

LENGTH OF LIFE IN DAYS

	MEDIAN	10TH PERCENTILE	MAXIMUM
Group 1	701	822	941
Group 2	1057	1226	1296
Group 3	808	918	1040
Group 4	941	1177	1299

□ ad libitum
▨ 60% of ad libitum intake

Age (months)

Figure 4. Influence of different times of initiation and durations of DR on longevity characteristics. From Masoro (1988b).

Criteria for Evaluating DR Effects

The Gompertz analysis provides the best approach (Finch et al. 1990) if the number of animals being studied is large, i.e., well in excess of 100 animals for each dietary regimen. If fewer than 100 animals are being studied, the maximum length of life, or preferably the age of the 10th-percentile survivors, is the best index, provided that the pitfalls discussed by Finch et al. (1990) are taken into account. With regard to criteria, however, it is important to recognize that DR does much more than influence longevity. Indeed, these other effects (retardation of age-related changes in physiological processes and of age-associated diseases) may be more important than life prolongation with respect to animal models used in toxicologic studies.

DR maintains a broad spectrum of physiological processes in a youthful state even at advanced ages. These effects range from gene expression to learning a maze. Many of these anti-aging actions on have been summarized in tabular form in a recent review article (Masoro 1993). If a DR regimen to be used for a toxicologic study differs in severity of restriction or age of initiation from the regimens that have been used in gerontologic studies, then it is desirable, if not necessary, to determine its physiological effects before the start of the toxicologic study.

The influence of a DR regimen on spontaneously occurring disease processes is probably of more importance to toxicology. DR involving a 40% restriction of food intake has been found to prevent or delay the

Table 2. Influence of dietary restriction (DR) with and without restriction of protein on the severity of nephropathy at the time of spontaneous death of male Fischer 344 rats

Dietary regimen	Number of rats examined	% of rats with lesions of following grade[a]					
		0	1	2	3	4	E
Ad libitum	182	0	4	14	14	23	45
Dietary restriction (DR, 40% reduction in energy and protein intake)	145	6	72	15	6	0	1
DR (40% reduction in energy but not protein intake)	63	2	41	33	16	2	6

[a]Severity of lesions increases from Grade 0 (no lesions) to Grade E (end-stage lesions). Only rats with Grades 4 or E lesions show evidence of kidney failure.
From Masoro et al. (1989).

occurrence of most age-associated disease processes (Masoro 1993). However, the extent to which a disease process is influenced by DR varies among species and strains of species and among disease processes within a strain or species. For example, in the male F344 rat, nephropathy sufficient to cause kidney failure is almost totally prevented by DR (40% reduction in food intake), and for this protection a reduction in protein intake is not a major factor (Table 2). The prevalence of many tumors at the time of spontaneous death is not reduced by DR in these rats, however (Shimokawa et al. 1993a), which of course may relate to the fact that DR results in death at much older ages. To investigate this possibility, data were analyzed by use of conditional survival curves by the Kaplan-Meier method (Gross and Clark 1975). This analysis revealed that DR delayed the age of fatal neoplastic disease occurrence (Shimokawa et al. 1991). Moreover, in the case of leukemia, which showed a higher prevalence at the time of death in DR than in AL rats, further analyses revealed that DR delayed the age of occurrence but not the rate of progression of this neoplastic disease (Shimokawa et al. 1993b). Thus, in the male F344 rat, DR retards neoplastic disease, but much less markedly than it influences nephropathy.

Mechanism of Anti-Aging Action of DR

A long-held premise in biological gerontology is that primary aging processes underlie most of the aging phenotype (Schneider 1987). The prevailing view is that DR slows aging by modulating one or more of

Table 3. Loss of body weight by male Fischer 344 rats during 48 hours following implantation of a jugular canula[a]

Age (months)	Ad libitum (AL) rats % body weight lost	Diet-restricted (DR) rats % body weight lost
3–4	4.8	1.8
17–18	11.2	3.2

[a]This is a summary table of a study yet to be published; for the dietary procedures used, see Yu et al. (1982).

these putative primary aging processes. For example, the action of DR has been linked to the glycation theory of aging (Masoro et al. 1992) and to the oxidative damage hypothesis of aging (Yu 1993).

Another possibility is that DR does not act specifically on primary aging processes but protects against damaging agents in general. In this view, aging processes are considered to be low-intensity, long-term damaging agents, with the actions of oxygen metabolites and of reducing sugars merely being specific examples.

DR has indeed been found to protect rodents from acute stressors. An example is our experience with the response of AL and DR male F344 rats to the surgical implantation of a jugular canula. Both young and older DR rats lost a lower percentage of body weight in response to this surgical procedure than did AL rats of a similar age (Table 3). Klebanov et al. (1993) studied the effect of DR on the response of young male BALB/C mice to the injection of carrageenan into the foot pad; DR was found to attenuate the inflammatory reaction. Further, Heydari et al. (1993) reported that DR 20-month-old male F344 rats were much more resistant to hyperthermic stress than were AL animals of the same age.

The adrenal cortical glucocorticoid system may provide a possible mechanism by which DR puts the rodent into a protective mode. Sabatino et al. (1991) found that DR results in daily afternoon peak concentrations of plasma free corticosterone in male F344 rats that are higher than in AL rats. The circadian pattern of plasma free corticosterone concentration is presented in Figure 5 for AL and DR male F344 rats in the 15- to 19-month age range; similar findings were obtained at all age ranges. Corticosterone is a double-edged sword—if its concentration is too low, the animal is unable to respond effectively to stress (Munck et al. 1984), and if it is too high, damaging effects occur (Sapolsky et al. 1986). Our hypothesis is that DR sets plasma free corticosterone concentration at optimal levels, thereby providing maximal protection without causing damage. This hypothesis is worthy of further study.

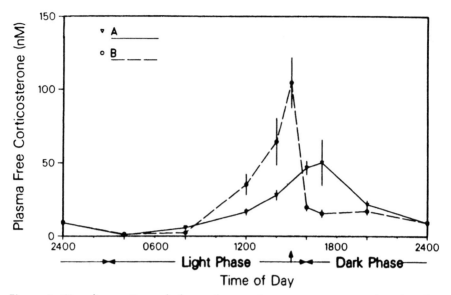

Figure 5. Circadian pattern of plasma free corticosterone concentrations in ad libitum fed (AL, Group A) and diet-restricted (DR, Group B) male F344 rats in the age range of 15–19 months. A total of 21 rats was used for each group. The study was of longitudinal design, with each rat having its circadian pattern measured at 6 month intervals from young adulthood until death. The vertical arrow on the x axis indicates when the Group B rats received the daily allotment of food. The findings at other age ranges were similar to those reported for the 15–19 month age interval. From Sabatino et al. (1991).

The heat shock protein system may constitute another potential general mechanism underlying the anti-aging actions of DR. This class of proteins is believed to protect animals from the adverse action of many stressors (Lindquist 1986). Heydari et al. (1993) found that the ability of hepatocytes of male F344 rats to express heat shock protein 70 (hsp 70) after a mild heat stress is enhanced by DR.

These reports and others suggest that, if DR has a general protective action in rodents, it would be expected to protect these animals from toxic agents. If, indeed, DR has this effect, is its usefulness for toxicologic studies compromised?

References

Finch CE, Pike MC, Witten M (1990) Slow mortality rate accelerations during aging in some animals approximate that of humans. Science 249:902
Gross AJ, Clark UA (1975) Survival distributions: reliability applications in the biomedical sciences. Wiley, New York

Heydari AR, Wu B, Takahashi R, et al. (1993) Expression of heat shock protein 70 is altered by age and diet at the level of transcription. Mol Cell Biol 13:2909

Iwasaki K, Gleiser CA, Masoro EJ, et al. (1988) Influence of restriction of individual dietary components on longevity and age-related disease in Fischer rats: the fat component and the mineral component. J Gerontol Biol Sci 43:B13

Klebanov S, Diais S, Stavinoha W, et al. (1993) Food restriction, elevated corticosterone and an attenuated inflammatory response to carrageenan in male BALB/c mice. Gerontologist 33(I):96

Lindquist S (1986) The heat shock response. Annu Rev Biochem 55:1151

Masoro EJ (1988a) Food restriction in rodents: an evaluation of its role in the study of aging. J. Gerontol Biol Sci 43:B59

Masoro EJ (1988b) Extension of life span. In Bianchi L, Holt P, James OFW, Butler RN (eds), Aging in liver and gastro-intestinal tract. MTP Press Ltd, Lancaster, UK, pp 49–58

Masoro EJ (1992) The role of animal models in meeting the gerontologic challenge of the 21st century. Gerontologist 32:627

Masoro EJ (1993) Dietary restriction and aging. J Am Geriatr Soc 41:994

Masoro EJ, Iwasaki K, Gleiser CA, et al. (1989) Dietary modulation of the progression of nephropathy in aging rats: an evaluation of the importance of protein. Am J Clin Nutr 49:1217

Masoro EJ, McCarter RJM, Katz MS, McMahan CA (1992) Dietary restriction alters characteristics of glucose fuel use. J Gerontol Biol Sci 47:B202

McCay C, Crowell M, Maynard L (1935) The effect of retarded growth upon the length of life and upon ultimate size. J Nutr 10:63

Munck A, Guyre PM, Holbrook NJ (1984) Physiological functions of glucocorticoids in stress and their relation to pharmacological actions. Endocr Rev 5:25

Sabatino F, Masoro EJ, McMahan CA, Kuhn RW (1991) Assessment of the role of the glucocorticoid system in aging processes and in the action of food restriction. J Gerontol Biol Sci 46:B171

Sapolsky RM, Krey LC, McEwen BS (1986) The neuroendocrinology of stress and aging: the glucocorticoid cascade hypothesis. Endocr Rev 7:284

Schneider EL (1987) Theories of aging, a perspective. In Warner HR, Butler RN, Sprott RL, Schneider EL (eds), Modern biological theories of aging. Raven Press, New York, pp 1–4

Shimokawa I, Higami Y, Hubbard GB, et al. (1993a) Diet and the suitability of the male Fischer 344 rat as a model for aging research. J Gerontol Biol Sci 48:B27

Shimokawa I, Yu BP, Higami Y, et al. (1993b) Dietary restriction retards onset but not progression of leukemia in male F344 rats. J Gerontol Biol Sci 48:B68

Shimokawa I, Yu BP, Masoro EJ (1991) Influence of diet on fatal neoplastic disease in male Fischer 344 rats. J Gerontol Biol Sci 46:B228

Weindruch R, Walford RL (1982) Dietary restriction in mice beginning at 1 year of age: effect on life span and spontaneous cancer. Science 215:1415

Weindruch R, Walford R (1988) The retardation of aging and disease by dietary restriction. Charles C. Thomas, Springfield, IL

Yu BP (1993) Oxidative damage by free radicals and lipid peroxidation in aging. In Yu BP (ed), Free radicals and aging. CRC Press, Boca Raton, FL, pp 57–88

Yu BP, Masoro EJ, McMahan CA (1985) Nutritional influences on aging of Fischer 344 rats. I. Physical, metabolic and longevity characteristics. J Gerontol 40:657–670

Yu BP, Masoro EJ, Murata I, et al. (1982) Life span study of SPF Fischer 344 male rats fed ad libitum or restricted diets: longevity, growth, lean body mass, and disease. J Gerontol 37:130

Husbandry Procedures Other Than Dietary Restriction for Lowering Body Weight and Tumor/Disease Rates in Fischer 344 Rats

G.N. Rao
National Institute of Environmental Health Sciences

Introduction

In recent years, the survival of most strains/stocks of rats commonly used for two-year carcinogenicity studies has been decreasing and is as low as 20% in some stocks. This decreased survival is a serious concern to researchers and regulatory agencies involved in evaluating the safety and carcinogenic potential of drugs, food additives, pesticides, and other chemicals. Contributing causes for mortality include anterior pituitary tumors in both sexes, nephropathy in males and mammary tumors in females of most strains/stocks of rats, and leukemia in Fischer 344 (F344) rats (Haseman et al. 1993). Dietary restriction (DR) for rats not subjected to any chemical treatment appears attractive because it lowers body weight, decreases the severity of nephropathy, decreases incidences of anterior pituitary and mammary tumors, and increases survival at two years (Witt et al. 1991).

Body Weight, Disease, and Dietary Restriction

The adult body weight of most strains/stocks of rats commonly used for toxicity and carcinogenicity studies increased by 20–40% during the last 20 years (Rao et al. 1990). There were no major changes in rodent diets used for long-term studies during the last 15 years, but the adult body weights of most strains/stocks of rats are still increasing. This increasing body weight has been accompanied by increasing incidences of anterior pituitary tumors and mammary tumors (Rao et al. 1987, 1990) and is possibly increasing the severity of nephropathy and decreasing survival in two-year studies. The major reason for this gradual increase in adult body weight year after year may be the intentional or inadvertent selection of breeding stock for faster growth and early reproduction (see Selection of Breeding Stock, below) to satisfy user preference for larger rodents at low cost (Rao 1993). Since rodents have had more than 50 generations elapse during the last 20 years (equivalent to approximately 1000 human years), even a 0.5% average increase in body weight per generation will lead to a large increase of adult body weight in 20 years.

Dietary restriction to lower the adult body weight is a treatment of the symptom but not the cause. A 30% DR may appear appropriate now, but as the body weights of future generations of rats increase further, a 40%, 50%, or greater degree of DR may be necessary to achieve appropriate reduction in adult body weight and to prevent decreased survival. Dietary restriction decreases the carcinogenicity of all major model chemical carcinogens (Hart 1993), especially those requiring metabolic activation (Pollard and Luckert 1985). This suggests that DR will decrease the sensitivity of the long-term carcinogenicity study by inhibiting or delaying the chemical carcinogenesis process, which is highly desirable for decreasing or delaying cancer risk in humans. Since the objective of a chemical carcinogenicity study in rodents is to assess the carcinogenic potential of the chemical, the effects of DR compromise the objective of such studies by inhibiting or delaying carcinogenic effects of test chemicals. At the doses fed to rodents, many chemicals may influence the energy expenditure of animals either by biochemical or behavior changes (e.g., stimulants, sedatives, antibiotics, anabolic agents). When fed ad libitum (AL), rats in chemical treatment groups may adjust caloric intake to compensate for chemical effects. However, DR may disproportionately influence the physiological processes and adult body weights of animals in chemical treatment groups as compared with the control, which will complicate interpretation of study results. Dietary restriction, either by limiting the quantity of the diet provided daily or by limiting the duration of feeding in the light (day) period (for convenience of technical personnel), will lead to consumption of most if not of all the food during the rodents' resting phase. Since rodents are nocturnal, feeding during the light period will result in disruption of their inherent circadian rhythm with a shift to a diurnal cycle and the associated changes in endocrine function and physiological processes (Duffy et al. 1989). The effects of modifying energy intake and the inherent nocturnal circadian rhythm on the toxic and carcinogenic responses to a chemical may be extremely complex, and the mechanisms involved in altering the effects will probably be difficult to identify and interpret.

Dietary restriction may require individual caging, which may not be an appropriate husbandry procedure for the well-being of most strains/stocks of rats (see Behavior Modification, below). Individual caging and daily feeding of a specified amount of diet will also be labor-intensive and expensive. Procedures associated with daily feeding of a specified amount may be prone to a high rate of errors (Rao and Huff 1991) and may be unmanageable in long-term chemical carcinogenicity studies. Errors of feeding (or not feeding) may be of little consequence to control groups but may introduce major complications in toxic responses of chemical treatment groups (especially for those animals in high-dose groups) and may compromise the quality of multiyear studies. Furthermore, if DR is used to assess the carcinogenic potential of a chemical, the same degree of DR should also be used for assessing pharmacokinetics, metabolism, toxicity, and other parameters with that chemical. This would cause a substantial and unjustifiable increase in effort for evaluation of chemicals in rodents. Dietary restriction is not a procedure for long-term or permanent resolution of problems associated with increasing adult body weight of rodents.

Permanent Resolution of High Body Weight-Associated Effects

Recommendations for resolving problems associated with high body weight (BW) include 1) selecting slower-growing breeding stock in the production colonies to lower the adult BW of the progeny, 2) modifying diets to delay the severity of chronic diseases and delay the development of spontaneous neoplasms, and 3) modifying feeding behavior by group-caging and making food available only during the night, the normal eating period for rodents.

Selection of Breeding Stock

Groups of rats of a given age will have a narrow-to-broad weight range depending on strain, stock, source, age, and other variables. To determine the influence of BW at 7 weeks of age on the growth pattern in 2-year studies, diet control groups of male F344 rats from the same production colony in five 2-year National Toxicology Program (NTP) studies started during a 9-month period and maintained under the standard conditions of the NTP Statement of Work were evaluated. From each control group of

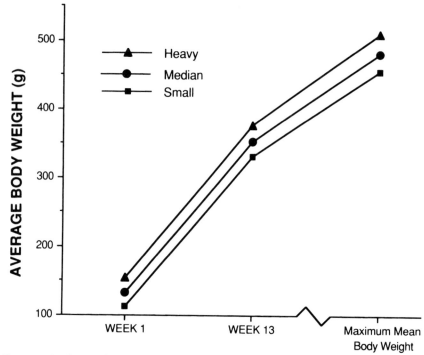

Figure 1. Body weight patterns of heavy, median, and small subgroups of control groups of male F344 rats in five studies. Corresponding subgroups of the five studies were combined.

50, three subgroups of 5 were identified: those with the lowest BW, BW close to the median, and highest BW. The average BW of corresponding subgroups of the five control groups at 7 and 20 weeks of age and maximum adult BW body weight were determined; the growth patterns are given in Figure 1. Results of this study indicated that the small male F344 rats remained small and the heavy rats remained heavy throughout the course of the studies, indicating that, in any selected population of rats, there will be smaller and slower-growing and heavier and probably faster-growing animals. Since larger rats of 8–10 weeks of age breed better and produce more and larger litters, the producers may be selecting heavier animals as breeders for future generations, leading to increasing BW generation after generation. Using the same procedure, one can lower the BW by selecting smaller and slower-growing rats as breeders for future generations. Smaller and slower-growing breeding stock may take an extra week or two to produce the first litter, and litter size could be smaller, resulting in an increase in the cost of rats. However, this increase could be

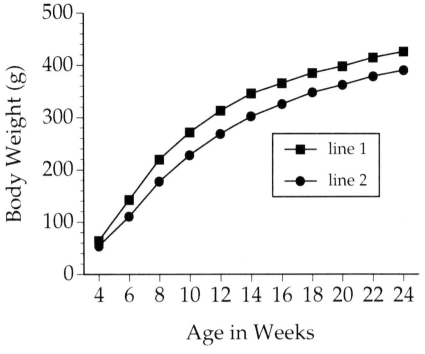

Figure 2. Growth patterns of two lines of male F344 rats from one production colony.

a small fraction of the cost/effort of using DR procedures. Figure 2 shows growth patterns (with mean BWs of groups) of two lines of male F344 rats from one production colony. Housing, care, management procedures, and health status of both lines of rats were same, and both lines were fed NIH-31 nonpurified diet ad libitum. Line 1 represents the male rats from an ongoing breeding colony produced by following the breeding (breeder selection) procedures in use for more than 10 years. Rats of line 2 were the progeny of selected breeders (from the same breeding colony as line 1) with average to less-than-average BWs over five generations. The progeny of the selected breeders (line 2) have an approximately 10% lower BW than the animals depicted by line 1. Pursuing the selection of smaller and slower-growing rats as breeding stock for several generations, a slower-growing smaller line of rat strains/stocks could be established. Since BW without DR appears to be highly associated with mammary and pituitary tumor incidence rates (Figures 3 and 4), breeding colonies producing smaller and slower-growing rats should decrease most of the problems associated with high adult BW.

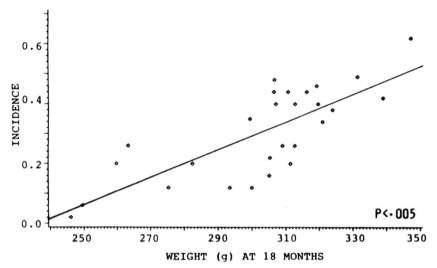

Figure 3. Association of body weight and incidence of mammary tumors in control groups of female F344 rats.

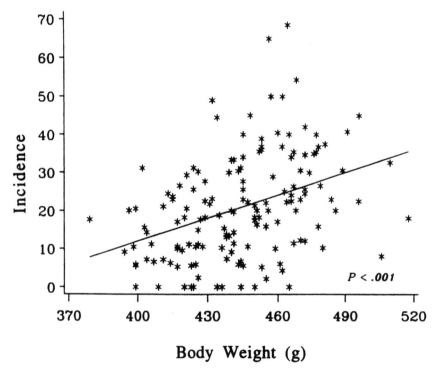

Figure 4. Association of body weight and incidence of anterior pituitary tumors in control groups of male F344 rats.

Diet Modification

Decreasing protein consumption by approximately 30% (by feeding a 15% protein nonpurified diet rather than a 23% protein nonpurified diet) markedly decreased the severity of nephropathy in F344 rats in a two-year study, without a significant decrease in BW as shown in Table 1 (Rao et al. 1993). Thus, decreasing the protein content of the diet to a level adequate for growth (~15%) will decrease the severity of nephropathy and nephropathy-associated mortality without dietary restriction.

Corn oil gavage at 5 mL/kg BW increased the calories from fat by 250% and resulted in an increase in adult BW of male but not female F344 rats (Rao and Haseman 1993). However, corn oil gavage, even with increased BW, decreased the incidence or delayed the development of leukemia and increased the survival of male but not female F344 rats as shown in Table 2 (Rao and Haseman 1993, National Toxicology Program 1994). Incidences of anterior pituitary and mammary tumors were associated with BW and not with corn oil gavage. The incidences of pancreatic acinar cell tumors appear to be influenced by a combination of BW and corn oil intake (Rao

Table 1. Effect of dietary protein concentration on urinary parameters and severity of nephropathy in F344 rats

	Male		Female	
Protein in diet (%)	15	23	15	23
Maximum body weight (g)	504	516	353	361
Protein consumption (g/rat/day)	2.9	4.3	2.0	2.8
Urine volume (mL/rat/day)	15	21	11	16
Urinary protein (mg/dl)	455	1439	173	349
Severity of nephropathy[*]	1.3	2.8	1.0	1.5

[*]Minimal = 1, mild = 2, moderate = 3, marked = 4 (see Rao et. al. [1993] for details).

Table 2. Influence of corn oil gavage on body weight, survival, and prevalences of selected tumors in F344 rats fed NIH-07 diet[a]

	Male		Female	
	Diet[b] Control	Corn Oil[c] Gavage	Diet[b] Control	Corn Oil[c] Gavage
Body weight (g)	478	504	348	330
Survival at 106 weeks	63	71	71	73
Leukemia	49	21	25	23
Anterior pituitary tumors	27	32	51	46
Mammary tumors	3	5	41	35
Pancreatic acinar cell tumors	1.2	7	0.4	0.8

[a]Except for body weight, all values are expressed as %.
[b]31 control groups with 50 rats/group.
[c]45 control groups with 50 rats/group.

and Haseman 1993). These results indicate that diets with higher fat content (>5%), especially corn oil types of fat, may decrease/delay leukemia and increase survival of male F344 rats. The mechanism(s) for this decrease/delay is not known. However, increasing the fat content of the diet will increase caloric density and may increase BW gain and incidences of BW-associated tumors.

Table 3. Influence of dietary fat and fiber on body weight, survival and prevalences of selected tumors—I[a]

	Male[b]		Female[b]	
	NIH-07	NTP-90	NIH-07	NTP-90
Maximum body weight (g)	523	499	349	324
Survival (at 110 wks)	48	52[c]	48	62
Leukemia				
Incidence	48	35	32	32
As cause of death	30	22	17	13
Mammary tumors				
Incidence	12	13	53	53
As cause of death	3	2	13	7
Adrenal Pheochromocytomas				
Incidence	32	18	3	3
Adrenal medullary hyperplasia incidence	17	5	11	2
Mesothelioma				
Incidence	2	8	0	2
As cause of death	0	7	0	0

NIH-07 diet contains ~5% fat, ~3.5% crude fiber, and ~23% protein. NTP-90 diet contains ~7% fat and ~9% crude fiber, and ~15% protein.
[a]Except for body weight, all values are expressed as %.
[b]60 rats/group.
[c]Due to a high incidence of mesothelioma-associated mortality, the survival was not significantly higher than in the NIH-07 diet group. However, a repeat study (see Table 4) with higher fat and fiber did not show higher incidence of mesothelioma, indicating that mesothelioma incidence in this study was probably due to chance.

Table 4. Influence of dietary fat and fiber on body weight, survival and cause of death—II

	Male[a]		Female[a]	
	NIH-07	NTP-91	NIH-07	NTP-91
Maximum body weight (g)	542	511	353	338
Survival at 110 weeks (%)	40	53	53	65
Cause of death or moribund				
sacrifice—mesothelioma (%)	2	1	0	0

NIH-07 diet contains ~5% fat, ~3.5% crude fiber, and ~23% protein. NTP-91 diet contains ~8.5% fat ~14% crude fiber, and ~15% protein.
[a]60 rats/group.

To determine the beneficial effects of higher dietary fat and fiber, experimental nonpurified diets (NTP-90 and NTP-91) with lower protein and higher fat and fiber content (Tables 3 and 4) than the NIH-07 diet were evaluated by 2-year studies. The ingredients of the experimental diets were similar to those of the NIH-07 diet (Rao et al. 1993), except that the fat content was increased by corn oil and the fiber content was increased by purified cellulose and oat hulls (Rao 1994). Housing, care, and management procedures of studies with the experimental diets in comparison with the NIH-07 diet were the same as described elsewhere (Rao et al. 1993). All diets were fed ad libitum.

Increasing the fiber content of the diet to compensate for increased calories due to higher fat content in the diet lowered the adult BW, decreased the incidence of leukemia in males, and increased the survival of male and female F344 rats as shown in Tables 3 and 4 (Rao 1994). Higher fat or fiber content in the diet (or a combination of the two) decreased the incidence of pheochromocytomas in males, decreased the incidences of adrenal medullary hyperplasia in both sexes (Table 3), delayed development of mammary tumors in females (Figure 5,) and decreased the number of female F344 rats sacrificed before 2 years due to mammary tumors (Table 3). The mechanism(s) responsible for the decrease in pheochromocytomas

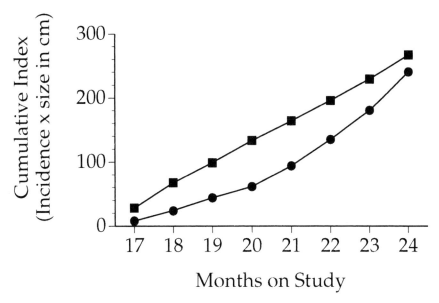

Figure 5. Effect of NTP-91 (●), a higher fat and fiber diet (see Table 4) than the NIH 07 diet (■) on the incidence/development of mammary tumors from 17 to 24 months of the study (18.5 to 25.5 months of age) in female F344 rats.

by higher dietary fat or fiber or by a combination of the two is not known. As indicated by human studies (Rose 1990), fiber may protect against breast cancer or mammary tumors by influencing estrogen metabolism. The beneficial effect appears to be due to estrogens binding to dietary fiber and increasing estrogen excretion through the feces. One study with an approximately 9% crude fiber diet (NTP-90) indicated that increased fiber consumption may have increased the incidence of mesothelioma and associated mortality in male F344 rats as shown in Table 3 (Rao 1994). However, a second study (NTP-91) with an approximately 14% crude fiber diet did not show an increased incidence of mesothelioma-associated mortality (Table 4), indicating that the increase observed in the first study was probably due to chance.

Behavior Modification

Review of the NTP historical data indicated that survival of individually caged F344 rats in diet-control groups without corn oil gavage was markedly lower for males and females when compared to the diet control groups of F344 rats caged as groups of five during the same period (Table 5). Survival of male Wistar rats housed four or five per cage was higher than when they were housed two or three per cage (Aldridge et al., ICI Pharmaceuticals, Ltd., and French and Williams, National Institute on Aging, unpublished data). Thus, group housing and associated social interaction appears to enhance the well-being and increase survival and life span of rats. Limited information (NTP) indicates that making food available from about 4 p.m. to about 8 a.m. instead of for 24 hours/day lowered the adult BW of male F344 rats by about 4% ($p < 0.04$) at about 15 months of age. Similar findings were reported with multiple-housed rodents by others (Tucker 1987). Thus, group housing and making food available only at night appear to increase survival in both sexes and lower adult BW of male rats.

Summary

In recent years the survival of most stocks/strains of rats commonly used for 2-year studies has been decreasing. There has been a marked increase in adult BWs of most strains/stocks of rats during the last 20 years. Increasing BW is strongly associated with decreasing survival and with increased incidences of mammary and anterior pituitary tumors. Dietary restriction appears to be attractive for revising these trends because it lowers BW and decreases incidences of BW-associated tumors. A major

Table 5. Survival and body weights of F344 rats housed one per cage or five per cage

	Males		Females	
	One/cage[a]	Five/cage[b]	One/cage[a]	Five/cage[b]
Survival at 110 weeks (%)	18.8 ± 7.0[**]	51.0 ± 8.5	48.4 ± 7.9[**]	62.8 ± 7.3
Maximum mean body weight (g)	478.8 ± 8.3	479.9 ± 16.0	326.4 ± 13.5[*]	348.0 ± 14.8

Values are mean ± SD.
[a]Five control groups with 50 rats/group.
[b]18 control groups with 50 rats/group.
[*]$p < 0.01$ (by Student's t-test).
[**]$p < 0.001$.

reason for increase in BW appears to be selection of breeding stock for faster growth and early reproduction. Dietary restriction to lower the adult BW treats the symptom but not the cause. Furthermore, DR decreases the carcinogenicity of all major model chemical carcinogens and thus will decrease the sensitivity of the long-term carcinogenicity study. In addition, procedures associated with DR could be cumbersome, labor-intensive, expensive, and unmanageable in long-term studies. Recommendations for permanent resolution of the high BW-associated effects include 1) selecting slower-growing breeding stock in production colonies to lower the BWs of the progeny for use in toxicity and carcinogenicity studies; 2) modifying diets to contain less protein to decrease the severity of nephropathy, increase fiber content to delay mammary tumor development, and increase fat, along with fiber, to decrease or delay the onset of leukemia in F344 rats; and 3) modifying behavior by group-caging and making food available only during the night, the normal eating period for rodents.

References

Duffy PH, Feuers RJ, Leaky JA, et al. (1989) Effects of chronic caloric restriction on physiological variables related to energy metabolism in the male Fischer 344 rat. Mech Ageing Dev 48:117

Hart RW (1993) Metabolic and biochemical effects of reduced caloric intake in rodents. Proceedings of the toxicology forum, annual winter meeting, February 1993. Toxicology Forum, Washington, DC, p 246

Haseman JK, Eustis SL, Ward JM (1993) Contributing causes of death in rats and utilization of this information in the statistical evaluation of tumor data. In Mohr U, Capen C, Dungworth D (eds), Pathology of aging animals, vol 2. ILSI Monographs. Springer-Verlag, New York, pp 629–638

National Toxicology Program (1994) Comparative toxicology studies of corn oil, safflower oil, and tricaprylin in male F344/N rats as vehicles for gavage. NTP Technical Report No. 426, NIH Publication No. 93-3157. National Toxicology Program, Research Triangle Park, NC

Pollard M, Luckert EH (1985) Tumorigenic effects of direct- and indirect-acting chemical carcinogens in rats on a restricted diet. J Natl Cancer Inst 74:1347

Rao GN (1993) Carcinogenesis and diet restriction. Environ Health Perspect 101:219

Rao GN (1994) Diet for Fischer-344 rats in long-term studies. Environ Health Perspect 102: 314

Rao GN, Edmondson J, Elwell MR (1993) Influence of dietary protein concentration on severity of nephropathy in Fischer-344 (F-344/N) rats. Toxicol Pathol 21:353

Rao GN, Haseman JK (1993) Influence of corn oil and diet on body weight, survival, and tumor incidences in F344/N rats. Nutr Cancer 19:21

Rao GN, Haseman JK, Grumbein S, et al. (1990) Growth, body weight, survival and tumor trends in F344/N rats during an eleven-year period. Toxicol Pathol 18:61

Rao GN, Huff J (1991) Letters to the editor. Fundam Appl Toxicol 16:617

Rao GN, Piegorsch WW, Haseman JK (1987) Influence of body weight on the incidence of spontaneous tumors in rats and mice of long-term studies. Am J Clin Nutr 45:252

Rose DP (1990) Dietary fiber and breast cancer. Nutr Cancer 13:1

Tucker MJ (1987) Factors influencing carcinogenicity testing in rodents. Human Toxicol 6:107

Witt WM, Sheldon WG, Thurman JD (1991) Pathological endpoints in dietary restricted rodents—Fischer 344 rats and B6C3F1 mice. In Fishbein L (ed), Biological effects of dietary restriction. Springer-Verlag, New York, p 73

The Sensitivity of the NTP Bioassay for Carcinogen Hazard Evaluation Can Be Modulated by Dietary Restriction

Frank W. Kari and Kamal M. Abdo
National Institute of Environmental Health Sciences

Introduction

It is well documented that dietary restriction (DR) with concomitant body weight reduction significantly increases longevity and decreases the incidence of background, chemical-, physical-, and biological-induced tumors in rats and mice (Tannenbaum 1940, Ross and Bras 1973, Pollard et al. 1984, Gross and Dreyfuss 1984). The interrelationships between body weights, survival rates, and tumor incidences suggest that practical benefits as well as problematic confounding factors may be introduced when feed intake and/or body weights are intentionally or unintentionally altered in toxicity and carcinogenicity bioassays for chemical hazard identification.

Typically, chronic exposure studies involve feeding rodents ad libitum (AL) while exposing them to several levels of chemical for up to two years. In DR paradigms the amount of food presented to control and exposed animals is restricted by amounts such that their body weights are decreased relative to that of AL animals. Since DR fosters leaner animals that live longer than more obese animals (Yu et al. 1985, Maeda et al. 1985), experiments conducted under DR may permit higher survival rates, thereby allowing more opportunities for chemical exposure and more time for pathologies to develop. It follows that these influences might enhance

This paper describes preliminary findings from four separate bioassays conducted by the National Toxicology Program. These studies have not yet been peer-reviewed, the data must be considered preliminary, and our findings are presented herein without identifying the chemicals. Complete reports of these four studies will be published following public peer review (anticipated spring 1995).

statistical power and increase the ability to resolve chemical effects in toxicity and carcinogenesis studies.

These potentially beneficial influences could be confounded by the propensity of DR to increase survival and generally decrease tumor incidence. A chemical observed to be toxic or carcinogenic in AL animals might not produce the same responses in DR or otherwise lighter animals. Such discordance could create interpretive challenges when comparing tumor outcomes both within and between studies. For example, observations collected for over a decade from the National Toxicology Program (NTP) show that some background tumor rates have comigrated with increasing body weights (Haseman and Rao 1992, Haseman 1993). Theoretically, comparisons between otherwise identical studies conducted several years apart could yield disparate results influenced primarily by the body weight of the animals.

Chemical-associated body weight depression in 13-week toxicity studies is routinely used, in conjunction with other factors, in selecting exposures for 2-year toxicity and carcinogenicity studies. By design, some top exposure levels selected for toxicity studies cause body weight depression. Comparisons within an experiment where there are exposure-related decreases in feed intake or body weight are potentially confounded by the influence of body weight on survival and disease processes. Since fewer tumors would be expected in the lighter animals compared to comparably dosed heavier animals, carcinogenic activity might be underestimated.

To begin assessing the merits and limitations associated with conducting bioassays under alternative dietary regimens, studies were undertaken to compare outcomes when four chemicals were evaluated under typical NTP bioassay conditions as well as by protocols employing DR. Specifically, experiments were designed 1) to evaluate the effect of DR (as an approximately 20% reduction in food intake relative to AL animals) on the sensitivity of the bioassay to detect chemically induced chronic toxicity and carcinogenicity; and 2) to evaluate the effect of weight matched control groups on the sensitivity of the bioassays. The results show a marked difference in bioassay outcome depending on the choice of protocol.

Materials and Methods

Experimental Design

Studies were designed to compare the toxicity and carcinogenicity of four chemicals evaluated under four different protocols. In each protocol, the effects of the chemical were assessed by a comparison between a group

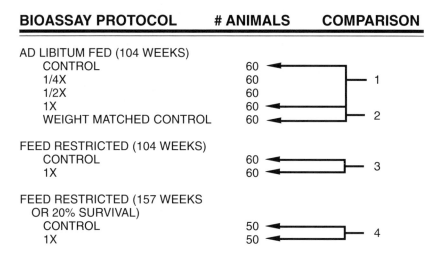

Figure 1. General design of feed restriction studies.

exposed to a single level of the chemical and a nonexposed control group (Figure 1). The exposure levels were chosen based on outcomes from 13-week prechronic evaluations conducted under AL conditions.

For the core bioassay, AL animals (60/group) were allotted to an unexposed control and three exposure groups (nominally 1/4X, 1/2X, and 1X) for 104 weeks. The comparison between the control group and the top exposure group (1X) was used to represent the outcome of the bioassay under AL conditions (Figure 1, comparison 1).

In a second experiment, outcomes from the top exposure AL group were instead compared with the weight-matched unexposed control group (Figure 1, comparison 2). In this group, the daily feed allotment was restricted such that their average body weights were matched to those of the top-dosed AL group (see Diets and Feeding, below).

Additionally, two groups of 60 animals (control and chemical-exposed) were both offered identical quantities of food, in amounts limited so that the unexposed control group would attain mean body weights approximately 80% that of the AL control group in protocol 1. Animals assigned to this DR paradigm were sacrificed at 104 weeks (Figure 1, comparison #3).

Since DR rats are expected to live longer than their AL counterparts, concurrent evaluations could result in comparisons at disproportionate times in their respective life spans, thereby masking age-dependent effects. Therefore, a protocol was employed to see if an additional year of exposure to

CHEMICAL	SPECIES*/GENDER	ROUTE OF EXPOSURE
A	RATS (M & F)	DOSED FEED
B	RATS (M) & MICE (M)	CORN OIL GAVAGE
C	MICE (M & F)	WATER GAVAGE
D	RATS (M & F)	DOSED FEED — (in utero, neonatal, and chronic exposure duration)

*Fischer 344/N
B6C3F1

Figure 2: Use of feed restriction protocols with different chemicals.

the chemicals would influence the tumor profile of DR animals. The diets of two groups (control and chemical-exposed) of 50 animals each were restricted, as in Figure 1, comparison 3, but the study was terminated either after 3 years or when survival in either group was reduced to 20% (Figure 1, comparison 4).

Chemical Exposures

The above general design was used to evaluate four different chemicals (nominally A, B, C, and D) according to the scheme in Figure 2. Dietary restriction treatments were not imposed across all chemical exposure groups of the four gender/species combinations typically used by the NTP. However, male and female rats and mice and the two principal modes of oral exposure are represented in these studies. Chemical A was administered to male and female rats as dosed feed, chemical B was given to male rats and male mice by corn oil gavage, chemical C was given to both sexes of mice by water gavage, and chemical D was administered to both sexes of rats as dosed feed; the exposure duration for chemical D ranged from in utero through adult stages of life.

The design for the experiment with chemical D is similar to that outlined for chemical A except that 70 animals per sex were used and the exposure began in utero. Female Fischer 344 (F344) rats (5 weeks of age) were maintained as described until they reached 10–12 weeks of age. They were then housed in groups of two and received NIH-07 diet with 0, 1/4X, 1/2X, or 1X ppm of Chemical D for 2 weeks before cohabitation (one male for two females). When pregnancy was ascertained, the females were then housed individually and maintained on their respective diet during pregnancy and until weaning of pups. The litters were culled to four males and four females on day 4 postpartum. During week 4 (between days 28 and 35 days postpartum), two males and two females were selected at random from each litter until 70 males and females per treatment group were amassed. These rats were administered the same level of the chemical as their respective dams. The animals were killed for evaluation after a single termination point of 129 weeks for males and 131 weeks for females.

For the dosed feed studies, a single concentration of the chemical (1X mg/kg diet) was blended in NIH-07 mash-type diet and was offered to the animals in feed hoppers in amounts described below. For gavage studies, a single concentration of chemical (1X mg/mL) was mixed in corn oil or water, and this mixture was given to animals at a dose of 5 (rats) or 10 (mice) mL/kg of body weight. Thus, for both feed and gavage exposures, AL and DR animals all received comparable masses of a given chemical on a body weight basis.

Animals and Care

Male and female F344 rats and B6C3F1 mice (5–6 weeks of age) were supplied by commercial vendors. Rats were segregated by sex and were housed (5 per cage) in solid-bottom polycarbonate cages placed on stainless steel racks (Lab Products, Maywood, NJ) with Sani-Chips (P.J. Murphy Forest Product Corp., Montville, NJ). Mice were individually housed. All animals were quarantined for two weeks before study to ascertain absence of disease and parasites. Animals were housed in temperature- and humidity-controlled rooms with 12-hour light-dark cycles (6 a.m. to 6 p.m. light) and 10 air changes per hour. Filtered city water was provided AL through an automatic watering system.

Diets and Feeding

NIH-07 open formula diet (4.0 kcal/g; Zeigler Bros. Inc., Gardners, PA) was offered either AL or in restricted quantities. When the test chemical was administered by dosed feed (chemicals A and D), the feed was a mash formulation; when chemicals were administered by gavage (chemicals B and C), the feed was processed precisely into 1-gram pellets, thereby facilitating presentation of restricted amounts by negating the requirement for weighing.

For the AL core bioassay, the control and three exposure groups (nominally 1/4X, 1/2X, and 1X) were allowed unlimited access to feed. Apparent feed intake (disappearance from feed hoppers, cage averages) and individual animal body weights were determined weekly for the first 13 weeks and monthly thereafter. Spillage was not determined, and therefore values reported for feed intake represent uncorrected indices of actual consumption.

The daily feed allotment for the "weight-matched" unexposed control group was restricted so that their average body weights would match those of the top-exposure (1X) AL group (Figure 1, comparison 2). The first two weeks of a study were used to establish baseline averages for body weights and feed intakes; consequently DR was not imposed during this period. Thereafter, daily food offerings to the weight-matched group were determined from the ratio of the mean body weight of exposed animals to that of the AL controls multiplied by the feed intake of the AL controls. For example, if, at a given weighing session, the top-dose animals weighed 90% of the AL controls, the desired amount of feed to be presented to each of the weight-matched animals for the following week would be 90% of the average AL intake of the unexposed animals during the previous week. The new weekly target value was then divided by 7 (days/week), multiplied by the number of animals in a given cage, and rounded to the nearest gram. The resultant mass of food was put in the cage feed hoppers daily. Corrections for body weight changes, if necessary, were made weekly for the first 13 weeks and monthly thereafter. Corrections for changes in the number of rats per cage due to mortality were made as they occurred.

In an analogous manner, control and exposed DR groups destined for sacrifice at either 104 or 157 weeks (Figure 1, comparisons 3 and 4) were both offered identical quantities of food constantly targeted to 80% that of the feed consumed by the AL unexposed group (Figure 1, comparison 1). This portion of the study was started 2 weeks later than the AL core experiment so that baseline values for body weights and feed intakes could be established.

For brevity, body weight at one year is used in this preliminary report as an indicator of animal growth. By this time, the animals generally will have had sufficient time to be influenced by the chemical but would not yet be greatly affected by morbidity and mortality (Turturro et al. 1993).

Clinical Examination and Pathology

Rats were observed twice daily for signs of morbidity, mortality, or toxicity. Moribund animals and those that survived until their scheduled termination were euthanized with CO_2 and necropsied. For each animal, 42 tissues were collected for routine histopathological examination, plus any tissues with grossly visible lesions. Tissues were preserved in 10% neutral buffered formalin, embedded in paraffin, sectioned at 5–6 mm, and stained with hematoxylin and eosin (Boorman et al. 1985)

Statistical Analysis

Differences in survival were analyzed by life table methods (Cox 1972, Tarone 1975). Tumor incidence data were analyzed by the logistic regression method described by Dinse and Haseman (1986). All reported p-values for tumor data are based on one-sided tests; those for survival are based on two-sided tests.

Results

Effects of Feed Restriction on Two Dosed Feed Studies

Chemical A. Chemical A was fed to male and female rats at levels of 0, 1/4X, 1/2X, and 1X mg/kg feed. Throughout the 104-week feeding studies there were chemical-associated decreases in mean body weights in both genders as exemplified by the 1-year body weights shown in Table 1. Control males averaged 451 grams and exposed animals were successively lighter, with the top-exposure males weighing 414 grams, approximately 92% that of unexposed controls. Weight-matched controls were 399 grams, about 97% of the target value. The mean body weight for AL female controls was 269 grams, and that of the three exposed groups were 270, 258, and 214 grams. The weight of this top-exposure group was 80% that of the AL controls, which was reasonably approximated by the weight-matched group weighing 199 grams after 12 months of study.

The 2-year survival rates (Table 1) of AL male rats exposed to the three levels of chemical A were statistically equivalent to that of their AL controls: control, 56%; 1/4X, 40%; 1/2X, 44%; and 1X, 44%. Corresponding values for females were 50%, 59%, 58%, and 59%. For both sexes, survival rates for weight-matched controls were significantly higher than for the AL controls: 68% versus 56% for males and 86% versus 50% for females.

Under the AL feeding protocol, the incidence of pancreatic acinar cell adenomas or carcinomas (combined) was increased in the male rats. The incidence in the control and the 1/4X, 1/2X, and 1X groups of males were 3/60 (5%); 2/59 (3%); 3/60 (5%); and 11/60 (18%), respectively. The incidence in the top-exposure group exceeded both that of the concurrent control group (5% versus 18%; $p < 0.01$) as well as the overall NTP historical incidence rate (4.8%) for this tumor. The incidence of pancreatic acinar cell hyperplasia increased with dose ($p < 0.05$). The incidence in the 1X group, 11/60 (18%), was higher than in the control, which was 5/60 (8%). No increase in the incidence of any tumor was observed in females rats administered chemical A under AL conditions.

Table 1. Feed consumption, 1-year body weights, and survival of male and female Fischer 344 (F344) rats in chronic dosed feed studies with two chemicals.

	Chemical A					
	Males			Females		
Bioassay protocol	Feed consumption (g/day/rat)	Body weight (g)/% of AL controls)	Survival rate (%)	Feed consumption (g/day/rat)	Body weight (g)/% of AL controls)	Survival rate (%)
AL (104 weeks)						
Control	16	451/100	56	10.8	269/100	50
1/4X	15.8	451/100	40	11.2	270/100	59
1/2X	15.1	437/97	44	11	258/96	58
1X	15.3	414/92	44	9.9	214/80	59
Weight-matched controls	14.1	399/89	68	8.5	199/74	86
DR (104 weeks)						
Control	13.6	372/82	70	8.5	207/77	71
1X	13.7	354/78	62	8.7	192/72	71
DR (137 weeks)						
Control	13.2	372/82	21	8.8	207/77	20
1X	13.3	354/78	27	8.6	192/72	22

	Chemical D					
	Males			Females		
Bioassay protocol	Feed consumption (g/day/rat)	Body weight (g)/% of AL controls)	Survival rate (%)	Feed consumption (g/day/rat)	Body weight (g)/% of AL controls)	Survival rate (%)
AL (121 weeks)						
Control	16.7	468/100	13	11.1	264/100	22
1/4X	16.6	465/100	12	11.7	263/100	30
1/2X	16.9	457/98	1	11.6	247/94	38
1X	16.6	430/92	23	11.1	236/88	43
Weight-matched controls	14.2	416/89	20	9.2	227/90	37
DR (121 weeks)						
Control	14.1	386/82	22	8.8	204/77	30
1X	13.9	391/84	45	8.8	204/77	42
DR (129 weeks)						
Control	13.1	386/82	17	9	204/77	30
1X	12.9	391/84	37	8.9	204/77	40

Table 2. Feed consumption, body weights, and survival estimates in gavage studies with chemical B conducted under various protocols

Chemical B

	Male mice			Male rats		
Bioassay protocol	Feed consumption (g/day/rat)	Body weight (g)/% of AL controls)	Survival rate (%)	Feed consumption (g/day/rat)	Body weight (g)/% of AL controls)	Survival rate (%)
AL (104 weeks)						
Control	5.3	52/100	84	13.8	469/100	70
1/4X	6.2	52/100	82	13.2	458/98	68
1/2X	6.4	50/96	82	13.3	462/99	68
1X	6.4	43/83	92	13.9	451/96	54
Weight-matched control	4.4	44/85	81	13.1	462/99	65
DR (104 weeks)						
Control	3.8	42/81	84	8.5	383/82	69
1X	3.9	34/65	88	8.5	360/77	82
DR (male mice 157 weeks, male rats 130 weeks)						
Control	3.7	42/81	40	8.2	383/82	21
1X	3.8	34/65	68	8.1	360/77	49

Chemical C

	Male mice			Female mice		
Bioassay protocol	Feed consumption (g/day/rat)	Body weight (g)/% of AL controls)	Survival rate (%)	Feed consumption (g/day/rat)	Body weight (g)/% of AL controls)	Survival rate (%)
AL (104 weeks)						
Control	5.4	50/100	81	5.6	51/100	67
1/25X	5.4	49/98	78	5.6	50/98	73
1/5X	5.6	47/94	82	5.6	46/90	77
1/X	5.8	40/80	83	5.7	39/76	76
Weight-matched control	4.4	42/84	85	4.5	40/78	74
DR (104 weeks)						
Control	4.3	36/72	98	4.3	33/65	94
1X	4.3	32/64	96	4.3	30/59	90
DR (157 weeks)						
Control	4.2	36/72	56	4.2	33/65	40
1X	4.2	32/64	74	4.2	30/59	39

[a]Kaplan-Meirer survival probabilities.

When the weight-matched control groups were used to evaluate the effect of the top-dose (1X) exposures, there were statistically significant ($p < 0.02$) increases in the incidence of pancreatic acinar cell tumors—1/60 (2%) versus 11/60 (18%)—and in the incidence of leukemia in exposed males—15/60 (25%) versus 29/60 (48%)—and leukemia in females—11/60 (18%) versus 19/60 (32%). Additionally, the incidence of urinary bladder papilloma (5%) in the 1X group of female rats was significantly increased compared to 0% incidence in their weight-matched counterparts. The 5% rate in the top-exposure group is considerably greater than the overall NTP historical control rate (0.22%). The increase in urinary bladder papillomas was accompanied by an increase in transitional cell hyperplasia at this site

Under the DR protocol, body weights of the control and exposed (1X) male rats averaged 372 and 354 grams after 12 months; these were 82% and 78% of AL control values. Corresponding weights for the DR female rats were 207 and 192 grams, which were 77% and 72% of AL control values. Compared to DR controls, the top exposure of chemical A was associated with 5–8% reductions in body weights of both male and female rats. For both genders, survival rates after 104 weeks for DR control and exposed rats were higher than for AL controls—70% and 62% versus 56% for males, and 70% and 76% versus 50% for females.

Under the DR protocol, no increase in any tumors could be attributed to chemical exposure in male rats after either 2 or 3 years. In females, the incidence of urinary bladder papillomas or carcinomas (combined) as well as transitional cell hyperplasia in the 1X restricted group was no greater than in the DR control at 2 years, but was after 3 years (2% versus 12%). This observation suggests that DR either delayed the onset of this tumor or that an additional year of chemical insult was required to cause the appearance of this tumor.

Chemical D. Exposure to chemical D began in utero and continued for 129 weeks (males) or 131 weeks (females). As seen in Table 1, mean body weights of control AL male rats at 1 year were 468 grams and decreased in an exposure-related manner whereby top-exposed males achieved a body weight of 431 grams, about 92% that of unexposed controls. The weight-matched males weighed 416 grams (approximately 97% of target value). One-year body weights of AL control female rats averaged 264 grams, while the average body weights for groups consuming chemical D were 263 grams (1/4X), 247 grams (1/2X), and 236 grams (1X). The average weight of the top-exposure group was 88% that of the controls. The average body weight of the weight-matched females was 227 grams, approximately 96% of the target value.

Table 3. Summary of carcinogenic effects in chronic bioassays by various protocols

Chemical route	Sex/ species	Tumor site	Comparison 1 AL	Comparison 2 Weight-matched	Comparison 3 DR (2 years)	Comparison 4 DR (3 years)
			Tumor incidence %			
A/feed	Male rats	Pancreas	5 vs. 18*	0 vs. 18*	0 vs. 0	0 vs. 4
		Leukemia	52 vs. 48	2 vs. 48*	30 vs. 45	52 vs. 55
	Female rats	Leukemia	37 vs. 32	18 vs. 32*	27 vs. 32	40 vs. 48
		Urinary bladder	2 vs. 5	0 vs. 5*	0 vs. 3	2 vs. 12*
D/feed	Male rats	Preputial gland	6 vs. 11	0 vs. 11*	1 vs. 2	NA
	Female rats	Clitoral gland	18 vs. 23	10 vs. 23*	7 vs. 25*	NA
		Liver	0 vs. 7*	0 vs. 7*	3 vs. 0	NA
B/gavage	Male mice	Liver	43 vs. 75*	25 vs. 75*	31 vs. 15*	46 vs. 38
		Adrenal cortex	10 vs. 8	0 vs. 8*	0 vs. 0	8 vs. 6
	Male Rats	Urinary bladder	0 vs. 8*	0 vs. 8*	0 vs. 0	0 vs. 0
C/gavage	Male and female mice		No carcinogenic effects under any protocol			

*Significantly different ($p < 0.05$) by logistic regression analysis.

Survival rates of AL rats at 121 weeks were 13% (control), 12% (1/4X), 1% (1/2X), and 23% (1X), for males and 22% (control), 30% (1/4X), 38% (1/2X), and 43% (1X) for females. Survival rates for weight-matched controls were 20% for males and 37% for females (Table 1).

Under the AL feeding protocol, exposure to chemical D was associated with increased incidences of liver tumors in female rats (7%) compared to the AL control group (0%) and the overall NTP historical incidence rate (0.67%). No significant increase in lesions could be attributed to chemical administration in male rats under AL feeding. When the weight-matched unexposed rats were used as controls (0%), increased liver tumors were again observed in the exposed AL female rats (7%). The incidence of clitoral gland tumors was also significantly increased: 10% of controls versus 23% of females exposed to chemical D. Employing weight-matched controls (0%) for comparisons with exposed male rats revealed a statistically increased incidence of preputial gland tumors in the exposed males (11%).

Body weights of the DR control and exposed (1X) rats at 1 year were 386 and 391 grams for males; for females, both groups averaged 204 grams. Thus, in both sexes, administration of chemical D had negligible effects on the body weights of the DR rats after 1 year of exposure. Survival rates for DR control and exposed groups were increased relative to the AL control: 22% and 45% versus 13% for males, and 30% and 41% versus 22% for females (Table 1).

In the DR experiment, there was a significant increase in clitoral gland tumors (7% in the unexposed group and 25% in the exposed group) No other chemical-associated lesions were observed.

Effects of Feed Restriction on Two Chemicals Evaluated by Gavage

Mice studies with chemical B. Chemical B was given to groups of 60 AL male mice by corn oil gavage at levels of 0, 1/4X, 1/2X, and 1X mg/ kg body weight. At one year, respective body weights were 52, 52, 50, and 43 grams (Table 2). After 104 weeks, survival rates in these respective groups were 84%, 82%, 82% and 92%. The weight-matched vehicle control group weighed 44 grams at one year, and the survival rate for this group was 81%.

Under AL conditions, 104 weeks of exposure to chemical B was associated with increases of hepatic adenomas or carcinomas (unexposed, 26/ 60; 1/4X, 40/60; 1/2X, 40/60; and 1X, 45/60; $p < 0.05$). Thus, as seen in Table 3, the AL unexposed and exposed mice yielded liver tumor incidences of 43% (26/60) and 75% (45/60), while a comparison between the weight-matched group and the top-dosed AL group yielded liver tumor incidences of 25% (15/60) and 75% (45/60; $p < 0.001$), respectively.

Under the DR protocol, the control and top-exposure groups weighed 42 and 34 grams at 1 year and had respective survival rates of 84% and 88% after 2 years. In contrast to the findings of the first two protocols, the incidence of liver tumors in these DR animals was significantly decreased in the chemical-exposed group—31% versus 15% (19/62 and 9/60; $p < 0.001$)—after 2 years—and were comparable—46% and 38% (22/48 and 19/50)—after 3 years.

Rat studies with chemical B. Ad libitum male rats were given chemical B by corn oil gavage at doses of 0, 1/4X, 1/2X, and 1X mg/kg body weight. After 1 year body weights for these groups averaged 469, 458, 462 and 451 grams. Two-year survivals were 70%, 68%, 68%, and 54%. Since there was negligible body weight depression (approximately 2–3% relative to AL controls in the top-exposure group) throughout the study, no adjustments were made to the weight-matched group, thereby yielding a redundant control group.

Under AL conditions there was a significant increase in urinary bladder hyperplasia (0/60, 1/60, 16/60, and 41/60) and papillomas (0/60, 0/60, 2/ 60, and 5/60) in the control and the three exposure groups. Thus, as shown in Table 3, unexposed and top-exposure AL rats had papilloma incidences of 0% and 8%. Similarly, comparison with the weight-matched control group (to which no restriction was imposed) yielded identical values.

In the DR groups, body weights at one year were 383 and 360 grams for the control and top-exposure groups, their respective 2-year survival rates were 69% and 82%. There were no urinary bladder papillomas after 2 or 3 years in any of these DR groups, nor was there evidence of a chemical-associated increase of any other neoplastic lesions.

Mice studies with chemical C. Groups of 60 male and female B6C3F1 AL mice were administered chemical C by water gavage at levels of 0, 1/25, 1/5, and 1X mg/kg body weight. Mean body weights of the males at 1 year averaged 50, 49, 47, and 40 grams, while corresponding values for the females were 51, 50, 46, and 39 grams. The weight-matched male and female mice averaged 42 and 40 grams. At 1 year, the respective body weights of the DR control and exposed male mice were 36 and 32 grams; for females, the body weights were 33 and 30 grams. No differences in survival were observed between any groups of male or female mice. No chemical-related increases in tumor incidences were observed in male or female mice regardless of feeding method (AL, weight-matched, or restricted), suggesting that chemical C is not carcinogenic in this species.

Incidences of liver tumors (and a variety of nonneoplastic lesions) in the restricted controls, the pair-weighted controls, and chemical-treated mice were depressed compared to AL controls, suggesting that the decreases are related to body weight depression.

Discussion

These studies were undertaken to compare outcomes when four chemicals were evaluated under typical NTP bioassay conditions as well as by protocols employing DR. Our results generally confirm numerous observations of others that DR increases survival rates and decreases incidence of neoplastic and nonneoplastic lesions at a variety of sites in both control and chemical-exposed animals. Furthermore, they demonstrate that the ability of bioassays to detect carcinogenic responses was altered by DR. As summarized in Table 3, three of the four chemicals studied were found to cause neoplastic lesions at four sites when evaluated under standard AL conditions. The target organs include the urinary bladder and pancreas in male rats and the livers in female rats and in male mice. Under the DR protocols, however, none of these tissues was detected as a target site for carcinogenesis after 2 or 3 years. Instead, two alternate sites, the clitoral gland (chemical D) and the urinary bladder (chemical A) were detected in female DR rats after 2 and 3 years, respectively. When the top-exposure group of AL animals was compared to the weight-matched control group, 10 sites were identified as targets of carcinogenesis. These included all four sites identified under the AL protocol, both sites identified under the DR protocol, and an additional four sites that were not identified under the other two protocols.

The increased survival of DR animals afforded more opportunities for dosing and additional time for tumor development; in no case, however, did this result in an enhanced ability to detect the carcinogenic response after 2 years. In fact, not a single tumor response associated with chemical exposure under AL conditions was identified under DR conditions after 2 or 3 years of exposure. Considered together, these results suggest that attributes associated with body weight and/or feed intake are stronger determinants of carcinogenesis than survival rates or duration of exposure to these chemical carcinogens.

It is noteworthy that the exposure levels used in these studies were based on the outcomes of 13-week prechronic evaluations conducted under AL conditions. Since DR causes a variety of pleiotropic responses that affect metabolism, distribution, and disposition of xenobiotics, it is probable that the minimally toxic doses (MTDs) will be altered under DR conditions. It remains to be seen what the outcomes would be if the chemicals were evaluated at exposure levels allowing comparable blood levels under AL and DR regimens.

Recognition that body weight is a covariant for at least some tumor outcomes can be used to enhance the sensitivity and refine the interpretation of these animal studies. This is exemplified by use of concurrent pair-weighted controls to estimate the effect of body weight changes on the "background" component of tumors in the chemical-dosed groups. The rationale for this comparison acknowledges that body weight changes alone, without chemical exposure, are accompanied by altered expression of a variety of pathological lesions and therefore presumes it reasonable to parse the confounding influences of body weight from other chemical-related effects. As summarized in Table 3, this practice had the effect of making four additional comparisons statistically significant with respect to chemical causation.

Although generalizations from this limited data should be viewed cautiously, they are very much relevant to our typical NTP experience. Tumors sites affected in these studies include rat and mouse livers and rat urinary bladder. These sites are among the most frequently affected sites in NTP chemical carcinogenesis studies (Huff et al. 1991). Furthermore, the degree of body weight restriction that we targeted, approximately 20%, is not greatly different from that seen in routine studies. Recent NTP experience (Technical Reports 370–420*) shows that in about half (48%) of the studies, the highest exposure levels were associated with body weight depressions ranging from 5–15% (Haseman, this volume). Since the NTP does not employ pair-weighting, however, it is clear that the bioassay is not currently being interpreted to its maximum sensitivity.

Inasmuch as rodent toxicity studies are designed to identify potential health risks, the intentional inclusion of experimental manipulations that desensitize this assay to an unknown, arbitrary, and variable degree must be viewed with caution. The preliminary findings from the experiments described here suggest that the sensitivity of bioassays to detect chemical-induced carcinogenesis can be modulated. The experimental design of our studies preclude determining whether there is a threshold level of DR that will increase longevity while having no impact on the sensitivity of tumor detection. However, our results show that the sensitivity is not enhanced due to increased survival rates. Regardless of whether the chemical evaluation is determined under DR or AL conditions, body weight changes should be expected in chemical-exposed animals, and further consideration should be given to using pair-weighted controls to help interpret these studies.

The observation that body weight/obesity may be a common risk factor in both the outcome of animal bioassays and the etiology and prognosis of a variety of human diseases (Kritchevsky 1993) enhances confidence that animal bioassays may be sensitive to host-chemical interactions relevant to humans outcomes. Opportunities are widespread for studying the role of diet and other physiological factors in modifying individual risk factors associated with chemical exposures.

References

Boorman G, Montgomery Jr CA, Hardisty J, et al. (1985) Quality assurance in pathology for rodent toxicology and carcinogenicity tests. In Milman H and Weisberger E (eds), Handbook of carcinogen testing. Noyes Publications, Park Ridge, NJ, pp 345–357

Cox DR (1972) Regression models and lifetables. J R Stat Soc B34:187–220

Dinse GE, Haseman JK (1986) Logistic regression analysis of incidental-tumor data from animal carcinogenicity experiments. Fundam Appl Toxicol 6:44–52

---

*Copies of NTP Technical Reports are available from the National Technical Information Service, U.S. Dept. of Commerce, 5285 Port Royal Road, Springfield, VA 22161.

Gross L, Dreyfuss Y (1984) Reduction of the incidence of radiation-induced tumors in rats after restriction of food intake. Proc Natl Acad Sci U S A 81:7596

Haseman JK (1993) The value and limitation of a large source of control animal pathology data. In McAuslane JAN, Parkinson C, Lumley CE (eds), Computerized control animal pathology databases: will they be used? Center for Medicine Research, London, pp 11–16

Haseman JK, Rao GN (1992) Effects of corn oil, time-related change, and inter-laboratory variability on tumor occurrence in control Fischer 344 (F344/N) rats. Toxicol Pathol 20:52

Huff J, Cirvello J, Haseman J, Bucher J (1991) Chemicals associated with site-specific neoplasia in 1394 long-term carcinogenesis experiments in laboratory rodents. Environ Health Perspect 93:247–270

Kritchevsky D (1993) Colorectal cancer: the role of dietary fat and caloric restriction. Mutat Res 290:63–70

Maeda H, Gleiser C, Masoro E, et al. (1985) Nutritional influences on aging of fischer 344 rats: II. pathology. J Gerontol 40:671–678

Pollard M, Luckert PH, Pan GY (1984) Inhibition of intestinal tumorigenesis in methylazoxymethanol-treated rats by dietary restriction. Cancer Treat Rep 68:405

Ross MH, Bras G (1973) Lasting influence of protein under- and over-nutrition on spontaneous tumor prevelence in the rat. J Nutr 103:944

Tannenbaum A (1940) The initiation and growth of tumors. Introduction. I. Effects of underfeeding. Am J Cancer 38:335

Tarone RE (1975) Tests for trend in life table analysis. Biometrika 62:679–682.

Tarturro A, Duffy P, Hart R (1993) Modulation of toxicity by diet and dietary macronutrient restriction. Mutat Res 295: 151–164

Yu B, Masoro E, McMahan A (1985) Nutritional influences on aging of fisher 344 rats. 1. physical, metabolic, and longevity characteristics. J Gerontol 40:657–670

The Effect of Caloric Modulation on Toxicity Studies

Angelo Turturro, Peter Duffy, and Ronald W. Hart
National Center for Toxicological Research

Introduction

Reliable, consistent estimation of the adverse effects of chemical and physical agents is key to both their rational regulation and to development of risk management approaches for solving practical problems associated with their use (Turturro and Hart 1987). A cornerstone in this process is the long-term toxicity test used to evaluate carcinogenic potential over an animal's "lifetime" (Interagency Staff Group 1986, Hart and Turturro, this volume). The most controlled long-term toxicity tests are those performed by the National Toxicology Program (NTP) (NTP 1983), which have evaluated almost 500 agents. In an effort to provide a consistent, reproducible test, the NTP has standardized test animals and a number of environmental variables, including feed type and animal maintenance. However, despite these controls, studies have observed a gradual rise in average body weight (BW) in B6C3F1 mice (Haseman et al. 1994) and an increase in average BW and plummeting survival at 24 months on test in Fischer 344 (F344) rats (the tests are conducted for 24 months, with the animals placed on test at approximately 5–6 weeks of age) (Rao et al. 1990). Additionally, there has been an increase in the incidences of a number of common spontaneous tumors in both mice and rats (Rao et al, 1990, Haseman et al. 1994). There also is such a wide variability in background tumor incidences in control animals that use of the historical control tumor database, which is useful for placing the potential induction of tumors by an agent into context, has become problematic (Roe 1994).

Experiments using dietary restriction (DR) have repeatedly suggested that one of the most significant modulators of spontaneous tumorigenesis is caloric consumption (Albanes 1987, Allaben et al. 1990, Turturro and

Dietary Restriction

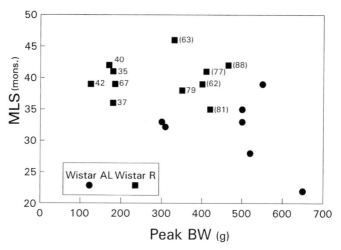

Figure 1. Peak body weight, maximal life span, and percentage of calorie or body weight restriction in male Wistar rats. Peak body weight (BW) and maximal life span (MLS) (average life span of cohort's longest lived decile) are shown for studies using AL (filled circle) and DR (filled squares) Wistar male rats. The numbers in parentheses denote DR performed by maintaining animals so that they attain the specified percentage of the BW of the AL controls. The other numbers near the squares denote DR performed by feeding the specified percentage of calories consumed by the AL controls. Note that there is significant variability in the BW of AL controls and that, as a result, studies that were reported to have restricted calories to 37% and 67% of that of AL controls resulted in animals with similar BW. There are some studies in which the AL control animals were smaller than the restricted groups in other experiments. (Data from Weindruch and Walford 1988.)

Hart 1992). Endocrine tumors are most sensitive to the effects of DR (Thurman et al. 1994), with almost every tumor type affected given the appropriate DR paradigm (Weindruch and Walford 1988, Turturro et al. 1994). Since uncontrolled food consumption resulting from ad libitum (AL) feeding may result in different consumption in different experiments, this factor may contribute to the variability in the spontaneous tumor incidence seen in the chronic bioassays. To evaluate this, we looked at a biomarker of dietary consumption in control animals in the most recent chronic bioassays, i.e., animal BWs at various times on test, and correlated this marker to cumulative incidences of some of the commonly seen tumors.

Methods

Previous work has shown that there is wide variation in food consumption in toxicity tests done in Sprague-Dawley (SD) rats (Keenan and Soper, this volume). This variability is reflected by the wide range (500–1000 g) in the average BW of these animals (based on cohorts of at least 50 animals). Average BW of F344 rats (Rao et al. 1990) and B6C3F1 mice (e.g., 30–60 g average) (Turturro et al. 1993, Haseman et al. 1994) also have varied significantly between experiments. In the Wistar rat (Figure 1), another rat often used in toxicity studies, male peak BW can vary from 300–600 g using AL animals. Also shown in Figure 1, which illustrates the relationship between life span and the peak BW of some previous studies using DR, is that the variability sometimes results in experiments in which animals subjected to DR are larger than the animals used as AL controls in other studies. This makes using percentage of DR as a common metric to compare experiments problematic.

A better metric appears to be BW at different times on test. The magnitudes of growth curves (estimated by BW at various ages) are directly correlated to shortening the time to onset of many chronic diseases, including cancer (McCay et al. 1935, 1943, Berg and Simms 1960, Ross 1976, Carroll 1975, Turturro and Hart 1992, Turturro et al. 1993, 1994) and nephropathy (Berg and Simms 1960). Increases in BW growth curves have been associated with increased mortality in rodents (Rose 1991, Turturro and Hart 1991). Also, when the effects of BW growth and food consumption per se can be separated, the effect of DR on tumors appears more related to BW than to food consumption. For instance, Holloszy and Smith (1986) reported that animals kept under chilled conditions (resulting in increased food consumption but lowered BW) developed few tumors, i.e., at levels consistent with their slowed growth rather than elevated food consumption. Based on these data and the practical problem that food consumption is not often reported in chronic studies although BW growth curves almost always are, BW at various times on test, rather than food consumption, were the biomarkers used in this to evaluate the effect of caloric modulation on disease and survival.

In an analysis of the pathological status of the F344 rat (Thurman et al. 1994) and the B6C3F1 mouse (Sheldon et al., in press) at various ages under either AL or DR feeding regimens, diseases such as pituitary tumors in rats and liver tumors in mice begin to appear after 18 months of age. Since these diseases could alter animal BW directly (e.g., tumors could change BW), BW was not evaluated after 12 months on test in order to reduce the chance of confounding any relationship by the effects of evident disease on BW. Data from some of the most recent bioassays (listed in Tables 1 and 2) were evaluated, with the BW at 2 months on test (BW2), 6 months (BW6), the average of BWs from 6 months to 12 months (BWAV), and 12 months (BW12) estimated by using the average weight

Table 1. Bioassays in B6C3F1 mice used in these analyses

Study/Type				
TR434/i	TR387/f	TR342/c	TR327/c	TR316/c
TR419/f	TR385/i	TR341/f	TR326/i	TR315/f
TR412/f	TR366/g	TR339/c	TR325/f	TR314/i
TR410/i	TR365/f	TR337/f	TR324/f	TR312/c
TR407/f	TR363/i	TR336/c	TR323/c	TR312/c
TR406/c	TR362/o	TR334/c	TR322/f	TR311/i
TR403/g	TR360/c	TR333/f	TR321/c	TR310/o
TR401/c	TR354/c	TR332/c	TR320/f	TR310/o
TR396/c	TR353/f	TR330/c	TR319/c	TR309/f
TR392/o	TR352/g	TR329/i	TR318/c	TR308/c
TR391/c	TR347/c	TR328/g	TR317/g	

i = inhalation studies, c = corn oil gavage studies, g = water gavage studies, f = feed studies, o = other, miscellaneous studies, TR = National Toxicology Program (NTP) Technical Report of an NTP bioassay (e.g., National Toxicology Program 1983). Evaluations were made using the untreated control animals in these studies. Repeated study numbers indicate that the study contained multiple controls and was used more than once. The reports are available from NTP Central Data Management, NIEHS, P.O. Box 12233, MD A0-01, Research Triangle Park, NC 27709.

Table 2. Bioassays in Fischer 344/N rats used in these analyses

Study/Type				
TR419/f	TR391/c	TR342/c	TR327/c	TR313/f
TR412/f	TR387/f	TR341/f	TR323/c	TR313/f
TR407/f	TR366/g	TR339/c	TR322/f	TR312/c
TR406/c	TR365/f	TR337/f	TR321/c	TR311/i
TR405/o	TR363/i	TR336/c	TR320/f	TR309/f
TR403/g	TR362/o	TR334/c	TR319/c	TR308/c
TR401/c	TR360/c	TR333/f	TR318/c	TR307/f
TR399/c	TR354/c	TR332/c	TR317/g	TR306/i
TR397/o	TR353/f	TR330/c	TR316/c	TR305/c
TR396/g	TR352/g	TR329/i	TR315/f	
TR392/o	TR347/c	TR328/g	TR314/i	

i = inhalation studies, c = corn oil gavage studies, g = water gavage studies, f = feed studies, o = other, miscellaneous studies, TR = National Toxicology Program (NTP) Technical Report of an NTP bioassay (e.g., National Toxicology Program 1983). Evaluations were made using the untreated control animals in these studies. Repeated study numbers indicate that the study contained multiple controls and was used more than once. The reports are available from NTP Central Data Management, NIEHS, P.O. Box 12233, MD A0-01, Research Triangle Park, NC 27709.

of the animals on test (usually 50 or 60 animals) at the week on test closest to the target time on test. Tumor incidences were obtained from the adjusted rate (i.e., using the Kaplan-Meier adjustment for intercurrent mortality [Kaplan and Meier 1958]) in the statistical analysis of primary neoplasms reported in each bioassay. Also reported in each bioassay are detailed methods of how tissues are evaluated for tumors. Although all common tumors (>10% incidence on average) were considered (i.e., total liver, lung, and lymphatic tumors in male mice), the major focus here will

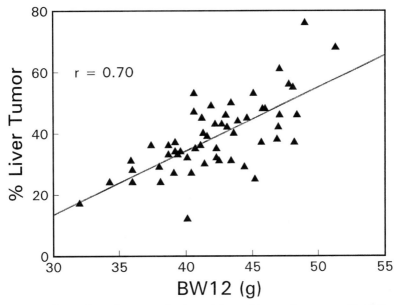

Figure 2. Relationship of average body weight at 12 months on test (BW12) and total liver tumor incidence at 24 months on test for control male B6C3F1 mice. Data are from the studies listed in Table 1; r is the correlation coefficient.

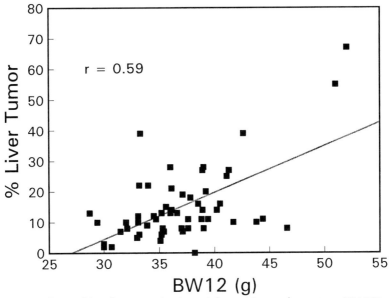

Figure 3. Relationship of average body weight at 12 months on test (BW12) and total liver tumor incidence at 24 months on test for control female B6C3F1 mice. Data are from studies listed in Table 1; r is the correlation coefficient.

be on total liver tumor incidence. In addition, BW at 13 weeks (BW13W) on test in the chronic portion of the NTP study was estimated for comparison to the BW for the control animals at the end of the relevant subchronic test, i.e, BW at 90 days (BW90D).

Results

Variability and Liver Tumors

The best BW biomarker for liver tumor incidence in male mice is BW12 (Figure 2), with less correlation for BW6 and BWAV, and no correlation for BW2 (Turturro and Hart 1994). BW12 has a correlation coefficient of 0.70 for male mice (Figure 2) and 0.59 for female mice (Figure 3). There is a wide, almost twofold variation in BW12 under the conditions of the bioassay. Because this indicates significant variability in growth curves in different bioassays, and because growth has such an impact on survival and tumorigenesis, these data suggest that the lack of control of this factor, in large part as a result of AL feeding, is significantly contributing to variability in the bioassay.

Study Type

The NTP database consists of different types of studies based on route of administration (e.g., inhalation, feed, corn oil gavage, etc.). Each of these study types treats control animals differently. These differing protocols result in differences in average tumor incidences for mice (Table 3) and rats (Table 4). Some of these differences have been characterized previously (Rao et al. 1990, Haseman et al. 1994). Separating studies as to type results in improved correlations of BW to tumor outcome. For instance, in inhalation studies, variability in BW12 accounts for up to 85% of the variability in liver tumor incidence in male mice (Figure 4), and up to 70% in female mice (Figure 5). Similar results with different levels of correlation occur for other tumor types in both mice and rats, with the best BW biomarker depending on tumor type, sex, etc. (e.g., BW2 for survival in male mice, BW2 for leukemia in rats, etc.).

Single and Group Housing

Some the clearest relationships seen between BW biomarkers and tumor incidences in mice are found in inhalation studies. One characteristic of inhalation studies is that the animals are almost always singly housed in inhalation chambers, so that any dietary modulation of BW and spontaneous tumorigenesis is not confounded by social interactions among the

Table 3. Average body weight at 12 months on test (BW12) and incidences of tumors for different study types for B6C3F1 mice

		i	c	g	f
Males	BW12 (g)	40.1	44.0	43.1	40.9
		Tumor incidence (%)[a]			
	Liver tumor	39.6	41.4	38.8	33.2
	Lymphoma	8.7	16.7	15.6	10.0
	Pituitary tumor[b]	0.5	1.1	2.6	0.3
Females	BW12 (g)	34.6	36.2	38.0	38.1
		Tumor incidence (%)[a]			
	Liver tumor	18.6	12.2	10.6	15.5
	Lymphoma	21.1	39.1	43.6	37.2
	Pituitary tumor[b]	23.2	33.3	23.6	20.0

i = inhalation studies (n = 8), c = corn oil gavage studies (n = 21), g = water gavage studies (n = 5), f = feed studies (n = 17). Data are from studies listed in Table 1.
[a]Total incidences after 24 months on test.
[b]Anterior pituitary tumor incidence.

Table 4. Average body weight at 12 months on test (BW12) and incidences of tumors for different study types for Fischer 344/N rats

		i	c	g	f
Males	BW12 (g)	460	486	470	460
		Tumor incidence (%)[a]			
	Mammary tumor	1.8	7.1	12.0	7.6
	Pancreatic acinar tumor	0	10.6	1.2	2.2
	Pancreatic islet adenoma	14.4	11.0	7.3	9.2
	Anterior pituitary tumor	63.2	43.8	59.5	38.8
	Mononuclear leukemia cell	65.8	27.1	59.5	56.7
Females	BW12 (g)	282	269	277	274
		Tumor incidence (%)[a]			
	Mammary tumor	26.4	42.4	54.3	43.8
	Pancreatic islet adenoma	4.7	1.3	0	2.9
	Anterior pituitary tumor	69.4	58.0	57.6	60.5
	Mononuclear leukemia cell	48.2	28.6	33.0	28.7

i = inhalation studies (n = 5), c = corn oil gavage studies (n = 22), g = water gavage studies (n = 6), f = feed studies (n = 15). Data are from studies listed in Table 1.
[a]Total incidences after 24 months on test.

animals. Some feed and corn oil gavage studies also used singly housed males to prevent fighting. When compared to similar studies that used group housing, it appears that, similar to the effects of different study types, group housing resulted in an alteration in study parameters, especially survival and incidence of lymphoproliferative disease (Table 5). The ability of housing conditions to confound the relationship between BW and survival is illustrated in Figure 6. The clear inverse relationship of BW2 and

Dietary Restriction

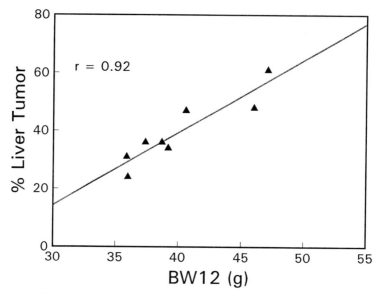

Figure 4. Relationship of average body weight at 12 months on test (BW12) and total liver tumor incidence at 24 months on test for inhalation study control male B6C3F1 mice. Data are from the mouse inhalation studies listed in Table 1; r is the correlation coefficient.

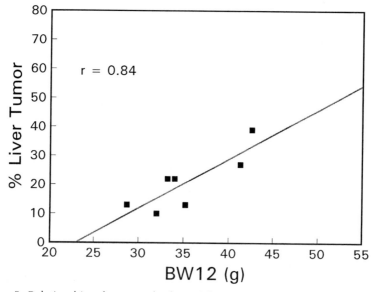

Figure 5. Relationship of average body weight at 12 months on test (BW12) and total liver tumor incidence at 24 months on test for inhalation study control female B6C3F1 mice. Data are from the mouse inhalation studies listed in Table 1; r is the correlation coefficient.

survival is obscured by studies in which the animals are group housed. In fact, there seems to be a positive relationship between BW2 and survival in group-housed animals. Although the mechanism for this is not known, one possibility is that an increased BW2 indicates that the dominant animals in a cage are larger and better able to survive the fighting that occurs with group-housed males. Similarly, if only studies that singly housed male mice are considered, variability in BWAV accounted for 87% of the variability of liver tumor incidence (Figure 7). This relationship is confounded when studies using group housing are also considered.

Table 5. Average survival and lymphoma incidence in single and group-housed male B6C3F1 mice[a]

	Single-housed	Group-housed
Survival %	64.6 ± 9	84 ± 10
Lymphoma incidence %	15 ± 5	7 ± 7

[a]Data are from same bioassays as in Figures 6 and 7, with $n = 9$ studies using singly housed and 12 studies using group-housed mice.

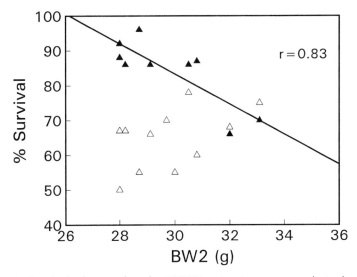

Figure 6. Survival of control male B6C3F1 mice in group- and singly housed conditions related to body weight at 2 months on test (BW2). Correlation coefficient (r) and regression line are for singly housed males (filled triangles). The group-housed animals (open triangles) are very different and appear to have opposite slope. Data are from noninhalation studies using single animal housing, i.e., reports TR419, TR407, TR406, TR396, TR392, TR387, TR385, TR363, and TR309 (Table 1), and are compared to data from similar studies using group-housed animals, i.e., reports TR410, TR401, TR391, TR365, TR360, TR354, TR353, TR341, TR327, TR323, TR321, and TR319 (Table 1).

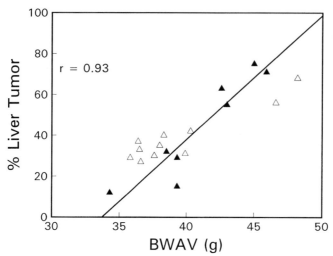

Figure 7. Liver tumor incidence in control male B6C3F1 mice in group- and singly housed conditions related to BWAV. Correlation coefficient (r) and regression line are for singly housed males (filled triangles). The group-housed animals (open triangles) are very different and appear to have opposite slope. Data are from noninhalation studies using single animal housing, i.e., reports TR419, TR407, TR406, TR396, TR392, TR387, TR385, TR363, and TR309 (Table 1), and are compared to data from similar studies using group-housed animals, i.e., reports TR410, TR401, TR391, TR365, TR360, TR354, TR353, TR341, TR327, TR323, TR321, and TR319 (Table 1).

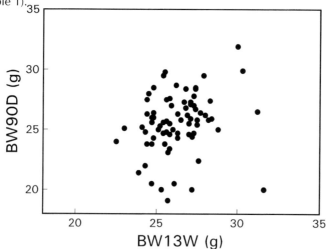

Figure 8. Correlation of body weight in control female B6C3F1 mice at the end of a subchronic study and the body weight at the same time on test during a chronic study. BW90D is the body weight at the end of a "90-day" subchronic study, BW13W is the body weight at the equivalent time (usually 13 weeks) in the chronic bioassay. There is no correlation significant at the $p < 0.05$ level. Data are from studies listed in Table 1.

Subchronic and Chronic Assays

Subchronic tests are used in almost all bioassays to determine which doses are used in the chronic studies. As shown for female mice (Figure 8), there is no significant ($p < 0.05$) correlation between the BW of animals at the end of a subchronic study (BW90D) and the BW of animals at an equivalent age in the chronic study upon which it was based (BW13W). Additionally, chronic experiments that use singly housed animals often have their dose levels determined by subchronic experiments that use group-housed animals. For instance, all of the noninhalation experiments illustrated in Figure 6 that had subchronic studies used group-housed mice. Since, as noted above, agent toxicity can be modulated by the animal growth curves, and housing conditions can modify the effect of growth curves, these data suggest that the dose levels determined for some chronic experiments may have been confounded by these parameters and that control of BW and housing conditions in a subchronic study is necessary for appropriate chronic study dose selection.

Intra-experiment Variability

It should be emphasized that the BWs used above are averages of 50–60 animals. In addition to inter-experiment variability, there is intra-experiment variability. This is illustrated in Figure 9, which shows the variability of BW12 for male mice in a typical chronic bioassay at the National Center for Toxicological Research (NCTR), which uses AL feeding and group housing (National Center for Toxicological Research 1991). Also illustrated in this figure is the distribution of BW12 in the same laboratory using dietary control (DC) (i.e., feeding the same amount of food to every animal) and single housing. In this experiment, the level to which DC is maintained results in a lower mean BW12, and the mean is adjusted to have the same mean as the AL study to assist in comparison.

Conclusions

Variability in growth curves and tumor incidences of common tumors in the NTP bioassays are quite significant. The variability in tumor incidences makes the historical tumor database almost useless. However, if the growth curves are used properly, the range is much smaller. For instance, using Figure 4, the range at BW12 of 40 g is approximately 36–40% of liver tumor incidence for inhalation studies, rather than the 20–60% range that exists for the entire BW range. Thus, a modified BW-adjusted historical tumor database can be used to evaluate whether a study is in the historical range or whether something has occurred in a particular experiment that affects the confidence in interpreting the outcome.

The effects of the variability in growth curves on the estimation of toxicity are complex. An important assumption in using linear low-dose extrapolation to estimate risk is that a toxicant interacts with some mechanism important for spontaneous tumorigenesis. If this is true, and the mechanisms of spontaneous tumorigenesis are influenced in some complicated manner by BW, there may be a complicated, nonlinear response of induced tumorigenesis to BW and growth curve variability. This is consistent with the data of Kritschevsky and Klurfield (1987), who found that DR inhibited the ability of 7,12 dimethylbenz(a)anthracene to induce mammary tumors in SD rats in a fairly linear fashion until a certain level (20% less than AL) was reached; any greater restriction totally inhibited this toxic action. Growth may act as a promoter of toxicant-induced disease, which would vary with different rates of growth.

In addition, a number of compounds alter growth rates, sometimes significantly. From the data shown, small differences in growth biomarkers could suggest large differences in predicted spontaneous tumor incidence. This impacts directly on risk estimation because the spontaneous tumor incidences are used to determine the number of excess tumors induced by an agent. Although problematic because of the potential nonlinear relationship between growth curves and induced toxicity noted above, adjusting the expected tumor incidence for common tumors to account for any toxicant-induced retardation or acceleration of growth (e.g., if a toxicant increases BW_{12}, the appropriate control may be animals with elevated body weight and a higher liver tumor incidence than in the concurrent control) may help interpret the effects of toxicants, especially the lowering of the incidences of observed common tumors by agents expected to be toxic.

Also, as noted above, survival is decreasing in F344 rats, which is causing concern. Survival in SD rats is so decreased that the ability of a study to detect a toxicant effect because of early death and evident pathology may be compromised (Keenan and Soper, this volume).

Thus, as a result of the effects of differential growth, it appears that chronic toxicity tests may have differential sensitivity to toxicants (e.g., if animals in an experiment have slow growth, it may be much more difficult to elicit tumors than if there is rapid growth), leading to an inconsistent test. Toxicant-induced changes in growth curves complicate the interpretation of effects; in addition, the presence of significant levels of spontaneous disease resulting from a rapid growth rate may decrease the ability to detect an agent-induced adverse effect.

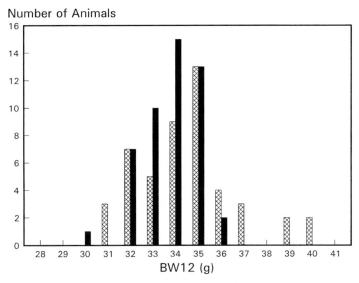

Figure 9. Distribution of body weights at 12 months on test (BW12) in a dietary controlled (DC) study using singly housed mice (solid bars) (Sheldon et al., in press) and a typical chronic study using ad-libitum-fed, group-housed mice (hatched bars) (NCTR 1991). Each experiment employed approximately 50 animals per group. Mean of the DC study was adjusted to coincide with the mean of the other study to assist comparison. Note the high percentage of high weight males in the typical chronic study.

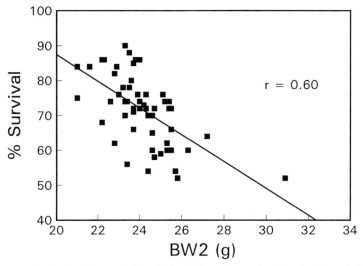

Figure 10. Survival of control female B6C3F1 mice related to body weight at 2 months on test (BW2). Data are from studies listed in Table 1; r is the correlation coefficient.

It appears that, in order to effect consistent testing protocols, to avoid the complications of agent-induced growth effects, and to keep survival and pathology within reasonable limits, there is a need to define a target growth curve for each study. As important as this is for the most controlled chronic bioassays (the NTP studies), it is even more important for the many bioassays done under less controlled conditions. These have been found to have at least similar, if not greater, variability than the NTP studies (Keenan and Soper, this volume). This growth curve would most simply be generated by controlling diet in singly housed animals.

With regard to criteria for this default curve, there should be a reasonable background incidence of common tumors so that agents can interact with spontaneous processes. However, the level should not be so high that it confounds the observation either by low survival or high disease rates. After deciding what background tumor incidences are desirable, one can estimate BW2, BW6, and BW12 based on the relationships demonstrated above (e.g., in Figure 4). The following constructions of default curves usually use the study type with the best correlation between BW and tumor incidence (or survival) on the premise that the relationship exists in all of the experiments but is obscured by other, uncontrolled factors.

For male mice, using Figure 4 (because inhalation studies use singly house animals), a desired background liver tumor incidence of 20% is equivalent to a BW12 of 35 g. Similarly, using Figure 6, 80% survival in males, using the singly housed animal data, results in a BW2 of 30 g. For female mice, using Figure 5, a liver tumor incidence of 20% results in a BW12 of 35 g, while a BW2 of 23 g will result in a survival rate of 80% (Figure 10).

In male rats, a leukemia incidence of 45% indicates the appropriate BW2 (Turturro and Hart 1994), while a survival rate of 75% indicates the BW12 (Figure 11). In females, a pituitary tumor incidence of 35% determines the BW6 (Figure 12), while survival of 75% determines the BW12 (Figure 13). The results of interpolating intermediate points from the patterns of normal growth curves are shown in Figures 14 and 15. These growth curves also correlate with a lymphoma rate of 10% in female mice, mammary tumor and leukemia rates of 20% each in female rats, and pituitary tumor incidence of 30% in male rats (data not shown).

These are suggestions for default growth curves; their applicability should be evaluated on a case-by-case basis.

Since dietary modulation has been shown to increase survival (Turturro and Hart, 1991) and resistance to a number of toxicants (Duffy et al., this volume), it is important that the preliminary studies that define the doses

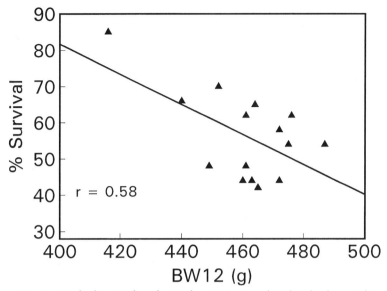

Figure 11. Survival of control male Fischer 344 rats related to body weight at 12 months on test (BW12). Data are from the feed studies listed in Table 2; r is the correlation coefficient.

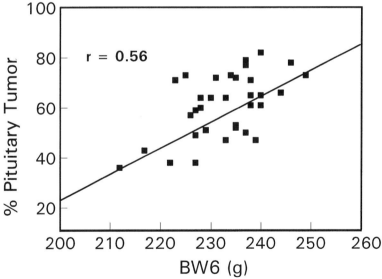

Figure 12. Relationship of average body weight (in grams) at 6 months on test (BW6) and total pituitary tumor incidence at 24 months on test for non-corn oil gavage studies in female Fischer 344 rats. Data are from the studies listed in Table 2, excluding the corn oil gavage studies, which have a greater propensity toward lower incidences of pituitary tumor than the other studies (see Table 3); r is the correlation coefficient.

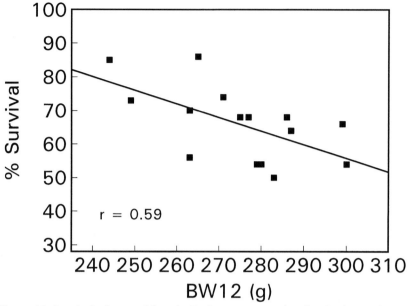

Figure 13. Survival of control female Fischer 344 rats related to body weight at 12 months on test (BW12). Data are from the feed studies listed in Table 2; *r* is the correlation coefficient.

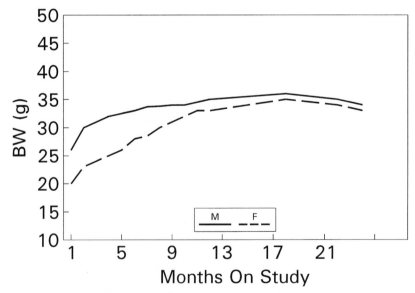

Figure 14. Proposed growth curve for male (solid) and female (dashed) B6C3F1 mice.

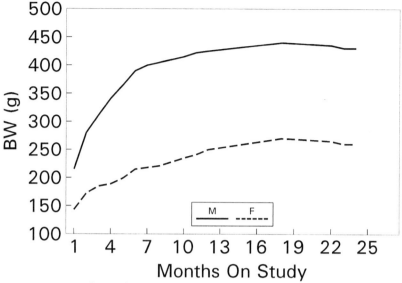

Figure 15. Proposed growth curve for male (solid) and female (dashed) Fischer 344/N rats.

for the chronic studies be performed using the same growth curves. With a decreased background tumor incidence, increased dose, and longer life span, it is probable that use of DC will increase the sensitivity of the bioassay as well as greatly reduce uncontrolled growth variability, leading to a consistent, reproducible test.

Acknowledgments

The authors wish to acknowledge the generous support of the National Institute of Aging/National Center for Toxicological Research Interagency Agreement on the Project on Caloric Restriction for this effort.

References

Albanes D (1987) Caloric intake, body weight and cancer. Food Res Inter 26:289

Allaben W, Chou M, Pegram R, et al. (1990) Modulation of toxicity and carcinogenicity by caloric restriction. Korean J Toxicol 6:167

Berg B, Simms H (1960) Nutrition and Longevity in the rat. 3: longevity and onset of disease with different levels of food intake. J Nutr 71:255

Carroll KK (1975) Experimental evidence of dietary factors and hormonal-dependent cancer. Cancer Res 35:3374

Haseman J, Bourbina J, Eustis S (1994) Effect of individual housing and other experimental design factors on tumor incidence in B6C3F1 mice. Fundam Appl Toxicol 23:44

Holloszy J, Smith E (1986) Longevity of cold-exposed rats: A re-evaluation of the "rate-of-living" theory. J Appl Physiol 61:1656

Interagency Staff Group (1986) Chemical carcinogens: a review of the science and associated principles. Environ Health Perspect 67:201

Kaplan EL, Meier P (1958) Nonparametric estimation from incomplete observations. J Amer Stat Assoc 53:457

Kritchevsky D, Klurfield D (1987) Caloric effects in experimental mammary tumorigenesis. Amer J Clin Nutr 45:236

McCay C, Crowell M, Maynard L (1935) The effect of retarded growth upon the length of lifespan and upon the ultimate body size. J Nutr 10:63

McCay C, Sperling G, Barnes L (1943) Growth, ageing, chronic diseases and lifespan in rats. Arch Biochem Biophys 2:469

National Center for Toxicological Research (1991) Pyrilamine: 104 week chronic dose study in rats and 104 week chronic dose study in mice. NCTR Technical Report 408 and 409. National Center for Toxicological Research, Jefferson, AR

NTP (National Toxicology Program) (1983) NTP Technical Report on the carcinogenesis bioassay of L-ascorbic acid (Vitamin C) (CAS No. 50-81-7) in F344/N rats and B6C3F1 mice (feed study). National Institutes of Health, Bethesda, MD

Rao G, Haseman J, et al. (1990) Growth, body weight, survival and tumor trends in F344/N rats during an eleven-year period. Toxicol Pathol 18:61

Roe FJC (1994) Historical histopathological control data for laboratory rodents: Valuable treasure or worthless trash? Lab Anim 28:148

Rose MR (1991) The evolutionary biology of aging. Oxford University Press, Oxford

Ross M (1976) Nutrition and longevity in experimental animals. In Winick M (ed), Nutrition and Aging. J. Wiley and Sons, New York, pp 23–41.

Sheldon W, Bucci T, Hart R, Turturro A (in press) Age-related neoplasia in a lifetime study of ad libitum and food restricted B6C3F1 mice. Toxicol Pathol

Thurman JD, Bucci T, Hart R, Turturro A (1994) Survival, body weight, and spontaneous neoplasms in ad libitum-fed and food-restricted Fischer 344 rats. Toxicol Pathol 22:1

Turturro A, Blank K, Murasko D, Hart R (1994) Mechanisms of caloric restriction effecting aging and disease. Ann N Y Acad Sci 719:159

Turturro A, Duffy P, Hart R (1993) Modulation of toxicity by diet and dietary macronutrient restriction. Mutat Res 295:151

Turturro A, Hart R (1987) Quantifying risk and accuracy in risk assessment: the process and its role in risk management problem solving. Med Oncol Tumor Pharmacother 4:125

Turturro A, Hart R (1991) Longevity-assurance mechanisms and caloric restriction. Ann N Y Acad Sci 621:363

Turturro A, Hart R (1992) Dietary alteration in the rate of cancer and aging. Exp Gerontol 27:583

Turturro A, Hart R (1994) Modulation of toxicity by diet: implications for response at low level exposures. In Calabrese E (ed), Biological effects of low level exposures: dose-response relationships. Lewis Publishers, Boca Raton, FL, pp 143–152

Weindruch R, Walford R (1988) Retardation of aging and disease by dietary restriction. C.C. Thomas, Springfield, IL

The Effects of Ad Libitum Overfeeding and Moderate Dietary Restriction on Sprague-Dawley Rat Survival, Spontaneous Carcinogenesis, Chronic Disease, and the Toxicologic Response to Pharmaceuticals

Kevin P. Keenan and Keith A. Soper
Merck Research Laboratories

Introduction

Laboratory rat survival in 2-year carcinogenicity studies has been declining over the past three decades throughout the pharmaceutical and chemical industry (Lang 1989, 1990, 1991, Burek 1990, Rao et al. 1990, Roe 1991, Roe et al. 1991, Jordan 1992, Keenan et al. 1992, 1995b, McMartin et al. 1992, Nohynek et al. 1993). This decline has been seen in all rat strains, including the Sprague-Dawley (SD) and the Fischer 344 (F344) rats, the most commonly used rats in toxicity and carcinogenicity studies (Lang 1990, Rao et al. 1990, Haseman and Rao 1992). This decline in survival has caused some regulatory agencies to question the adequacy of exposure of rats on carcinogenicity studies that result in less than 50% survival, i.e., 25 animals alive per group, at the end of the 2-year period (Burek 1990, Lang 1991, Jordan 1992).

Dietary or food restriction is a well-established method of extending the life span of rodents. It has been known for several decades that the common practice of unlimited ad libitum (AL) feeding of nutritionally rich, high-energy diets is overfeeding and has many negative effects on physiologic and toxicologic endpoints and results in poor survival

Data and information presented in this review chapter have been published as cited, presented as meeting abstracts or posters, or have been submitted or are being prepared for journal publication.

compared with the beneficial effects of moderate caloric restriction (Weindruch and Walford 1988, Finch 1990, Fishbein 1991, Masoro 1991a,b, Masoro and McCarter 1991, Masora et al. 1991, Roe 1991, Roe et al. 1991, Keenan et al. 1992, 1994c,d, Yu 1993). The beneficial effects of dietary (caloric) restriction have been documented in studies of aging and senescence in invertebrates, rodents, and nonrodent vertebrates such as fish, birds, and other mammals, including humans (Weindruch and Walford 1988, Finch 1990, Fishbein 1991, Masoro 1991a, Masoro and McCarter 1991). The most successful caloric restriction regimens provide essential nutrients at adequate amounts but restrict caloric intake to 30–70% below AL food consumption levels. The beneficial effects on longevity appear to depend primarily on caloric restriction, since the specific restriction of fat, protein, minerals, or other nutritional components without caloric restriction does not increase the overall long-term survival or maximum species-specific life span. A chronic 30–40% restriction of energy intake without essential nutrient deficiency lowers the incidence and/or delays the onset of most spontaneous and induced tumors, reduces the severity and/or onset of most spontaneous degenerative diseases such as nephropathy and cardiomyopathy of rats, and extends the average and maximal life span of rodents (Weindruch and Walford 1988, Finch 1990, Fishbein 1991, Masoro 1991a,b, Masoro and McCarter 1991, Masora et al. 1991, Roe 1991, Roe et al. 1991, Keenan et al. 1992, 1994c,d, Yu 1993).

On the basis of the National Cancer Institute's 1976 recommendations and the National Toxicology Program's (NTP's) 1984 recommendations, an international consensus developed that the duration of a rat carcinogenicity study should be 24 months. The Food and Drug Administration's Redbook (1982) states that the group size should be 50 rats/sex/group and that at least 25 of these rats, or 50%, should survive until 24 months. This guideline is widely accepted internationally. The SD rat is the principal stock used by the pharmaceutical industry in the United States. However, rat survival has been declining in all commercially available rat stocks and strains, including the relatively long-lived F344 rat (Lang 1990, Rao et al. 1990). The NTP has reported F344 rat 2-year survival as low as 30% in males and 50% in females (Dr. G. A. Boorman, personal communication).

Although both genetic and environmental factors are involved, rat survival can be improved by simple caloric restriction (Weindruch and Walford 1988, Finch 1990, Fishbein 1991, Masoro 1991a,b, Masoro and McCarter 1991, Masora et al. 1991, Roe 1991, Roe et al. 1991, Keenan et al. 1992, 1994c,d, Yu 1993). Our laboratory has undertaken long-term studies of dietary restriction (DR) with the SD rat and has shown that this method will improve survival, lower the incidence and/or severity of chronic diseases associated with AL overfeeding, and thereby improve the rat as a model to test human safety of candidate pharmaceuticals. Our laboratory has also studied a modified diet with lowered protein, fat, and energy content and increased fiber content and found no survival benefit if this diet was fed AL (Keenan et al. 1992, 1994a,b, 1995a,b). These data indicate that SD rats should be maintained by DR (caloric restriction) by providing approximately 60–65% of true AL food consumption per day of a standardized diet (Keenan et al. 1992, 1994a,b,c,d).

The need for a controlled feeding method is further illustrated when we examine the relationship between 104-week survival and daily food consumption data from other laboratories using the same SD rat stock (Sprague-Dawley, Crl:CD®(SD)BR, Charles River Laboratories), single-housed and fed the same feed (Purina Certified Rodent Chow 5002). Data on 58 SD control groups from 2-year carcinogenicity studies that had been started during the 1980s indicated considerable variability in survival and food consumption between laboratories and within the same laboratory. Survival ranged from 7.0% to 73.0% for males and 29.0% to 68.0% for females. Daily food consumption (g/day) ranged from 21.7–32.3 g/day for males and 16.1–24.9 g/day for females (Keenan et al. 1994a,b). This variability in food consumption is likely due to differences in feeders, caging and husbandry methods used in different laboratories. It is apparent from these data that AL food consumption varies widely from laboratory to laboratory. Therefore, any consideration of modification of AL food consumption by a given percentage must be understood in the context that AL varies greatly depending on specific laboratory conditions, even with the same stock of rat and the same feed. For example, in the studies from our laboratory restricted SD male rats were given approximately 21.5 g of food per day, which is a 35% reduction of our true AL food consumption level for SD males of 33 g/day. In a laboratory where SD male rat AL food consumption is 23 g/day, this amount is less than a 10% DR.

The data from these 58 studies show a statistically significant relationship ($p < 0.001$) between the percentage survival and the average food consumption in Charles River SD rats. For male SD rats, the correlation by linear regression (Snedecor and Cochran 1989) between the percentage survival and the average food consumption was $r = -0.48$; for females, the correlation was $r = -0.47$ (Keenan et al. 1994a). The greater the food consumption per day, the greater the likelihood of poor survival by 104 weeks (Keenan et al. 1994a,b). These data show that food consumption and survival are tightly correlated but quite variable from laboratory to laboratory. Therefore, it is the amount of feed provided per day that appears critical in the long-term survival and health of the animals rather than a given percentage (20–40%) of the extremely variable AL levels reported by different laboratories using the same SD rat stock and identical diets. It should be noted that only rodents (rats and mice) are maintained by AL feeding. Other laboratory animals (i.e., dogs and primates) are fed measured amounts of feed, and it is considered poor veterinary practice to do otherwise.

The scientific motivation for requiring 50% survival at the end of a rat carcinogenicity study is twofold. First, a substantial number of rats should be exposed to the test compound for the full 2 years of the study. Second, the ability of the study to detect a treatment effect is a working definition of adequate statistical sensitivity or power of the bioassay. For example, a decrease in survival from 50% to 20% causes a decrease in statistical sensitivity, especially for late-occurring tumors. We estimate that increasing the total sample size from 50 to 75 rats/sex/group offsets some of this decrease for most tumor types if survival does not decline further, but at a cost of a nearly 50% increase in animal use, work load, manpower, and time to complete the bioassay (Keenan et al. 1995b). Trying to solve the problem of declining survival and the potential loss of statistical sensitivity by simply increasing group size has disadvantages in addition to the increased time and expense of the study. Increasing the group size to 75 rats/sex/group will not offset the loss of statistical power if the 24-month survival declines to 10% or less. Also, problems from treatment-related mortality are exacerbated in a study completed with a low control survival. The preferable solution to the loss of statistical power is to increase the 2-year survival to nearly or above 50%, since this increases the total time the animals are exposed to the test compound and increases the sensitivity of the bioassay to distinguish a true treatment effect from concurrent controls.

The Effects of Ad Libitum Overfeeding and Moderate Dietary Restriction on Rat Carcinogenesis and Age-related Diseases

A 106-week carcinogenicity study in SD rats was conducted to determine the effects of different diets and feeding regimens on survival and on the incidence and severity of spontaneous degenerative disease and tumors in SD rats. A 52-week interim necropsy was performed where approximately 10 rats/sex/group were sampled. The original group size was 70 rats/sex/group for each of five dietary regimens. The rats were 36 days of age at the initiation of the study with the males weighing 115–175 g and the females weighing 91–156 g. These SD rats (Crl:CD® (SD) BR) were obtained from Charles River Laboratories, Inc., Raleigh, NC, and were housed in individual stainless steel wire cages in environmentally controlled, clean air rooms with a 12-hour light-dark cycle.

The experimental groups were designed to compare two different diets obtained from Dr. G.C. Ballam, Purina Mills, Inc., St. Louis, MO as well as moderate DR. The diets and the dietary restrictions were as follows:

a) Purina 5002 Certified Rodent Chow fed AL (5002 AL) as pellets. (This diet contains approximately 21.4% protein, 5.7% fat, and 4.1% crude fiber and has a calculated metabolizable energy value of 3.07 kcal/g.)

b) Purina 5002 Certified Rodent Chow fed AL for approximately 6.5 hours per day (5002 6.5h DR) during the light cycle.

c) Purina 5002 Certified Rodent Chow given in measured amounts (5002 DR) at approximately 65% of adult AL food consumption (approximately 16 g/day for females and 21.5 g/day for males).

d) Purina 5002-9 Certified diet fed AL (5002-9 AL) as extruded pellets. (This diet contains approximately 13.6% protein, 4.6% fat, and 15.7% crude fiber and has a calculated metabolizable energy value of 2.36 kcal/g.)

e) Purina 5002-9 Certified DR (5002-9 DR) as a measured amount to provide approximately the same caloric intake as the animals fed regimen c (approximately 20.8 g/day for females and 28.8 g/day for males).

All rats underwent complete necropsy examinations. The rats that survived until the terminal necropsy were weighed, deeply anesthetized by ether inhalation, and killed by exsanguination. Numerous tissues were collected from all rats, including all gross lesions. Routine histologic sections from paraplast-embedded tissues were stained with hematoxylin and eosin for microscopic examination of all animals. In addition, 5-bromo-2'-deoxyuridine (BrdU) immunohistochemistry was used to stain selected target tissues from approximately 10 rats/sex/group that had been surgically implanted 1 week prior to necropsy with osmotic minipumps for the continuous delivery of BrdU for cell proliferation and stereology studies at the 52-week interim and terminal necropsy (Frank et al. 1993, Keenan et al. 1994a, 1995a).

Body Weight and Food Consumption

Anticipated changes in the rate of body weight gain associated with the restriction of caloric intake included significant decrements in all groups relative to the 5002 AL group. These changes were present after the first week, and between approximately weeks 1 and 10, the rate of change in weight gain was most dramatic. After 10 weeks, body weight gain began to level off.

During the second year several other changes were apparent. In females, the 5002 AL animals continued to gain weight, whereas weight gain in the 5002-9 AL group remained at the same level from 1 year on. In addition, although food consumption (and therefore caloric intake) was similar, females in the 5002 6.5h DR group appeared to continue to gain weight, whereas those animals in the 5002 DR group essentially reached a plateau after 25 weeks, which was maintained for the study duration. In males, the weight relationships among the various groups that were present after 52 weeks remained essentially the same for the remainder of the study.

Mean daily food consumption over 2 years is given in Table 1. All mean values were within 10% of the desired targets, which were determined by calculation from the median food consumption values of AL SD rats from this laboratory between 10 and 17 weeks of age. These AL target values (true AL food consumption) for the 5002 diet fed AL were 24.6 g/day for females and 33 g/day for males.

When the measured food consumption was calculated on the basis of grams of feed consumed per gram body weight, it was apparent that the amount of a given food (5002 or 5002-9) a rat consumed per gram of body weight was remarkably similar for the AL and DR groups. Indeed, after correction for food wastage, the actual kilocalories consumed per day was determined, revealing that the AL and DR groups had essentially the same caloric intake per gram body weight per day. In fact, AL rats fed either diet consumed the greatest number of grams and kilocalories per rat, but the same grams and kilocalories per gram body weight as rats restricted to 65–70% of AL food consumption (Table 1).

It was clear that early body weight gain and AL feeding were tightly correlated with survival and other biomarkers for aging as discussed below. Development had not been retarded by dietary (caloric) restriction, because brain weights measured at the 52-week interim and terminal necropsies were not different between the AL and DR rats, but body weights and other organ weights were generally higher in the AL animals, as was their percentage body fat by carcass analysis (Keenan et al. 1992, 1994b,c,d,). Organ weights were analyzed by the Tukey trend test (Tukey et al. 1985).

These data rule out hypotheses that food restriction acts by reducing the intake of calories or other nutrients per unit body mass, since the amount of grams or kilocalories consumed per gram body weight is very similar

Table 1. Sprague-Dawley rat food consumption and caloric intake over 2 years (weeks 1–103)

	Mean food consumption (g/day)	Per rat		Per gram body weight[a]	
		g/day[b]	kcal/day[c]	g/day	kcal/day
Females					
5002 AL	24.2	24.3	76.3	0.038	0.119
5002 6.5h DR	15.9	18.7	58.7	0.048	0.150
5002 DR	16.2	14.6	45.8	0.046	0.144
5002-9 AL	35.1	25.5	58.9	0.054	0.125
5002-9 DR	21.3	17.7	40.9	0.059	0.136
Males					
5002 AL	32.0	27.5	86.4	0.033	0.103
5002 6.5h DR	20.2	21.6	67.8	0.034	0.107
5002 DR	22.5	18.9	59.3	0.035	0.109
5002-9 AL	46.5	35.8	82.7	0.046	0.106
5002-9 DR	28.6	25.5	58.9	0.049	0.113

AL = ad libitum, DR = diet-restricted.
[a]Calculated from week 103 average food consumption (g/day) and average body weights.
[b]Week 103 food consumption (g/day) corrected for wastage (5002, 12%; 5002-9, 14%).
[c]5002 diet has 3.07 kcal/g and 5002-9 diet has 2.36 kcal/g metabolizable energy.

between the AL and DR groups. For a given feed, it appears that the rat's body mass adjusts to the reduced food intake in such a way that there is no decrease in caloric or other nutrient intake (grams or kilocalories) per unit body mass. This suggests that the conversion of calories to body mass is similar under both conditions and that metabolic rate, oxygen consumption and food utilization may be similar (Masoro 1991a,b, Masoro and McCarter 1991, Masoro et al. 1991, Yu 1993). The conclusion to be drawn from this observation is that the reduction in nutrient intake per rat rather than per-unit of body mass is the maneuver that prevents excessive growth, early body weight gain, early onset of spontaneous degenerative disease and tumors, and poor survival (Masoro 1991a, Keenan et al. 1992, 1994a,b, 1995a,b, Gumprecht et al. 1993).

Clinical Pathology and Tissue Biochemistry

Slight to moderately lower mean leukocyte counts were noted in the DR females and males. Corresponding decrements were noted in lymphocyte and in segmented neutrophil and monocyte counts for these groups. In general, these lower leukocyte counts represent values closest to the normal range for younger rats of this stock and indicate the better general health of the older DR rats at 52 and 104 weeks (Keenan et al. 1994b).

Serum biochemistry findings after 52 and 104 weeks demonstrated few differences between the AL and DR groups. In general, all the DR rats had moderate decreases in serum cholesterol, and triglycerides, and the DR females' glucose decreased relative to the AL females at 52 weeks (Gumprecht et al. 1993, Keenan et al. 1994b).

Special tissue biochemistry performed on liver samples indicated that DR rats appear to be better able to withstand spontaneous oxidative injury than AL rats. At 52 weeks, the mean hepatic glutathione (GSH) content of males from all DR groups was clearly higher (approximately 50%) than that of livers from either AL group. Although a similar difference in GSH was not observed in females, the hepatic malondialdehyde (MDA) content in the DR females (all groups) was significantly lower (24–45% of the 5002 AL group for 5002 DR rats and 33% of the 5002-9 AL group for 5002-9 DR rats) than that of the AL females at 52 weeks (Keenan et al. 1994b). Interestingly, after 104 weeks no differences between the various dietary regimens were observed in liver GSH or MDA. However, in rats of either sex or dietary regimen (AL or DR), there was a general decline (40–60%) in hepatic GSH content and moderate decrements in MDA levels (Keenan et al. 1994b, 1995a). The differences noted at 1 year indicate that livers of DR SD rats are better able to defend against oxidative injury, as has been shown as well in F344 rats (Yu 1993).

Measurement of liver enzymes involved in biotransformation included both phase I and phase II systems. Phase I–associated enzyme activities included cytochrome P450s characterized with specific substrates, total cytochrome P450 content, and NADPH cytochrome c reductase activity. Phase II–associated enzymes included GSH-S-transferase and GSH-peroxidase (Keenan et al 1994b).

In these assays there were only minor alterations in the P450 enzyme profiles, which were characterized as slight increases in activity, none of which would be expected to have a significant impact on the metabolic profile of administered xenobiotics. The largest change in any P450 activity after 52 weeks was a three-fold induction of aniline hydroxylase (ethanol-inducible form P450 IIE) in the 5002-9 DR females, which may have been associated with the presence of high plasma ketone levels as reflected by the presence of a higher incidence of urinary ketones seen in the 5002-9 DR rats fed this high-fiber diet (Keenan et al 1994b).

Similarly, there were no changes in cytochrome c reductase activity or in the activities of either GSH-S-transferase or GSH-peroxidase that could be attributed to the various diets or modes of restriction. There were clear decrements in the levels of these activities in all groups as a function of time, with the largest decreases (approximately 30–50%) occurring in NADPH cytochrome c reductase between 1 and 2 years (Keenan et al. 1994b).

Table 2. Sprague-Dawley rat 104-week survival[a]

	%[b]	First week ($p < 0.05$[c])	Average weeks of survival
Females			
5002 AL	36	—	86.4
5002 6.5h DR	41**	—	88.3**
5002 DR	62*	65	98.3*
5002-9 AL	24**	—	89.5**
5002-9 DR	81*	69	98.3*
Males			
5002 AL	7	—	80.1
5002 6.5h DR	57*	57	95.7*
5002 DR	74*	51	99.6*
5002-9 AL	32*	91	86.6*
5002-9 DR	75*	51	99.6*

AL = ad libitum, DR = diet-restricted.
[a]From all causes of mortality except the 52-week interim necropsy.
[b]Percentage survival (Kaplan-Meier) to end of week 104 (Keenan et al. 1994a,d).
[c]First week at which the difference in survival between each treatment group and the 5002 AL groups became significant ($p < 0.05$) and remained significantly ($p < 0.05$) different.
*$p \leq 0.001$
**$p > 0.10$ compared with the 5002 AL groups.

Survival and Average Time on Study

There was no difference in mortality between any of the dietary groups during the first 52 weeks of this study. However, over the course of the second year of the study, very obvious differences in survival became apparent. In general, the best survival was seen in those rats fed by DR through measurement. The differences in survival were highly statistically significant between the AL and DR groups (with the exception of the 5002 6.5h DR females) based on a log-rank test, and clearly represent a dramatic increase in survival as the result of dietary restriction (Table 2).

The average survival for the 5002 AL rats was 80.1 weeks for males and 86.4 weeks for females. By comparison, the 5002 DR rats survived 99.6 weeks for males and 98.3 weeks for females. This comparison clearly shows that moderate DR of the same diet by 30–35% results in significant differences in both long-term survival and the average number of weeks the DR-fed rats were on study (Table 2). Under these study conditions, such effects would increase the animals' exposure to a test compound.

The most common cause of death in rats fed either diet under AL or DR conditions was pituitary tumors. This was followed by mammary gland tumors in females and renal and cardiovascular disease in the AL males. The largest number of deaths of undetermined cause was seen in the 5002 AL males. The only unusual cause of death was seen in the 5002-9 rats.

Six male and three female 5002-9 AL rats and one 5002-9 DR female died or were killed owing to the effects of the high fiber content of the 5002-9 diet, which induced colonic impaction, dilation, and chronic colitis. Otherwise, both sexes of SD rats fed either diet AL (5002 or 5002-9) had an earlier onset of, and more severe, lesions and tumors than their DR counterparts (Keenan et al. 1994a,b, 1995a,b).

Organ and Tissue Changes

Differences between the terminal body weights of the 5002 AL and DR rats reflect the changes in body weight gains observed during the course of the study. The AL animals were the heaviest, which appears to reflect body fat composition as observed at the 52-week interim necropsy and the terminal necropsy (Keenan et al. 1992, 1994a,b, 1995a). In contrast, no difference in absolute brain weight was seen between the AL and DR rats of both sexes; thus, the percentage of brain weight is the most appropriate relative comparison to make when evaluating organ weights between the groups (Keenan et al. 1992, 1994a,b, 1995a). Compared with the 5002 AL group, absolute and relative (percentage brain weight) weights of spleen, heart, kidneys, liver, adrenals, lungs, thyroids, ovaries, and pituitaries were generally smaller and frequently significantly so in the 5002 DR group of both sexes. In most cases the lower organ weights correlated with a decrease in lesion onset, incidence, and severity (Keenan et al 1992, 1994a,b, 1995a,b, Gumprecht et al. 1993).

The morphologic changes observed in all of the tissues examined in both the AL and DR groups reflect neoplastic, proliferative, and degenerative processes commonly seen in control SD rats during the course of a 2-year carcinogenicity study. The overall incidence of both benign and malignant neoplasia was remarkably similar between both groups and was within the range of incidence seen in historical SD rat controls in both our laboratory and the laboratories of others (Lang 1992, McMartin et al. 1992, Keenan et al. 1995b). However, the 5002 AL rats tended to have earlier onset and larger neoplasms, and these tumors were more likely to be contributing factors to early mortality. The 5002 DR groups, although having a similar incidence of neoplasms, had a later onset and tended to have smaller tumors that were found incidentally at necropsy rather than resulting in mortality. In contrast to the incidence trends in neoplastic disease, the AL groups clearly had a higher incidence and greater severity of renal disease and cardiovascular disease compared with their DR counterparts. These lesions frequently contributed to or were the cause of death in the

Table 3. Summary of pathology and mortality by 2 years

	5002 AL		5002 DR	
	Females	Males	Females	Males
Number necropsied[a]	58	57	60	60
Number with FD or ES[b]	38	53	24	16
Neoplasms[c]	58 (36)[d]	46 (23)	58 (23)	49 (13)
Malignant neoplasms[c]	15 (6)	14 (3)	22 (6)	15 (6)
Benign neoplasms[c]	56 (30)	45 (20)	57 (17)	40 (7)
Renal disease[c]	35 (1)	57 (13)	3 (0)	22 (0)
Heart disease[c]	13 (0)	31 (6)	2 (0)	11 (0)

AL = ad libitum, DR = diet-restricted.
[a]Except interim necropsy (week 52).
[b]Number with fatal disease (FD = found dead, ES = early sacrifice when moribund).
[c]Number with diagnosis.
[d]Number in parenthesis = number of FD or ES with neoplasm or disease condition as cause of death.

Table 4. Summary of pituitary pathology by 2 years

	5002 AL		5002 DR	
	Females	Males	Females	Males
Number examined	58	57	60	60
Adenoma[a]	47 (21)[b]	38 (20)[b]	47 (15)[b]	28 (7)[b]
Carcinoma[a]	1 (1)[b]	0 (0)[b]	3 (3)[b]	0 (0)[b]
Focal hyperplasia[a]	8 (0.34)[c]	11 (0.44)[c]	8 (0.23)[c]	16 (0.65)[c]

AL = ad libitum, DR = diet-restricted.
[a]Number with diagnosis.
[b]Number in parenthesis = number with fatal tumors (cause of death/reason killed).
[c]Number in parenthesis = average grade of hyperplasia (grades 0–5).

5002 AL males. A summary of these general trends for the 5002 AL and 5002 DR rats is shown in Table 3.

Pituitary Tumors

The incidence of pituitary tumors was almost identical between both dietary groups, but the 5002 AL animals were more likely to have fatal tumors than were their 5002 DR counterparts (Table 4). A statistically significant decrease in the age-adjusted incidence (Peto's test) of pituitary tumors was observed in DR males and females compared with the AL groups (Peto et al. 1984, Keenan et al. 1994b, 1995b).

Mammary Gland Tumors

The overall incidence of benign and malignant mammary gland tumors was similar between the 5002 AL and the 5002 DR females (Table 5). However, the tumors observed in the 5002 AL females frequently were larger and contributed to the early death of these animals. Other proliferative and degenerative changes seen in the mammary glands of the 5002

Table 5. Summary of mammary gland pathology by 2 years

	5002 AL		5002 DR	
	Females	Males	Females	Males
Number examined	58	57	60	60
Adenocarcinoma[a]	9 (2)[b]	0	10 (1)[b]	0
Adenoma[a]	6 (0)[b]	1	6 (2)[b]	0
Fibroadenoma[a]	30 (8)[b]	2	25 (2)[b]	0
Fibroma[a]	1 (0)[b]	1	0 (0)[b]	1
Lobular hyperplasia	28 (1.16)[c]	6 (0.19)[c]	23 (0.80)[c]	7 (0.18)[c]
Galactocele[a]	20	7	14	6

AL = ad libitum, DR = diet-restricted.
[a]Number with diagnosis.
[b]Number in parenthesis = number of females with fatal tumors (cause of death/reason killed).
[c]Number in parenthesis = average grade of lobular hyperplasia (grades 0–5).

Table 6. Summary of mammary gland tumor volume and estimated doubling time

	5002 AL	5002 DR	*p* value
Number of females with tumors/total	39/70	33/70	0.016[a]
Time to 50% prevalence (weeks)	85	101	0.008
Time to 25% prevalence (weeks)	62	73	0.01
0–25 weeks after palpation			
Number with tumor	26	20	
Mean growth time (weeks)	11	12	0.95
Doubling time (weeks)	3.1	4.7	0.24
Average tumor volume (cm^3)	29	19	0.55
26+ weeks after palpation			
Number with tumor	13	13	
Mean growth time (weeks)	37	35	0.41
Doubling time (weeks)	12	5.2	0.41
Average tumor volume (cm^3)	308	68	0.003

AL = ad libitum, DR = diet-restricted
[a]Age-adjusted incidence for fibroadenoma only; not significant for other mammary gland tumors (adenoma, fibroma, adenocarcinoma).

AL rats were generally of a higher grade or greater severity than similar lesions observed in the 5002 DR rats. For females, the only tumor type other than pituitary tumors showing a statistically significant ($p < 0.05$) change in age-adjusted tumor incidence (Peto's test) were mammary tumors, particularly fibroadenomas (Keenan et al. 1995b). The incidence of these lesions is shown in Table 5 for the 5002 AL and the 5002 DR rats.

Table 5 does not fully indicate how DR alters mammary tumor onset time. Because mammary gland tumors can be palpated and measured when the animals are alive, an analysis of growth rate is possible with mammary tumors (all types), the most common palpable tumor in the SD rat. Our present analysis is limited to the 5002 AL and 5002 DR females (Table 6).

Tumor volume was defined as the total volume (by measurement) of all mammary gland tumors at necropsy. Growth time was defined as the number of weeks from first palpation of a mammary gland tumor until death from any cause including terminal necropsy. Average tumor volume was defined as the geometric mean (Snedecor and Cochran 1989) of total tumor volume for each animal with a mammary tumor. The doubling time (time for the tumor to double in size) was calculated. All p-values were two sided, with statistical significance set at 0.05. The 5002 AL and 5002 DR females were compared for total mammary tumor volume, estimated tumor growth rate, and tumor growth time (Keenan et al. 1995b).

Although both groups had similar numbers of tumor-bearing animals, females in the 5002 DR group had later initial palpation times on average (Table 6). The time to 50% or 25% prevalence estimated by Kaplan-Meier curves (Kaplan and Meier 1958, Keenan et al. 1995b) indicated that the time to tumor for the DR females was significantly longer than for the AL females. This indicated that DR females had a later average initial palpation time. In addition, females with longer times from initial palpation to death had smaller average growth rates, so the data were divided into two strata, 0–25 and ≥26 weeks of growth time (Keenan et al. 1995b). For females with less than 26 weeks of tumor growth time, no statistically significant ($p > 0.05$) differences were observed between the AL and DR groups, either in mean growth time, estimated doubling time, or average tumor volume. Females bearing tumors for 26 weeks or more, likewise, had no difference in average growth time or doubling time for AL and DR groups, but 5002 AL females with mammary tumors palpated 26 weeks or more before death had larger average tumor volumes compared with similar 5002 DR females ($p = 0.003$). The geometric mean tumor volume was 308 cm^3 in the 5002 AL group, compared with 68 cm^3 in the 5002 DR group. This difference remained statistically significant ($p = 0.005$) after adjustment for tumor growth time by analysis of covariance (Snedecor and Cochran 1989). This difference in tumor volume of the AL females reflects the great variability in individual tumor volumes and suggests that AL females can sustain larger tumors owing to their greater body size. These results show that moderate dietary (caloric) restriction causes a delay in the time of onset of mammary tumors but does not appear to affect mammary tumor growth rates after initial palpation (Keenan et al. 1995b).

Delaying the time of tumor onset or even decreasing the incidence at a given time will not decrease the ability of a 24-month bioassay conducted by moderate DR to detect a compound-related carcinogenic effect. Studies of the effect of caloric restriction on known direct-acting

carcinogen-induced tumorigenesis do show a lowered incidence and onset rate at a given time (Tucker 1979, Pollard and Luckert 1985, Albanes
1987, Weindruch and Walford 1988, Pollard et al. 1989, Weindruch 1989,
Fishbein 1991, Ruggeri 1991, Rogers et al. 1993, Kritchevsky 1993). Although many of these studies were terminated at 6 to 12 months, they
detected treatment-related tumors in the DR animals and clearly distinguished them from concurrent untreated controls. Thus, the ability of
these bioassays to detect a carcinogenic response at even early times
meets the working definition of adequate statistical sensitivity or power.
Compared with AL or overfed rats, more DR rats will live to 24 months
and thus more animals will be exposed to the test compound for the 2-
year period (Table 2). In our study, the average time on study for 5002
AL rats was 80.1 weeks for males and 86.4 weeks for females, whereas
5002 DR rats were on study for an average of 99.6 weeks for males and
98.3 weeks for females (Table 2). This difference was significant
($p < 0.001$) and means that the 5002 DR males would have 19.5 weeks
(5 months) and 5002 DR females would have 11.9 weeks (3 months) of
additional exposure to a test compound. This increased exposure would
increase the statistical sensitivity of the bioassay to detect a treatment-
related event (Keenan et al. 1994a,b).

   The sensitivity (power) of a rodent carcinogenicity study is strongly
affected by survival, the incidence of tumors at terminal sacrifice, the distribution of tumor onset times, and other factors. Diet can have a marked
effect on both survival and tumor onset times (Tucker 1979, Maeda et al.
1985, Pollard and Luckert 1985, Albanes 1987, Pollard et al. 1989,
Weindruch 1989, Masoro et al. 1991, Shimokawa et al. 1991, 1993,
Kritchevsky 1993, Rogers 1993, Keenan et al. 1995b). Moderate DR improves survival, which tends to increase assay sensitivity because early
deaths are reduced. However, DR also has the potential to delay average
time of tumor onset, which tends to decrease assay sensitivity. Experimental data from our 2-year rat study were used to assess the net effect of
moderate DR on assay sensitivity. We compared the mammary gland tumor data on the 5002 AL and 5002 DR females.

   To quantify the sensitivity, or power, of carcinogenicity studies to detect a treatment effect, five factors must be specified for each treatment
group (McKnight and Crowley 1984): 1) the number of animals studied,
2) the distribution over time of death in tumor-free animals, 3) the distribution of tumor onset over time in controls, 4) the effect of treatment on
survival and on tumor onset, and 5) for nonpalpable tumors, the effect of
tumor on subsequent survival. For palpable tumors, we do not need to be

concerned about survival subsequent to tumor onset because the response variable is time of initial palpation. For nonpalpable tumors, however, any lethality associated with tumor development can affect the statistical analysis (Kaplan and Meier 1958, Peto and Peto 1972, Peto et al. 1980, McKnight and Crowley 1984).

It is not possible to observe factors 3, 4, and 5 for nonpalpable tumors unless many interim sacrifices with large numbers of animals are used. For palpable tumors, however, time of tumor onset can be approximated by time of initial palpation, and we do not need to know the effect of tumor on subsequent survival. For this reason we have restricted our attention to palpable mammary gland tumors. Our estimates for the distribution of tumor onset (palpation) times are most accurate for common tumors, so we restricted our power calculations to mammary tumors (all types) in females.

Statistical power can be defined only for a specified dose-response effect (carcinogenic response as a function of dose) and depends on the number of treatment groups and doses selected. For simplicity, we assumed one control group and one treated group with 70 females per group, the same sample size used in our study. Although most carcinogenicity studies employ multiple dose groups, these calculations served to indicate the relative power of studies under AL or DR feeding regimens. Calculations were restricted to mammary tumors, assumed for simplicity to be palpable in all females before death. Our response variable was the week of initial tumor palpation.

In controls, the true distribution of tumor onset times was fixed as the observed distribution in the appropriate diet study group, either AL or DR. Under the null hypothesis of no treatment effect, the same distribution was used for the treatment group. Two alternative hypotheses were considered. Under the "constant effect" alternative, treatment was assumed to double the risk of mammary tumor onset during each week of follow-up throughout the study. Under the "late onset" alternative, treatment had the same effect on tumor incidence by terminal sacrifice, but most of the effect occurred late in the follow-up period (Keenan et al. 1995b).

Table 7 shows the estimated power of carcinogenicity studies to detect a variety of hypothetical treatment effects, and the impact of DR and AL feeding. If there was no true effect of treatment, the probability of a statistically significant result at $p < 0.01$ was estimated to be 0.6% under AL feeding and 0.2% under DR (false-positive rate). Neither probability was statistically different from the theoretical 1% level, and AL did not differ from DR ($p = 0.62$). If hypothetical treatment doubled the risk of

Table 7. Mammary gland tumor power calculations and probability of a statistically significant result ($p < = 0.01$)[a]

Female group	True effect of treatment		
	None[b]	Constant effect[c]	Late-onset effect[d]
5002 AL	0.6 ± 0.4%	80 ± 1.8%	59 ± 2.2%
5002 DR	0.2 ± 0.4%	80 ± 1.8%	69 ± 2.1%

[a]Each table entry is based on 500 computer-simulated experiments.
[b]False positive rate ± standard error; AL is not statistically different from DR ($p = 0.62$). Neither entry in this column is statistically different from the 1% level expected ($p > 0.05$).
[c]Hypothetical treatment doubles the risk of mammary tumor onset throughout the study. AL is not statistically different from DR ($p = 0.87$).
[d]The risk of mammary tumor by terminal sacrifice in treated rats is the same as in the previous column, but most of the treatment effect occurs late in follow-up. AL has a statistically significantly lower power compared with DR ($p = 0.001$).

mammary tumor onset throughout the study ("constant effect" in Table 7), the power was 80% under either AL or DR diets. It should be noted that the "treatment effect" under DR was less than that under AL. Because the risk of mammary tumor onset was generally less each week under DR compared with AL feeding, doubling the risk was a smaller increase under DR. However, if the treatment effect was not constant but occurred primarily late in follow-up ("late onset effect" in Table 7), the power under AL feeding was 59%, significantly less ($p = 0.001$) compared with the power of 69% under DR. This is not surprising, since DR animals have much better survival. In summary, these calculations support the position that moderate DR will not lower the sensitivity of carcinogenicity studies in rats (Keenan et al. 1995b).

Although some feel that restriction of reported AL caloric intake decreases the incidence of certain spontaneous neoplasms in rodents (Tucker 1979, Weindruch and Walford 1988, Pollard et al. 1989, Weindruch 1989, Fishbein 1991, Rogers et al. 1993), our studies and those of others (Maeda et al. 1985, Shimokawa et al. 1991, Keenan et al. 1995b) indicate that this is not true for all types of tumors when an evaluation is made at the time of spontaneous death or when the animals are maintained for a significant portion of their life span, as is the case in a 24-month carcinogenicity study. In lifetime studies of AL and DR rats, the percentage of F344 rats with neoplastic disease at the time of spontaneous death was greater in DR than in AL rats (Maeda et al. 1985). However, this finding may be due to the increased longevity of the DR rats. When theoretical survival curves were generated for these studies based on all neoplastic disease or specific neoplastic diseases (leukemia/lymphoma) as the cause of death, the results of the analysis of male F344 rats studied at the time of spontaneous death show that the effects of caloric restriction delayed death due to neoplastic disease in general (Shimokawa et al. 1991). Protein restriction and fat restriction without energy restriction had no influence on death owing to neoplastic disease. The main influence of dietary manipulation of F344 male rats fed AL indicates that energy restriction significantly influences the *onset time* of spontaneous tumors. Restriction of protein or fat without decreased energy intake was not associated with this effect. Energy restriction appeared to delay the occurrence of death due to leukemia/lymphoma in F344 male rats, but energy restriction has this action by delaying the onset of leukemia/lymphoma and not by decreasing the progression and severity of the disease after onset occurs (Shimokawa et al. 1991, 1993). These observations are consistent with our studies of SD rats that indicate that the time of onset for pituitary and mammary gland tumors is delayed by DR, but not the overall incidence in a 24-month study (Keenan et al. 1994a,b, 1995b).

Kidney

In contrast to the similar overall final incidence of neoplasia between the AL and DR groups, there were clear differences in the incidence and severity of chronic renal disease (CRD) observed in these animals. The 5002 AL males had the earliest onset of, highest incidence of, and most severe CRD, and this was a common cause of mortality (Table 3). Kidney tissue from BrdU-treated rats showed high labeling in areas of tubular basophilia, cellular infiltration, interstitial fibrosis, and glomerular sclerosis. The DR rats fed either diet (5002 or 5002-9) had a lower incidence and severity of most of the morphologic changes, and significant CRD was not a contributing factor to early death in the DR animals.

These observations were quantified to determine the glomerular area (μm^2), the glomerular sclerosis index, the tubulointerstitial index (%), and the tubular BrdU labeling index (Gumprecht et al. 1993, Keenan et al. 1995a). The data from the terminal necropsy are summarized with geometric means in Table 8 (Keenan et al. 1995a).

These quantitative data support the histological observation that the 5002 AL males have the most severe renal disease, followed by the 5002-9 AL males. In addition, these data and similar quantitative data collected on the 52-week interim necropsy animals indicate that AL feeding of either diet is associated with the early development of glomerular hypertrophy that correlates with the severity of glomerular sclerosis seen at the terminal necropsy (Gumprecht et al. 1993, Keenan et al. 1995a). These

Table 8. Summary of chronic renal disease (CRD) at 2 years[a]

	5002 AL	5002 DR	5002 6.5h DR	5002-9 AL	5002-9 DR
Females					
Glomerular area ($\mu m^2 \times 10^4$)	24.2	18.1	21.0	22.9	19.1
Glomerular sclerotic index	104.3	72.4	91.2	102.4	74.0
Tubulointerstitial index	16.5	11.8	12.6	11.5	10.5
Tubular BrdU labeling index	3.6	1.0	1.8	2.1	1.1
Males					
Glomerular area ($\mu m^2 \times 10^4$)	29.3	26.2	26.8	28.9	20.9
Glomerular sclerotic index	140.7	87.9	65.2	102.0	80.4
Tubulointerstitial index	21.6	11.5	12.7	18.9	9.3
Tubular BrdU labeling index	5.1	1.1	1.4	4.1	1.3

AL = ad libitum, DR = diet-restricted.
[a]From Keenan et al. (1995a).

data suggest that the pathogenesis of chronic renal disease in the rat is initiated by glomerular hypertrophy and that moderate DR prevents the early development of this change. Interestingly, the lower protein in the 5002-9 diet was of minimal benefit when fed AL but was of some benefit when fed on a restricted basis (Gumprechet et al. 1993, Keenan et al. 1995a).

The observations at the 52- and 106-week necropsies in this study support our hypothesis that the initial glomerular hypertrophy leads to glomerular sclerosis and, subsequently, to tubular and interstitial damage in the progression of CRD (Gumprecht et al. 1993, Keenan et al. 1994a,b, 1995a). Analysis of the dietary manipulations in this study provide strong evidence that caloric restriction, not protein restriction, is primarily responsible for decreases in the incidence and severity of glomerular lesions and delays the progression of spontaneous CRD. These data and results from related studies support our hypothesis and suggest that caloric restriction is most important in delaying the progression of glomerular sclerosis and nephron loss initiated by the early development of glomerular hypertrophy in AL overfed rats (Keenan et al. 1992, 1994a, 1995a, Gumprecht et al. 1993).

Table 9. Summary of cardiac pathology by 2 years[a]

	5002 AL		5002 DR	
	Females	Males	Females	Males
Number examined	58	57	60	60
Cardiomyopathy[b]	13 (0.33)[c]	31 (1.32)	2 (0.05)	11 (0.25)
Cellular infiltration[b]	33 (0.71)	45 (1.07)	11 (0.20)	28 (0.56)
Degeneration[b]	15 (0.36)	40 (1.35)	1 (0.02)	7 (0.12)
Fibrosis[b]	49 (1.36)	55 (2.35)	19 (0.33)	52 (1.22)

AL = ad libitum, DR = diet-restricted.
[a]From Keenan et al. (1995a).
[b]Number with diagnosis.
[c]Number in parenthesis = average grade of lesion (grades 0–5).

Heart

The incidence and severity of cardiomyopathy as seen morphologically as cellular infiltration, myocardial degeneration, and fibrosis were greater in the 5002 AL animals and were contributing factors to mortality in the 5002 AL males (Table 3). For purposes of this study, cellular infiltration, myocardial degeneration, and fibrosis were individually graded on all animals. When an individual manifests all of these changes, an overall grade was given under the category of cardiomyopathy. The incidence and average grade of these lesions are given in Table 9 for the 5002 AL and 5002 DR rats (Keenan et al. 1994b, 1995a).

Liver

Degenerative liver lesions were most evident and severe in the 5002 AL rats. Degenerative changes such as hepatocellular periportal vacuolation and telangiectasis were more evident and severe in the 5002 AL rats than in the DR animals, although most of the DR animals lived for a longer period than their AL counterparts. In animals showing hepatocellular periportal vacuolation, particularly the females, there was increased BrdU nuclear labeling of hepatocytes in this region. These and other changes correlated with the increased hepatic malondialdehyde content and the decreased hepatic glutathione content observed in the 5002 AL rats at 52 weeks (Keenan et al. 1994a, 1995a).

Bile duct hyperplasia occurred at a similar incidence in both the AL and DR rats, but was of slightly greater relative severity in the AL animals. Basophilic and eosinophilic altered hepatocellular foci (AHF) were seen in all groups with a similar incidence and severity. The overall incidence of liver tumors was similar between the AL and the DR groups; however, the 5002 DR animals' median survival was more than 104 weeks

Table 10. Summary of liver pathology by 2 years[a]

	5002 AL		5002 DR	
	Females	Males	Females	Males
Number examined	58	57	60	60
Hepatocellular adenoma[b]	1	2	3	2
Hepatocellular carcinoma[b]	1	4	0	2
Bile duct adenoma[b]	0	0	0	1
Bile duct carcinoma[b]	0	1	0	0
Basophilic AHF[b,c]	31 (1.05)	32 (0.75)	39 (0.93)	32 (0.57)
Eosinophilic AHF[b,c]	19 (0.530)	13 (0.39)	21 (0.52)	20 (0.52)
Bile duct hyperplasia[b]	52 (1.60)	52 (1.86)	46 (1.12)	56 (1.38)
Cellular infiltration[b]	54 (1.44)	53 (1.42)	47 (0.82)	52 (1.05)
Periportal vacuolation[b]	41 (1.69)	23 (0.72)	8 (0.15)	11 (0.20)
Telangiectasis[b]	25 (0.83)	26 (0.77)	21 (0.57)	18 (0.40)

AL = ad libitum, DR = diet-restricted.
[a]From Keenan et al. (1995a).
[b]Number with diagnosis.
[c]AHF = altered hepatocellular foci.
[d]Number in parenthesis = average grade of lesion (grades 0–5).

compared with the median survival of 85–90 weeks of their AL counterparts (Keenan et al. 1994a). The incidence and average grade of these liver lesions are given in Table 10 for the 5002 AL and 5002 DR rats (Keenan et al. 1995a).

Pancreas

The AL rats of both sexes, in general, had the highest incidence and severity of proliferative and degenerative changes in their pancreas, particularly islet fibrosis and acinar atrophy. In contrast, the overall incidence of islet cell tumors was similar between AL and DR rats. The incidence of islet cell carcinoma was higher in the 5002 AL males, suggesting an earlier onset of these tumors (Keenan et al. 1995b). The diagnosis of islet carcinoma was used only when the tumor cells had invaded the capsule of the tumor (Keenan et al. 1995b).

The remaining tumors, proliferative lesions, and degenerative changes observed in this study generally indicated that the 5002 AL group had a higher incidence and/or severity of degenerative and proliferative lesions. However, the overall incidence of benign and malignant neoplasms was remarkably similar between the AL and DR groups (Table 3) (Keenan et al. 1995b).

Summary

Moderate dietary (caloric) restriction as practiced in this study did not have any adverse effects on the overall long-term health or lesion incidence of the DR animals. The overall incidence of benign and malignant neoplasia was very similar between the AL and the calorically restricted groups (Keenan et al. 1995b). Although the incidence was similar, the tumors observed in the AL animals were generally larger, of earlier onset, and more likely to be contributing factors to the early death of these animals. In contrast, degenerative disease, particularly renal and cardiovascular disease in the males, showed clear differences in both incidence and severity between the AL and calorically restricted groups, and serious renal and cardiovascular disease was frequently identified as a contributing factor to the early death of the AL males. The conclusion to be drawn from this study is that true AL feeding of laboratory rats is overfeeding with many adverse outcomes, including reduced survival.

These data are consistent with the observations of others over the past four decades that reducing dietary energy intake by caloric restriction will increase the maximum life span, retard age-related senescence and degeneration, and delay or prevent the appearance of age-related diseases (Weindruch and Walford 1988, Finch 1990, Fishbein 1991, Masoro 1991a,b, Masoro and McCarter 1991, Masoro et al. 1991, Roe 1991, Roe et al. 1991, Keenan et al. 1992, 1994a,b, 1995a,b, Yu 1993).

The results of this study underline the healthful action of moderate DR on SD rat longevity, spontaneous carcinogenesis, and age-related proliferative and degenerative lesions. A number of major hypotheses have been proposed for the beneficial effects of DR, but many have been ruled out by recent basic studies (Weindruch and Walford 1988, Finch 1990, Fishbein 1991, Masoro 1991a,b, Masoro and McCarter 1991, Masoro et al. 1991, Yu 1993, Keenan et al. 1994b). However, the effects induced by AL overfeeding or moderate DR involve the means by which rodents utilize fuel in more or less damaging ways over the course of their life spans. Most of the basic work in DR supports hypotheses of reducing the adverse metabolic effects of glucose and oxygen, both of which facilitate potentially toxic processes (Weindruch and Walford 1988, Finch 1990, Fishbein 1991, Masoro 1991a,b, Masoro and McCarter 1991, Masoro et al. 1991, Yu 1993). Results from our study support most of the widely held hypotheses that caloric restriction acts by modulating the characteristics, but not the rate, of fuel use in a way to prevent long-term damage through oxidative damage or glycation (Weindruch and Walford 1988, Masoro 1991a,b, Masoro and McCarter 1991, Masoro et al. 1991, Yu 1993). In contrast, our present methods of AL overfeeding rodents apparently accelerates primary aging processes that result in lower survival and the early onset and increased severity of age-related degenerative diseases and diet-related endocrine tumors. The beneficial effects of this moderate level of DR in improving longevity, preventing the development of chronic degenerative disease, and delaying the onset of diet-related endocrine tumors will result in the animals being exposed to test compounds for a longer period of time and will thus, improve the sensitivity of the bioassay to detect compound-specific chronic toxicity and carcinogenicity. This standardized, controlled method of moderate DR provides a given amount of food that is still within the range of current "AL" food consumption (Keenan et al. 1995a,b). Moderate DR does not adversely affect the rats' health, physiology, or metabolic profile and, thus, improves the rat carcinogenicity bioassay for the evaluation of human safety.

Effect of Dietary (Caloric) Restriction on the Toxicity of Pharmacological Agents

A 14-week study was conducted to compare the effects of dietary or caloric restriction versus AL feeding on the toxic response of SD rats to five different pharmacological agents administered at or near the maximum tolerated dose (MTD). The AL group was provided free access to Purina Certified Rodent Chow 5002 and the DR group received the same diet at approximately 21.5 g/day for males and 16.0 g/day for females as described above (Gumprecht et al. 1993, Keenan et al. 1994a, 1995a,b). Complete toxicologic pathology evaluations were conducted on all of the animals. Treatment groups consisted of 20/sex for both of the control groups (AL and DR) and 10/sex for each of the drug treatment groups (AL and DR), which included phenobarbital (100 mg/kg/day); clofibrate (500 mg/kg/day); L-647,318, a 3-hydroxy-3-methyl-glutaryl coenzyme A (HMG CoA) reductase inhibitor (150 mg/kg/day); cyclosporine A (25 mg/kg/day); and MK-0458, a dopamine agonist (4 mg/kg/day). All drugs were administered by oral gavage, and the doses selected were based on known MTD doses (Keenan et al. 1994b).

Although minor differences were seen in clinical signs, in all cases significant clinical, hematologic, and biochemical changes were detected in both the AL and DR groups given MTD doses of these compounds. Liver samples taken at necropsy from control, phenobarbital, clofibrate, and cyclosporine A treated groups were analyzed for liver microsomal cytochrome P450 content and drug-metabolizing enzyme activities. There were no significant differences in cytochrome P450 contents or other enzyme activities in the samples from control or treated AL or DR rats (Keenan et al. 1994b).

Liver samples from AL and DR rats from control and clofibrate-treated groups were analyzed for peroxisome fatty acyl CoA oxidase (FACO) activity. Dietary restriction had no effect on the basal activity of this enzyme in either gender. Both male and female AL and DR rats treated with clofibrate had marked increases in FACO activity. The magnitude of the change in male rats was greater than that seen in female rats under both dietary regimens (Keenan et al. 1994b).

The major objective of this toxicity study was to compare the effects of moderate DR versus AL overfeeding on the toxicologic response of five different pharmacologic agents administered to rats at or near the MTD for 14 weeks. With the very large doses administered, many of the anticipated manifestations of toxicity of these compounds were observed. While quantitative differences were seen in terms of clinical signs, individual biochemistry parameters, organ weights, and the severity of lesion morphologic change relative to the dietary regimen, the qualitative changes were very similar for each compound under AL and DR feeding as done in this laboratory. In general, organ weight changes and the morphologic response seen in AL and DR rats appeared proportional to their respective controls. The main morphologic changes are compared in Table 11 (Keenan et al. 1994b).

Considering the beneficial effects of long-term moderate DR, it is anticipated that the chronic toxicity induced by the classes of compounds tested in this study could be readily detected in animals maintained by moderate DR. Moreover, the beneficial effects of moderate DR in preventing long-term degenerative disease and delaying the onset of diet-related endocrine tumors would result in the animals being exposed to the test compound for a longer period of time (Table 2) and, thus, allowing a better assessment of chronic toxicity and carcinogenicity. For the pharmacologic classes of compounds tested in this study, MTDs and no-effect dose thresholds would be readily detectable with moderate DR (Keenan et al. 1994b).

Conclusion

While both genetic and environmental factors are involved, it is clear that low rat survival is closely correlated with AL overfeeding and can be improved by simple moderate dietary (caloric) restriction as practiced in the above studies (Keenan et al. 1992, 1994a,b, 1995a,b, Gumprecht et al. 1993). Our laboratory has undertaken studies of moderate dietary (caloric) restriction with the SD rat and has shown that this method will improve survival and lower the incidence and severity of diseases associated with

Table 11. Comparison of morphologic response to different compounds between ad libitum (AL) and dietary-restricted (DR) Sprague-Dawley rats

Compound	Morphologic change	Females		Males	
		AL	DR	AL	DR
Phenobarbital	Liver weight[a] (%)	148	149	125	147
(100 mg/kg/day)	Centrilobular hypertrophy[b]	3.3	3.3	3.8	3.6
Clofibrate	Liver weight[a] (%)	172	147	135	148
(500 mg/kg/day)	Diffuse hypertrophy[b]	2.7	2.7	2.7	2.9
	Bile duct hyperplasia[b]	1.0	0.7	0.8	0.5
L-647,318	Liver weight[a] (%)	100	105	90	99
(150 mg/kg/day)	Periportal atypia[b]	2.5	2.3	1.8	2.2
	Bile duct hyperplasia[b]	1.6	2.2	1.1	1.4
	Basophilic AHF[b,c]	0.5	0.9	0.0	0.1
	Eosinophilic AHF[b,c]	0.3	0.7	0.1	0.1
Cyclosporine A	Thymic weight[a]	60	87	33	63
(25 mg/kg/day)	Thymic depletion[b]	4.5	4.1	4.2	4.2
	Kidney weight[a]	115	110	121	111
	Nephropathy[b]	3.0	2.5	2.8	2.9
MK-0458	Pituitary weight[a]	69	71	87	91
(4 mg/kg/day)	Ovary weight[a]	214	165	—	—
	Corpora lutea retention[b]	2.5	2.7	—	—

[a]Percent change of relative organ weights compared with AL controls (relative organ weights were expressed as a percentage of brain weight).
[b]Mean histologic grade (0–5).
[c]AHF = altered hepatocellular foci.

AL overfeeding. Our laboratory also studied a modified diet of reduced protein and increased fiber and found no survival benefit if this diet was fed AL (Keenan et al. 1994a). A comparison of five reference pharmaceutical agents at the MTD doses failed to show significant differences in the toxicologic profiles seen in 14-week studies comparing AL and DR feeding (Keenan et al. 1994b).

These data are consistent with other observations over the past four decades that reducing energy intake by caloric restriction will increase the maximum life span, retard age-related senescence and degeneration, and delay or prevent the appearance of age-related diseases and tumors (Weindruch and Walford 1988, Finch 1990, Fishbein 1991, Masoro 1991a,b, Masoro and McCarter 1991, Masoro et al. 1991, Roe 1991, Roe et al. 1991, Keenan et al. 1992, 1995a,b, Yu 1993). Although the mechanisms underlying these effects are not completely understood, our data support several widely held hypotheses that caloric restriction acts by modulating the characteristics, but not the rate, of fuel use in such a fashion as to prevent long-term damage of such fuel use through oxidative damage or glycation (Weindruch and Walford 1988, Finch 1990, Fishbein 1991, Masoro 1991a, Masoro and McCarter 1991, Masoro et al. 1991, Yu 1993). It is apparent that caloric restriction is affecting survival primarily through its action on fuel utilization.

The studies conducted at Merck Research Laboratories demonstrate that unlimited access to food (so-called ad libitum feeding) is clearly overnutrition with many adverse outcomes, including premature death from the early onset of degenerative disease and cancer (Keenan et al. 1992, 1994a,b, 1995a,b, Gumprecht et al. 1993). Conversely, moderate food restriction, acting through the mechanisms of caloric restriction, delays the onset of these degenerative diseases and cancers, resulting in greater longevity of the animals under study. Data from Wistar rat food restriction studies by Roe (1991) and Roe et al. (1991) have already been used in risk assessment models of human overnutrition (Lutz and Schlatter 1992). Lutz and Schlatter (1992) concluded that the effect of overfeeding on cancer alone could possibly account for 60,000 out of every 1 million human deaths. Our studies of moderate DR further show that it will not adversely affect the health or metabolic profile of the animals, which will be exposed to a test compound for a much longer period than AL overfed rodents (Keenan et al. 1992, 1994a,c,d). Thus, moderate DR will result in a more appropriate rodent model for long-term toxicity and carcinogenicity studies to assess the human safety of candidate pharmaceuticals.

Acknowledgments

The authors wish to thank Mrs. B.J. Morgan for preparing this chapter. Data reviewed in this chapter have been provided by the following collaborating scientists: Drs. P.F. Smith (Searle Laboratories, Skokie, IL), G.C. Ballam (Purina Mills, Inc., St. Louis, MO), R.L. Clark (Rhone-Poulenc Rorer, Collegeville, PA), P. Lang (consulting toxicologist, Charles River Laboratories), G. Wolfe (Hazleton Laboratories, Vienna, VA), C.E. Cover (E.I. DuPont de Nemours and Co., Newark, DE), and R. Wang, J. DeLuca, S. Grossman, L. Gumprecht, P. Hertzog, and C.M. Hoe (Merck Research Laboratories, West Point, PA). The authors thank Drs. C.P. Peter, C.F. Hollander, M.J. van Zwieten, J.D. Burek, R.T. Robertson, and D.L. Bokelman of Merck Research Laboratories for advice and support of this work and Professor E.J. Masoro (University of Texas, San Antonio, TX) for his critical review of the design of the carcinogenesis study.

References

Albanes D (1987) Caloric intake, body weight, and cancer: a review. Nutr Cancer 9:199–217

Burek JD (1990) Survival experience with the Sprague-Dawley-derived rat. Transcript of 1990 Toxicology Forum Annual Meeting, Aspen, CO., pp 506–518

Finch CE (1990) Longevity, senescence and the genome. University of Chicago Press, Chicago

Fishbein L (ed) (1991) Biological effects of dietary restriction. Springer-Verlag, New York

Food and Drug Administration (1982) Toxicological principles for the safety assessment of direct food additives and color additives used in food. USDA, Center for Food Safety and Applied Nutrition, Washington, DC

Frank JD, Cartwright ME, Keenan KP (1993) Intensification of 5-bromo-2'-deoxyuridine immunohistochemistry with zinc buffered formalin postfixation and ABC elite solution. J Histotech 16(4):329–334

Gumprecht LA, Long CR, Soper KA, et al. (1993) The early effects of dietary restriction on the pathogenesis of chronic renal disease in Sprague-Dawley rats at 12 months. Toxicol Pathol 21(6):528–537

Haseman JK, Rao GR (1992) Effects of corn-oil, time-related changes and inter-laboratory variability on tumor occurrence in control Fischer 344 (F-344/N) rats. Toxicol Pathol 20(1):52–60

Jordan A (1992) FDA requirements for nonclinical testing of contraceptive steroids. Contraception 46:499–509

Kaplan EL, Meier P (1958) Nonparametric estimation from incomplete observations. J Am Stat Assoc 53:457–481

Keenan KP, Smith PF, Ballam GC, et al. (1992) The effect of diet and dietary optimization (caloric restriction) on rat survival in carcinogenicity studies—an industrial viewpoint. In McAuslane JAN, Lumley CF, Walker SR (eds), The Carcinogenicity Debate. (Centre for Medicines Research Workshop) Quay Publishing, Lancaster, England, pp 77–102

Keenan KP, Smith PF, Hertzog P, et al. (1994a) The effects of overfeeding and dietary restriction on Sprague-Dawley rat survival and early pathology biomarkers of aging. Toxicol Pathol 22(3):300–315

Keenan KP, Smith PF, Soper KA, (1994b) The effect of dietary (caloric) restriction on rat aging, survival, pathology and toxicology. In Mohr U, Dungworth DL, Capen CC (eds), Pathobiology of the aging rat, vol. 2. ILSI Press, Washington, DC, pp 609–628

Keenan KP, Soper KA, Hertzog PR, et al. (1995a) Diet, overfeeding and moderate dietary restriction in control Sprague-Dawley rats, II: Effects on age-related proliferative and degenerative lesions. Toxicol Pathol, in press

Keenan KP, Soper KA, Smith PF, et al. (1995b) Diet, overfeeding and moderate dietary restriction in control Sprague-Dawley rats, I: Effects on spontaneous neoplasms. Toxicol Pathol, in press

Kritchevsky D (1993) Energy restriction and carcinogenesis. Food Res Int 26:289–295

Lang PL (1989) Survival of Crd:CD®BR rats during chronic toxicology studies. Charles River Laboratories Reference Paper, pp 1–4

Lang PL (1990) Survival of CDF®(F-344)/CrlBR rats during chronic toxicology studies. Charles River Laboratories Reference Paper, pp 1–4

Lang PL (1991) Changes in life span of research animals leading to questions about validity of toxicologic studies. Chem Regul Reporter 14:1518–1520

Lang PL (1992) Spontaneous neoplastic lesions and selected non-neoplastic lesions in the Crl:CD®BR rat. Charles River Laboratories Reference Paper, pp 1–36

Lutz WK, Schlatter J (1992) Chemical carcinogens and overnutrition in diet-related cancer. Carcinogenesis 13(12):2211–2216

Maeda H, Gleiser CA, Masoro EJ, et al. (1985) Nutritional influences on aging of Fischer 344 rats, II: Pathology. J Gerontol 40:671–688

Masoro EJ (1991a) Biology of aging: facts, thoughts, and experimental approaches. Lab Invest 65:500–510

Masoro EJ (1991b) Use of rodents as models for the study of "normal aging": conceptual and practical issues. Neurobiol Aging 12:639–643

Masoro EJ, McCarter RJM (1991) Aging as a consequence of fuel utilization. Aging 3:117–128

Masoro EJ, Shimokawa I, Yu BP (1991) Retardation of the aging process in rats by food restriction. Ann N Y Acad Sci 621:337–352

McKnight B, Crowley J (1984) Tests for differences in tumor incidence based on animal carcinogenicity experiments. J Am Stat Assoc 79(307):639–648

McMartin DN, Sahota PS, Gunson DE, et al. (1992) Neoplasms and related proliferative lesions in control Sprague-Dawley rats from carcinogenicity studies: historical data and diagnostic considerations. Toxicol Pathol 20:212–225

Nohynek GJ, Longeart L, Geffray B, et al. (1993) Fat, frail and dying young: survival body weight and pathology of the Charles River Sprague-Dawley-derived rat prior to and since the introduction of the VAF® variant in 1988. Hum Exp Toxicol 12:87–98

Peto R, Peto J (1972) Asymptotically efficient rank invariant procedures. J Roy Stat Soc, ser A, 135:185–207

Peto R, Pike MC, Day NE, et al. (1980) Guidelines for simple, sensitive significance tests for carcinogenic effects in long-term animal experiments. In IARC monographs on the evaluation of carcinogenic risk of chemicals to humans: supplement of long-term and short-term screening assays for carcinogens: a critical appraisal. IARC, Lyon, France, pp 311–426

Pollard M, Luckert PH (1985) Tumorigenic effects of direct- and indirect-acting chemical carcinogens in rats on a restricted diet. J Natl Cancer Inst 74:1347–1349

Pollard M, Luckert PH, Snyder D (1989) Prevention of prostate cancer and liver tumors in L-W rats by moderate dietary restriction. Cancer 64:686–690

Rao GN, Hasemen JK, Grumbein S, et al. (1990) Growth, body weight, survival, and tumor trends in F-344/N rats during an eleven-year period. Toxicol Pathol 18(1):61–70

Roe FJC (1991) 1200-rat biosure study: design and overview of results. In Fishbein L (ed), Biological effects of dietary restriction. Springer-Verlag, New York, pp 287–304

Roe FJC, Lee PN, Conybeare G, et al. (1991) Risk of premature death and cancer predicted by body weight in early life. Hum Exp Toxicol 10:285–288

Rogers AE, Zeisel SH, Groopman J (1993) Diet and carcinogenesis. Carcinogenesis 14(11):2205–2217

Ruggeri BA (1991) The effects of caloric restriction on neoplasia and age-related degenerative processes. In Alfin-Slater RB, Kritchevsky D (eds), Human nutrition, 7: Cancer and nutrition. Plenum Press, New York, pp 187–210

Shimokawa I, Yu BP, Higami Y, et al. (1993) Dietary restriction retards onset but not progression of leukemia in male F344 rats. J Gerontol 48:B68–B73

Shimokawa I, Yu BP, Masoro EJ (1991) Influence of diet on fatal neoplastic disease in male Fischer 344 rats. J Gerontol 46:B228–232

Snedecor GW, Cochran WG (1989) Statistical methods, 8th ed. Iowa State University Press, Ames, Iowa, pp 149–176

Tucker MJ (1979) The effect of long-term food restriction on tumours in rodents. Int J Cancer 23:803–807

Tukey JW, Ciminera JL, Heyse JF (1985) Testing the statistical certainty of a response to increasing doses of a drug. Biometrics 41:259–301

Weindruch R (1989) Dietary restriction, tumors and aging in rodents. J Gerontol 44:67–71

Weindruch R, Walford RL (1988) The retardation of aging and disease by dietary restriction. Charles C Thomas, Springfield

Yu BP (ed) (1993) Free radicals in aging. CRC Press, Boca Raton, FL

The Effect of Dietary Restriction and Aging on the Physiological Response of Rodents to Drugs

Peter H. Duffy, Ritchie J. Feuers, James L. Pipkin, Thomas F. Berg, Julian E.A. Leakey, Angelo Turturro, and Ronald W. Hart
National Center for Toxicological Research

Introduction

Although relatively little is known about the long-term (or even the significant short-term) effects of dietary restriction (DR) in humans, a wide array of rodent studies has indicated that reduced energy (calorie) intake alters a number of key biological processes in such a way as to promote good health, decrease disease, and increase longevity (Maeda et al. 1985, Masoro 1988, McCay et al. 1935, Sarkar et al. 1982, Walford et al. 1974, Weindruch and Walford 1982). Numerous studies have shown that DR decreases the carcinogenic capacity of a number of well-studied carcinogens (Kritchevsky et al. 1984, Ruggeri et al. 1987). Additionally, in other studies, the relative toxicity and mortalities associated with certain prescription drugs were found to be significantly reduced by DR in an age-dependent fashion (Berg et al. 1994).

Nutrition studies suggest that it is important not only to investigate how DR modulates toxicity but also to determine how procedural and environmental factors, such as the frequency and method of feeding as well as the duration of DR (acute versus chronic), impact this variable (Duffy et al. 1989). Drug toxicity studies also support use of DR to increase the survival of rodents in the chronic bioassay while increasing the accuracy and decreasing the variability in drug-testing regimens (Berg et al. 1994). If such results are to be used to predict potential risk factors, to determine health hazards, and to design more effective drug-testing protocols, then additional studies are necessary to provide baseline data related to the effects of DR and age on drug and chemical toxicity.

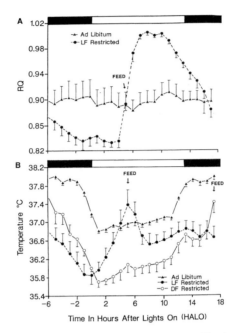

Figure 1. Circadian rhythms for respiratory quotient (A) and body temperature (B) for young male Fischer 344 rats (18–19 months of age) that were fed an ad libitum diet or a restricted diet during the light phase (LF) or the dark phase (DF) of the photoperiod cycle. Individual animals were measured for a 7-day interval; data are expressed as group mean values ± standard errors. Dark and light bars signify periods of lights-off and lights-on, respectively; HALO = hours after lights on (Duffy et al. 1989).

Physiologic Effects of Dietary Restriction by Daily Feeding

A variety of physiologic and behavior variables that relate to energy metabolism were monitored in DR rodents that received a daily ration of NIH-31 diet that was reduced in calories by 40% compared to ad libitum (AL) rodents (Duffy et al. 1989, Duffy et al. 1990a, 1990b, 1991). In these studies, rodents were individually housed and DR was started at 14 months of age. DR animals were given 40% less food than was consumed by AL animals, whereas AL animals were given more food than they could eat. DR animals were given food during the dark phase (5 hours after lights-off) when AL feeding was at its maximum level so that the circadian rhythms for DR and AL animals would be synchronized. The circadian rhythms for respiratory quotient (RQ) and body temperature in AL and DR rats (17 months) that were tested under light-feeding (LF) and dark-feeding (DF) regimes are compared in Figure 1. The effects of DR on

Table 1. Effects of diet and feeding time on physiologic and behavorial variables in old male B6C3F1 mice

Measurement	AL group Mean ± S.E.	LF restricted group Mean ± S.E.	% of AL	DF restricted group Mean ± S.E.	% of AL	AL vs. LF	AL vs. DF	LF vs. DF
Total food consumption (g)	5.21 ± 0.10	3.35 ± 0.02	64.3	3.41 ± 0.08	65.5	A***	B***	–****
Caloric consumption (kcal/g)	22.66 ± 0.43	14.57 ± 0.09	64.3	14.83 ± 0.35	65.5	A***	B***	–****
Total water consumption (g)	3.64 ± 0.38	5.15 ± 1.07	141.5	3.72 ± 0.07	102.2	A*	–****	C*
Number of feeding episodes	16.51 ± 0.69	2.80 ± 0.23	17.0	4.32 ± 0.37	25.6	A***	B***	C*
Number of drinking episodes	11.13 ± 1.28	9.03 ± 0.44	81.1	8.99 ± 0.67	80.8	–***	–***	–***
Average body temperature (°C)	36.78 ± 0.08	35.54 ± 0.15	96.6	35.11 ± 0.18	95.5 A***	B**	–***	–***
Max–min body temperature (range) (°C)	37.98–35.81 (2.18)	38.15–32.24 (5.91)	271.1	37.52–32.93 (4.59)	210.6	A***	B***	–***
Average activity (pulse/hr)	10.54 ± 2.30	18.26 ± 1.71	173.2	26.50 ± 5.57	251.4	A*	B**	–***
Average O$_2$ consumption (g/LBM) (mL g^{-1} hr^{-1})	3.34 ± 0.16	3.44 ± 0.16	103.0	3.19 ± 0.06	95.5	–***	–***	–***
Max–min respiratory quotient (range)	0.95–0.86 (0.09)	0.99–0.80 (0.19)	211.0	1.01–0.77 (0.24)	267.0	A***	B***	–***

Results of Student's t-test analysis:
A = AL × LF restricted comparison (significant effect), B = AL × DF restricted comparison (significant effect), C = LF restricted × DF restricted comparison (significant effect) (adapted from Duffy et al. 1990b).
AL = ad libitum, LF = restricted group fed during light period, DF = restricted group fed during dark period, LBM = lean body mass, Age = 28 months.
* = P < 0.05.
** = P < 0.01.
*** = P < 0.001.
**** = P > 0.05.

various physiologic and behavior variables in old (28 months of age) male B6C3F1 mice are compared in Table 1. Individual animals were measured for 7 days, and data are expressed as group means (10 animals/group). These data indicate that factors such as DR and the time of feeding greatly alter the circadian rhythms for important physiologic variables (Duffy et al. 1989).

Metabolic output (oxygen consumption) per gram of lean body mass (LBM) is not affected by DR in either rats (Duffy et al. 1989) or mice (Duffy et al. 1990a). However, body temperature was significantly reduced by DR in both species with the magnitude of the reduction being greater in mice (Duffy et al. 1990a) than in rats (Duffy et al. 1989). The daily variations in RQ were increased by DR, indicating rapid substrate-dependent shifts in metabolic pathways from carbohydrate metabolism (immediately after feeding) to fatty acid metabolism (several hours before feeding). Similar results were found in other studies in which the level of key liver enzymes associated with intermediary metabolism (Feuers et al. 1989) and drug metabolism (Leakey et al. 1989) were altered by DR in a manner that would enhance metabolic efficiency.

The DR-induced reduction in body temperature observed in rodents appears to be highly correlated with an increased ability of cells to perform DNA repair (Lipman et al. 1989) and with decreased protooncogene expression (Nakamura et al. 1989). These observations may indicate that homeostatic control mechanisms at the whole-animal level can act to modulate key biochemical processes at the molecular level, including gene expression, which in turn may regulate processes related to aging, disease resistance, and drug toxicity.

Nutritional status (i.e., 40% DR) and age were also found to alter motor activity and cardiac performance. Spontaneous activity (Table 1) was increased by DR in mice (Duffy et al. 1990a), whereas DR ($p < 0.05$) significantly decreased heart rate in young 12-month-old rats (Figure 2). When treadmill running performance was monitored in AL and DR rats using a progressive ramping exercise regimen, both the running speed and endurance (total run time) were significantly ($p < 0.05$) greater in DR rats compared to AL rats (Figure 3). Feeding behavior was also dramatically affected by the nutritional state. The duration of food consumption was compressed by DR so that most of the food was consumed during the first few hours immediately after feeding. DR mice ate fewer meals (feeding episodes) but spent more time feeding per meal and consumed more food per meal than did their AL counterparts. Although food consumption was decreased by DR, water consumption in DR mice remained at or near the

Heart rate (beat/min)

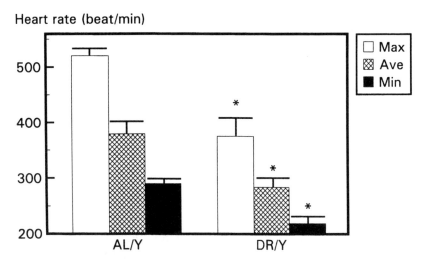

Figure 2. Effect of diet on heart rate in young AL and DR rats (12 months of age). Normal maximum, average, and minimum daily values ± standard errors are shown for rats under control conditions. A conventional statistical method (Student's t-test) was used to analyze the data. * = significant difference between AL and DR groups ($p < .05$).

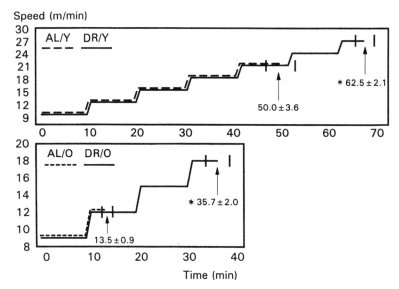

Figure 3. Effect of DR and age on treadmill running performance in male F344 × BNF₁ rats. Running speed and total run times ± standard errors are shown for young rats (12 months of age) and for old rats (36 months of age). A conventional statistical method (Student's t-test) was used to analyze the data. * = significant difference between the AL and DR groups ($p < 0.5$).

AL drinking level so that total water consumption exceeded food consumption in DR mice, and therefore water consumption per gram of body weight increased markedly. Data comparing food and water consumption variables in AL and DR mice are given in Table 1 (Duffy et al. 1990a).

The excessive drinking behavior exhibited by DR rodents (Duffy et al. 1989, Duffy et al. 1990a) is strikingly similar, if not identical, to the phenomenon of schedule-induced polypsia (SIP), a behavior pattern in which abnormal drinking activity is triggered by the intermittent feeding of small amounts of food. Evidence supporting the premise that DR effects are regulated by the central nervous system is that pituitary-adrenal hormones, such as corticosterone, which are altered by DR (Sabatino et al. 1991), have been shown to play an important role in the acquisition of SIP (Levine and Levine 1989). Recent data indicating that DR-induced SIP is regulated by the lateral hypothalamus (Winn et al. 1992), that lowered temperature and metabolism in DR rats during hypothermia (Duffy et al. 1989, 1990b) are regulated by the anterior hypothalamus (Lyman 1982, Beckman 1978), and that DR elevates the expression of stress proteins in the hypothalamus (Pipkin et al. 1994) all strongly suggest that many of the physiological effects of DR are a result of alterations in hypothalamic control function.

Physiologic Effects of Dietary Restriction by Alternate Day Feeding

Mice conditioned to chronic DR by an alternate day (FAD) feeding regimen were used to determine if rapid changes in the availability of metabolic substrates can alter physiologic functions under conditions of chronic DR (Duffy et al. 1994). In this study, mice were placed on a DR feeding regimen in which they were given a double ration (120% of AL) on day 1 of FAD, followed by fasting on day 2 of FAD. The total food consumption during the two days was 60% of AL. The range over which both body temperature and metabolism fluctuated was significantly greater in FAD mice than in DR mice that were conditioned to daily feeding (FD). The temperature, oxygen consumption, and RQ values observed on day 1 of FAD were similar to those found in AL mice, whereas the levels of these variables on day 2 of FAD were significantly lower than in FD mice (Duffy et al. 1994). Mice respond to the energy deficit imposed by FD by lowering body temperature and reducing LBM, but they do not alter metabolic rate (Duffy et al. 1990b). Conversely, FAD mice maintain a constant LBM while lowering metabolic rate and body temperature during fasting conditions (day 2 of FAD).

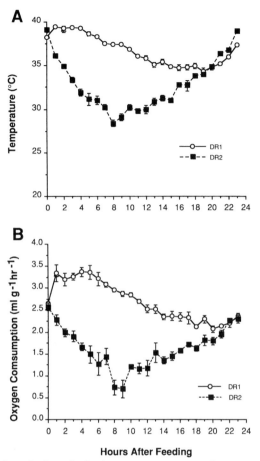

Figure 4. Circadian rhythms for body temperature (A) and oxygen metabolism (B) in old female B6C3F1 mice (28 months of age) are given for day 1 of alternate day dietary restriction when food was presented (DR1) and for day 2 of alternate day dietary restriction when no food was presented (DR2). Group mean values representing hourly averages ± standard errors are shown for the 24-hour interval (Duffy et al. 1994).

Experimental results obtained with FD and FAD feeding regimens suggest that rodents respond to a chronic feed/fast FAD regimen through a different mechanism than was previously observed in chronic FD (Duffy et al. 1990a). Circadian rhythm data for metabolism and body temperature in FAD mice (Figure 4) indicate that FAD led to the expression of two distinct biological rhythms (temporal oscillators) in the mice (Duffy et al. 1994). Measured parameters associated with metabolic output such as oxygen consumption, RQ, and food consumption were synchronized primarily to the 48-hour feeding regimen, while behavior variables such as motor activity and water consumption continued to be synchronized to the 24-hour photoperiod cycle rather than the feeding cycle (Duffy et al. 1994). The magnitude of the physiologic changes found in this study suggests that the effects of acute DR may differ significantly from those found in chronic DR.

Effect of Dietary Restriction and Age on the Cardiotoxicity of Isoproterenol

A study was conducted to determine the effect of DR and age on the toxicity of the drug isoproterenol (IPR) and to monitor physiological performance in rats under conditions of maximum metabolic output. In this experiment, the test subjects were individually housed and DR was implemented at 14 months of age. DR rats were given 40% less food than was consumed by their AL counterparts, whereas AL rats were given more food than they could eat. β-adrenergic agonists such as the catecholamine IPR have been used therapeutically in cardiac disorders to regulate (accelerate) heart rate and as antiasthmatics in the treatment of respiratory disease (Ferraiolo et al. 1987). High doses of IPR lead to maximum cardiac output, arrhythmias, and cardiac arrest (Sono et al. 1985). IPR is a useful drug to access cardiac fitness and the effects of heart disease because it produces cardiac effects (e.g., ischemia and myocardial infarction) that are similar to those resulting from occlusion of the coronary artery in humans.

Cardiac fitness as measured by heart rate, electrocardiogram (ECG), and body temperature was assessed in different age groups of male AL and DR rats. Physiological performance was measured in young (12-month-old) and old (36-month-old) AL and DR rats that were given a single intramuscular (IM) injection of IPR in doses ranging from .005 mg/kg LBM to 300 mg/kg LBM. A sample size of 10 AL rats and 10 DR rats were tested at each dose. The results of this study indicate that the dose of IPR necessary to elicit 50% mortality was between 10–20 mg/kg LBM in young AL rats and .005–.01 mg/kg LBM in old AL rats, whereas less than 10% mortality was found in young and old DR rats at a dose as high as 300 mg/kg LBM. Therefore, the observed differences in IPR-induced mortality were thus both age- and diet-dependent. Interestingly, IPR elevated the heart rate in both young AL and DR rats to a similar extent during the 24 hours after exposure, indicating that the efficacy of the drug was not affected by the nutritional state at that time. Supporting this conclusion is the observation that inversion of the T complex (ventricular repolarization) of the

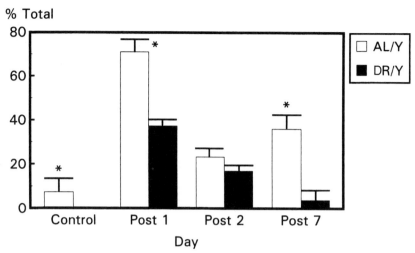

Figure 5. Effect of exposure to isoproterenol (10 mg/kg) on abnormal ECG waveforms in young AL and DR male F344 × BNF$_1$ rats (12 months of age). Percentage of abnormal waveforms ± standard errors are shown for the control day, together with days 1, 2, and 7 after injection. A conventional statistical method (Student's t-test) was used to analyze the data. * = significant difference between AL and DR groups. ($p < 0.5$). The value for the DR/Y control is 0% and is not shown on the graph.

ECG waveform occurred within minutes after IPR injection in both AL and DR rats, thereby suggesting that severe ischemia occurred in both groups. Apparently DR dramatically increased the fibrillation threshold of the heart, since long periods of drug-induced ischemia in DR rats did not cause arrhythmias and cardiac arrest.

The percentage of abnormal ECG waveforms (Figure 5) was lower in young DR rats than in AL rats during the control period and for up to 7 days past IPR exposure. Additionally, ECG variables, such as the amplitude of the R wave (ventricular contraction), P wave (atrial contraction), and T waves, as well as the QTE (duration of ventricular repolarization) and QRS (duration of ventricular depolarization) intervals changed adversely with age and after exposure to IPR. DR decreases the drug- and age-induced changes in these variables.

The rapid reduction in core temperature that occurs in DR rats immediately after IPR injection may protect the heart from damage, thereby increasing survival potential. Bonisch et al. (1979) reported that extraneuronal uptake of IPR and the efflux of IPR and its metabolites into heart tissue were reduced by lowering temperature. The fact that IPR-related cardiac arrhythmias are linked to calcium (Ca^{++}) overload and free radical damage to the sarcolemma and mitochondria suggests that DR may protect membranes by blocking Ca^{++} channels, thereby preventing cardiac arrest (Bonisch et al. 1979). Pretreatment with the drug verapamil, a known Ca^{++} channel blocker, before IPR exposure produced a similar antifibrillatory response as seen with DR (Will-Achuhab et al. 1986). Additionally, the expression of heat shock proteins (HSPs), such as HSP 70 and HSP 90, which have been reported to protect cells from damage resulting from changes in environmental temperatures and drug-related perturbations, were found to be increased by DR and decreased by the aging process (Pipkin et al. 1994). The rapid onset of hypothermia in DR rats following IPR injection may serve to increase the expression of HSPs, thereby decreasing mortality in DR rats.

Effect of Dietary Restriction and Age on the Toxicity of Ganciclovir

The effects of diet, age, and time of dose on mortality, behavior, and physiological function were monitored in mice injected with 400 mg/kg LBM of the drug ganciclovir (DHPG) (Berg et al. 1994). DHPG is a potent antiviral drug that is used therapeutically in immunosuppressed patients in the treatment of AIDS and in organ transplant recipients to prevent the occurrence of potentially lethal secondary infections. Unfortunately, the drug is extremely toxic, and exposure to DHPG must be limited.

Young (7–10-month-old) and middle-aged (19–22-month-old) female AL and DR mice were dosed with DHPG at 0, 6, 12, and 18 hours after lights-on (HALO) to determine mortality at different times of the day. The level of DR used in this study was 40%, and singly housed test subjects were placed on DR at 14 months of age. Mortality was 53% in AL young mice, 20% in AL middle-aged mice, 1.8% in DR middle-aged mice, and 1.7% in DR young mice. Therefore, as with IPR, both an age and a nutritional effect were observed. The toxicity of DHPG in AL mice was dependent on the time of dosing, with the greatest mortality occurring at 6 HALO (73%) and the lowest mortality occurring at 12 HALO (8%) (Berg et al. 1994). The fact that food consumption and activity declined in AL mice indicates an abnormal behavioral response that is highly correlated to the increased toxicity of DHPG in the nutritional group. DHPG-induced changes in physiological and behavioral performance in DR mice, characterized by increased temperature, decreased metabolic rate, and increased water consumption, may promote cytoprotective mechanisms such as the induction of stress proteins or the reduction of oxidative damage resulting from increases in catalase and other free radical scavengers.

Implications for the Design and Interpretation of Toxicity Studies

A number of procedural factors should be carefully analyzed and implemented to successfully incorporate DR into the experimental design of toxicity studies. First, chronobiological testing procedures that effectively synchronize the biological rhythm in DR and AL test groups may be important in the study of certain drugs. Differences in the activity period of AL and DR mice can be compensated for by adopting a night feeding regimen for DR animals, thereby ensuring that experimental variables in AL and DR rodents are sampled at equivalent phases of their biological cycles (Duffy et al. 1989, Duffy et al. 1990a). This appears to eliminate some confounding variables that can result from trying to interpret data from unsynchronized AL and DR animals.

Noninvasive physiological variables such as whole-body metabolism, temperature, motor activity, heart rate, and ECG are excellent biomarkers of aging, nutritional state, and toxicity and therefore should be used in drug toxicity studies (Duffy et al. 1989, 1990b). As described earlier, metabolism and temperature at the whole-animal level are highly correlated with, and in some cases may directly regulate, biochemical events at the molecular level, such as DNA damage and repair (Lipman et al. 1989), protooncogene expression (Nakamura et al. 1989), intermediary metabolism (Feuers et al. 1989), and drug metabolism (Leakey et al. 1989). All of these factors are known to modulate drug metabolism and toxicity. Noninvasive chronobiological and physiological markers can be used for "reconnaissance" to predict which variables will be affected by age, DR, and toxicity and also to pinpoint specific time intervals when maximum effects may occur. This scientific approach promotes the efficient and economical use of material, equipment, and animal resources.

Behavioral variables related to DR also have broad implications for toxicity studies. The excessive drinking (SIP) that is triggered by DR increases the rate of fluid turnover and may accelerate the elimination and metabolism of drugs, thereby decreasing their toxicity (Duffy et al. 1994). Increased motor activity in DR rodents with no increase in metabolic rate indicates greater metabolic efficiency (Duffy et al. 1989, 1990b) together with increased efficiency of drug-metabolizing enzymes (Leakey et al. 1989).

The results of the FAD study clearly demonstrate that the physiological effects of acute DR are significantly different from those associated with chronic DR (Duffy et al. 1994). Changes in those parameters related to temperature and metabolism can have a dramatic effect on drug and chemical toxicity, as well as drug efficacy and metabolism (Duffy et al. 1994). Therefore, in addition to the level of DR, factors such as the frequency and duration of feeding are important parameters that must be evaluated when designing toxicity studies and interpreting results. These types of protocol design factors need to be considered before DR can be applied to humans.

Nutrition and aging studies are highly relevant to government agencies that regulate the use of toxic substances. Recent drug (DHPG) studies revealed that the time of day at which drugs are administered has a profound effect on mortality and toxicity (Berg et al. 1994). These results are extremely useful in developing effective drug-dosing and -testing regimens that optimize the timing of drug delivery in a way that increases the efficacy of drugs while decreasing drug toxicity (Berg et al. 1994). The IPR studies presented here, together with the DHPG studies (Berg et al. 1994), conclusively demonstrate that age and nutritional state modulate drug toxicity as a consequence of aging and age-related disease. The elderly take a high percentage of prescription drugs compared with other age groups, and all segments of the human population, including the elderly, often eat low-calorie diets. Unfortunately, current drug-testing protocols in rodents do not adequately monitor the effects of age and nutrition, and consequently, additional studies are necessary to develop a database that can be used to accurately assess the interactions among these variables in normal as well as "high risk" animal populations.

At present, most chronic bioassay studies are plagued with the problem of low survival rate and a heterogeneity of response in AL control populations. Since previous studies (Duffy et al. 1989, 1990a) have shown that DR increases life span, a long-lived control population could be developed by maintaining DR rodents at a minimal level of food restriction (possibly 5–10%), thereby increasing longevity. The use of DR to maintain control populations at a constant predetermined body weight or (LBM) may prove to be an effective method for ensuring that control groups from different studies or animal colonies are comparable and consistent. The prudent and careful use of DR will increase the sensitivity and accuracy of food and drug toxicity studies and is a potential strategy for significantly improving the health of test animals.

References

Beckman AL (1978) Hypothalamic and midbrain function during hibernation. In Veale WL, Lederis K (eds), Current studies of hypothalamid function. Basel, Karger, pp 29–43

Berg TF, Breen PJ, Feuers RJ, et al. (1994) Acute toxicity of ganciclovir: effect of dietary restriction and chronobiology. Food Chem Toxicol 32(1):45–50

Bonisch H, Uhlig W, Frendelenburg U (1979) Temperature sensitivity of the extraneuronal uptake and metabolism of isoprenaline in the perfused rat heart. Pharmacology 306:229–239

Duffy PH, Feuers RJ, Leakey JA, et al. (1989) Effect of chronic caloric restriction on physiological variables related to energy metabolism in the male Fischer 344 rat. Mech Ageing Dev 48:117–133

Duffy PH, Feuers RJ, Nakamura KD, et al. (1990b) Effect of chronic caloric restriction on the synchronization of various physiological measures in old female Fischer 344 rats. Chronobiol Int 7:113–124

Duffy PH, Feuers RJ, Hart RW (1990a) Effect of chronic caloric restriction on the circadian regulation of physiological and behavioral variables in old male B6C3F1 mice. Chronobiol Int 7:291–303

Duffy PH, Feuers RJ, Leakey JA, Hart RW (1991) Chronic caloric restriction in old female mice: changes in the circadian rhythms of physiological and behavioral variables. In Fishbein L (ed), Biological effects of dietary restriction. Springer-Verlag, New York, pp 245–263

Duffy PH, Feuers RJ, Pipkin JL, Hart RW (1994) Effect of chronic caloric restriction: physiological and behavioral response to alternate day feeding in old female B6C3F1 mice. Age 17:13–21

Ferraiolo BL, Halldin MM, Asscher Y, et al. (1987) Pharmacokinetics and excretion of unique beta-adrenergic agonists. Pharmacology 34:57–166

Feuers RJ, Duffy PH, Leakey JA, et al. (1989) Effect of chronic caloric restriction on hepatic enzymes of intermediary metabolism in the male Fischer 344 rat. Mech Ageing Dev 48:179–189

Kritchevsky D, Weber MM, Klurfeld DM (1984) Dietary fat versus caloric content in initiation and promotion of 7,12-dimethylbenz(a)anthracene-induced mammary tumorigenesis in rats. Cancer Res 44:3174–3177

Leakey JEA, Cunny HC, Bazare J, et al. (1989) Effects of aging and caloric restriction on hepatic drug metabolizing enzymes in the Fischer 344 rat. I: the cytochrome p-450 dependent monooxygenase system. Mech Ageing Dev 48:145–155

Levine R, Levine S (1989) Role of the pituitary-adrenal hormones in the acquisition of schedule-induced polydipsia. Behav Neurosci 103(3):621–637

Lipman JM, Turturro A, Hart RW (1989) The influence of dietary restriction on DNA repair in rodents: a preliminary study. Mech Ageing Dev 48:135–143

Lyman CP (1982) Entering hibernation. In Lyman C, Malan A, Wang L (eds), Hibernation and torpor in mammals and birds. Academic Press, San Diego, pp 37–53

Maeda H, Gleiser CA, Masoro EJ, et al. (1985) Nutritional influences on aging of Fischer 344 rats. II. pathology. J Gerontol 40:671–688

Masoro EJ (1988) Mini-review: food restriction in rodents: an evaluation of its role in the study of aging. J Gerontol 43:B59–B64

McCay CM, Crowell MF, Maynard LA (1935) The effect of retarded growth upon the length of life span and upon the ultimate body size. J Nutr 10:67–69

Nakamura KD, Duffy PH, Turturro A, Hart RW (1989) The effect of dietary restriction on MYC protooncogene expression in mice: a preliminary study. Mech Ageing Dev 48:199–205

Pipkin JL, Hinson WG, Feuers RJ, et al. (1994) The temporal relationships of synthesis and phosphorylation in stress proteins 70 and 90 in aged caloric restricted rats exposed to bleomycin. Aging Clin Exp Res 6:125–132

Ruggeri BA, Klurfeld DM, Kritchevsky D (1987) Biochemical alterations in 7,12-dimethylbenz[a]anthracene-induced mammary tumors from rats subjected to caloric restriction. Biochem Biophys Acta 929:239–246

Sabatino F, Masoro EJ, McMahan CA, Kuhn RW (1991) Assessment of the role of the glucocorticoid system in the aging processes and in the action of food restriction. J Gerontol 46:B171–B179

Sarkar NH, Fernandes G, Telang NT, et al. (1982) A low-calorie diet prevents the development of mammary tumors in C3H mice and reduces circulating prolactin level, murine mammary tumor virus expression, and proliferation of mammary alveolar cells. Proc Natl Acad Sci U S A 79:7758–7762

Sono K, Kurahashi K, Fujiwara M (1985) Changes in the incidence and duration of ventricular fibrillation in dependence on the extraneuronal accumulation of isoprenaline in the perfused rat heart. Arch Pharmacol 331:76–81

Walford RL, Liu RK, Gerbase-Delima M, et al. (1974) Long term dietary restriction in immune function in mice: Response to sheet red blood cells and to mitogenic agents. Mech Ageing Dev 2:447–454

Weindruch R, Walford RL (1982) Dietary restriction in mice beginning at one year of age: effect on life span and spontaneous cancer incidence. Science 215:1415–1418

Will-Achahab L, Krause EG, Schulze W, Bartel S (1986) Adrenergic regulation of the acute ischaemic myocardium: facts, interpretation and consequences. Cor Vasa 28(2):107–113

Winn P, Clark JM, Clark AJ, Parker GC (1992) NMDA lesions of lateral hypothalamus enhance the acquisition of schedule-induced polydipsia. Physiol Behav 52(6):1069–1075

Statistical Considerations in Long-Term Dietary Restriction Studies

Joseph K. Haseman

National Institute of Environmental Health Sciences

Introduction

Control Fischer 344 (F344) rats in National Toxicology Program (NTP) long-term rodent carcinogenicity studies are showing a steady time-related increase in body weight and a corresponding decrease in survival (Rao et al. 1987, 1990b, Haseman and Rao 1992, Haseman 1993). Untreated F344 rats are now approximately 15% heavier than they were a decade ago (Table 1), and 2-year survival rates of 50% or less for untreated male F344 rats now frequently occur. Such reduced survival may adversely affect the sensitivity of the rodent bioassay for detecting carcinogenic effects. It is unclear why body weights are increasing and survival is decreasing in control animals. Increased body weights may be due to the selective breeding of heavier animals, and time-related increased incidences of leukemia may be contributing to the reduced survival.

In the carcinogenicity studies carried out by the National Cancer Institute (NCI) during the 1970s, body weights in control B6C3F1 mice showed no time-related increases (Rao et al. 1990a). However, in later NTP studies, mice began to show body weight trends similar to those observed in F344 rats (Haseman 1992, Haseman et al. 1994). One possible explanation is the NTP protocol change from group housing to individual housing. Individually housed mice have shown greater body weights and higher tumor incidence than group-housed mice (Haseman et al. 1994). The NTP has now returned to the practice of group housing female mice but continues to individually house male mice because of fighting problems among group-housed males (Haseman et al. 1994).

Increased body weights become an issue because of the well-documented association between elevated body weight and increased tumor incidence,

Table 1. Maximum mean body weight and survival (at 100 weeks of age) in untreated control F344 rats in NCI/NTP Studies

Year of study start	Number of studies	Body weight (g)		Survival (%)	
		Males	Females	Males	Females
1971	3	408	307	85	83
1972	22	422	310	77	78
1973	42	441	313	82	82
1974	10	444	315	77	83
1975	4	466	318	66	81
1976	9	433	316	74	82
1977	15	447	321	75	79
1978	12	441	308	75	79
1979	11	452	319	76	84
1980	10	470	335	66	75
1981	12	477	347	62	69
1982	6	482	359	62	67
1983	7	482	346	56	70

Source: Haseman and Rao (1992), plus three additional 1983 studies.

a correlation that in the NTP historical control database is most notable for mammary gland fibroadenoma in female F344 rats (Rao et al. 1987, Haseman and Rao 1992, see also Figure 1) and liver tumors in B6C3F1 mice (see Figure 2).

Body Weight Reductions in NTP Studies

The association between tumor incidence and body weight is important because the top dose used in long-term rodent studies may decrease body

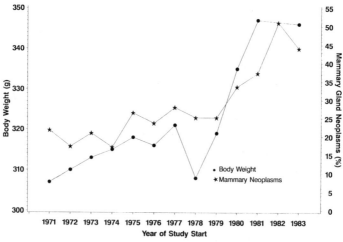

Figure 1. Time-related changes in body weight and mammary gland neoplasms in untreated female F344 rats from 2-year NCI/NTP studies.

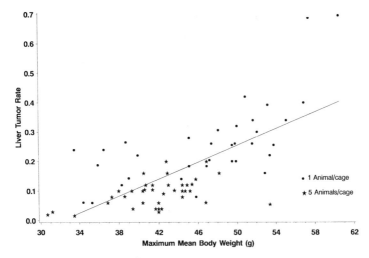

Figure 2. Association between liver tumor incidence and maximum mean body weight in control female B6C3F1 mice. The line represents the best fitting least squares linear regression based on data from both individually and group-housed animals.

weight, thereby possibly making it more difficult to detect carcinogenic effects. Table 2 shows the distribution of body weight reductions in the top-dose group for the chemicals evaluated by the NTP in Technical Reports 370–420. These body weight reductions refer to the mean body weight reduction in the top-dose group relative to controls over the course of the study. Approximately 40% of NTP studies produce average body weight decrements in the top-dose group in the 0–5% range. Although such small body weight changes may not be cause for concern, a substantial number of chemicals show average body weight reductions in the top-dose group that exceed 5%. For female mice in particular, body weight differences of 15% or more frequently occur (Table 2).

Table 2. Average differences in body weight between top-dose and control groups for 45 chemicals evaluated in NTP technical reports 370–420

Mean body weight relative to controls (%)	Male rats	Female rats	Male mice	Female mice	% carcinogens
0 to +5	3	5	9	6	43 (10/23)
−5 to 0	21	16	18	16	30 (21/71)
−10 to −5	13	11	7	10	46 (19/41)
−15 to −10	7	9	5	3	38 (9/24)
Decrease >15	1	4	6	10	33 (7/21)
Total	45	45	45	45	

Table 3. Example of the possible confounding effects of reduced body weight: incidence of liver tumors in female B6C3F1 mice given o-nitroanisole

Dose (ppm)	Survival (%)	Body weight reduction (%)[a]	Tumor rate (%)
0	76 (38/50)	—	34 (17/50)
666	52 (26/50)	2	44 (22/50)
2000	66 (33/50)	12	74 (37/50)[*]
6000	90 (45/50)	38	40 (20/50)

[a]Change relative to controls.
[*]$p < 0.001$ increase relative to controls.

Are these reduced body weights of sufficient magnitude to mask the detection of carcinogenic effects and produce false-negative outcomes? One way to approach this matter is to evaluate the association between rodent carcinogenicity and body weight changes in the top-dose group. Table 2 shows that the likelihood of a chemical being declared a rodent carcinogen is not correlated with overall body weight changes in the top-dose group relative to the concurrent control group. Alternatively, the mean body weight reduction in the top-dose group is 7% for both carcinogens and noncarcinogens and is 5% for those chemicals with "equivocal" evidence of carcinogenic activity, again no difference. These results suggest that there are relatively few (if any) rodent carcinogens that are undetected solely because of an association between tumor incidence and body weight.

Nevertheless, examination of those studies with extreme body weight reductions shows some interesting trends. For example, in the NTP o-nitroanisole study (NTP Technical Report 416), top-dose female mice reached a maximum mean body weight of only 27.2 g compared with 51.7 g in the untreated controls. The average body weight reduction in the top-dose group relative to controls over the course of the study was 38%. The NTP considered this chemical to provide some evidence of liver carcinogenicity in female mice; the liver tumor rates are summarized in Table 3. Importantly, increased liver tumor incidence was not observed in the top-dose group. Survival differences could not explain this lack of increase in liver tumors, since survival in the top-dose group actually exceeded that of the controls (Table 3). The NTP considered the increased liver tumor incidence in the mid-dose group to be chemically-related, and concluded that the lack of effect in the top-dose group was due to the reduced body weights. However, if the data from the mid-dose group were not available, this particular carcinogenic effect would have been missed altogether, since the liver tumor responses in the low- and high-dose female mouse groups were similar to controls.

Table 4. Incidences of mammary gland tumors in female rats for NTP studies with no survival differences and >10% decreases in body weight in the top-dose group

NTP technical report no.	Mammary gland tumors (%)		Mean body weight decrease in the top-dose group
	Control	Top dose	
304	46 (23/50)	14 (7/50)	17
321	40 (20/50)	2 (1/50)	14
333	32 (16/50)	10 (5/50)	22
357	60 (30/50)	12 (6/50)	16
368	20 (10/50)	4 (2/50)	15
387	42 (21/50)	4 (2/50)	21
389	33 (20/60)	14 (8/59)	17
395	48 (24/50)	10 (5/50)	12
404	36 (18/50)	8 (4/50)	26
406	44 (22/50)	12 (6/50)	11
407	46 (23/50)	24 (12/50)	12
408	30 (15/50)	4 (2/50)	11
409	58 (29/50)	18 (9/50)	12
439	30 (15/50)	10 (5/50)	16
442	46 (23/50)	48 (24/50)	10

A second example of the possible confounding effect of body weight on tumor incidence is illustrated in Table 4. With one exception, top-dose female F344 rats with normal survival rates and reduced (> 10%) mean body weights relative to controls have significantly lower incidences of mammary gland fibroadenoma than controls. The one exception is p-nitrobenzoic acid (Technical Report 442), for which the top-dose tumor incidence is similar to that of the control group (Table 4). One possible explanation for this finding is that p-nitrobenzoic acid causes mammary gland fibroadenoma, but this is not detected because the potential tumor increase is offset by a decrease related to the reduced body weight. On the other hand, since this chemical had the smallest body weight reduction of the Table 4 chemicals (and only a 3% body weight reduction during the first 13 weeks of the study), it is also possible that the body weight reduction was not of sufficient magnitude to alter the incidence of mammary gland fibroadenoma.

Concerns about the association between tumor incidence and body weight most often involve possible false-negative outcomes, since only rarely does the chemical under study actually increase body weight (Table 2). However, false-positive results could conceivably occur if the chemical under consideration causes a marked increase in body weight. Table 5 shows an example of a chemical that increases body weight and liver tumor incidence in male and female B6C3F1 mice. Although this particular chemical (N-methylolacrylamide) produced tumors at multiple sites in

Table 5. Incidence of liver tumors in top dose and control mice receiving N-methylolacrylamide

| Sex | Incidence of liver tumors(%) | | Body weight[a] top dose |
	Controls	Top dose	
M	24 (12/50)	52* (26/50)	+8
F	12 (6/50)	35* (17/49)	+13

*$p < 0.01$ vs. controls; considered by the NTP to be a chemically related effect.
[a]Mean body weight in top-dose group relative to controls during the course of the study, which used an ad libitum feeding protocol for all groups.

mice, there is at least the possibility that the increased body weights may have influenced the incidence of liver tumors in the top-dose group.

NTP Dietary Restriction Study: Statistical Issues

Because of concerns about the impact of body weight on the interpretation of rodent carcinogenicity studies, the NTP initiated a comprehensive dietary restriction (DR) study. This study included the standard bioassay design with three dosed groups and controls, fed ad libitum (AL) (Huff et al. 1988). However, the study used additional groups to study the impact of body weight differences. These included 1) a control group that was weight matched to the mean body weight of the high-dose AL group, and 2) food-restricted control and high-dose groups. The study design specified that food-restricted controls receive the amount of food necessary to maintain a body weight of approximately 80% of AL controls (food-restricted high-dose animals received this amount of food also).

One objective of the NTP study was to compare the various protocols with respect to sensitivity for detecting rodent carcinogenicity. Four different chemicals were examined, identified in this report as simply chemicals A, B, C, and D. More detailed information on experimental design is given elsewhere (Kari and Abdo, this volume). The results from the NTP DR study given in this report should be considered preliminary, since the pathology data have not yet been finalized and no formal evaluation of the data by the NTP has yet taken place. Data from these four studies will be published following public peer review, which is tentatively scheduled for June 20–21, 1995.

Dietary restriction studies introduce new statistical issues, including 1) determining the reproducibility of body weights with food restriction and/ or by weight matching, 2) evaluating the association between body weight and site-specific tumor incidence, and 3) comparing study protocols with regard to sensitivity for detecting rodent carcinogenicity. These issues are discussed in more detail below.

Table 6. Average percent body weight difference[a] from ad libitum (AL) controls in NTP dietary restriction studies

Sex/ species	Chemical[b]	Weight-matched control	High dose	Food-restricted control	Food-restricted high dose
MR	A	−5.4	−6.5	−14.3	−19.1
MR	B	−0.6	−2.8	−12.3	−19.0
MR	D	−8.0	−7.2	−13.0	−15.6
FR	A	−16.1	−16.9	−14.9	−26.0
FR	D	−13.9	−10.5	−14.9	−17.2
MM	B	−10.8	−12.0	−13.0	−29.4
MM	C	−17.1	−17.5	−26.9	−33.2
FM	C	−21.2	−17.9	−31.6	−35.5

MR = male rats, FR = female rats, MM = male mice, FM = female mice.
[a]Averaged over the course of the study.
[b]Chemical D was evaluated in a lifetime study; others were 2-year studies.

Table 6 compares the mean body weights of AL controls with those of the two other control groups (weight-matched and food restricted) and with the two high-dose (AL and food-restricted) groups. Table 6 shows that 1) with the exception of male rats given chemical B, body weights in AL high-dose animals were reduced 7–18% relative to AL controls; 2) the food restriction protocol produced relatively consistent body weight reductions for chemicals A, B, and D, but not for chemical C; 3) food-restricted high-dose animals were considerably lighter than either food-restricted controls or high-dose AL animals; and 4) the weight matching was reasonably successful in achieving similarity of the average body weights of high-dose and weight-matched control groups.

Comparison of the growth curves of high-dose and weight-matched control groups indicates that the two groups tended to show different body weight patterns over time. High-dose animals tended to show a steady increase in body weight, whereas weight-matched controls often showed an "up and down" body weight response over time, implying that the amount of food made available to them varied significantly from week to week as the experimenters attempted to match the body weights of the high-dose group. Figures 3 and 4 illustrate the most striking of these patterns and suggest that further improvements may be possible in the application of the weight-matching procedures.

Table 7 compares site-specific tumor rates of untreated male and female F344 rats with those of food-restricted controls. In this table, rates have been averaged over studies for simplicity, but the statistical evaluation (logistic regression analysis) compared the two control groups within

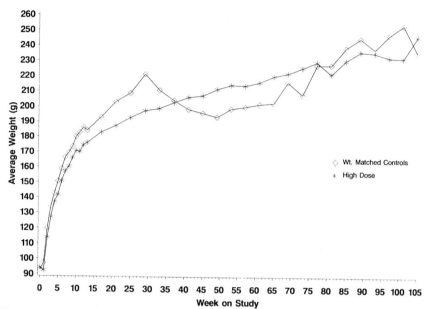

Figure 3. Two-year dosed feed study of chemical A in female rats.

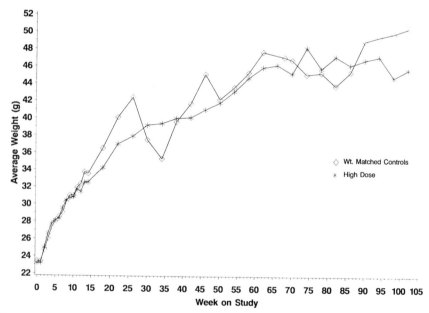

Figure 4. Two-year gavage study of chemical B in male mice.

Table 7. Comparison of survival and tumor rates (%) in ad libitum (AL) and food-restricted (FR) control F344 rats.

Tumor type	Males		Females	
	AL	FR	AL	FR
Adrenal gland pheochromocytoma	24 (38/160)	19 (30/159)	3 (3/109)	7 (7/106)
Thyroid gland c-cell tumors	9 (15/160)	10 (16/159)	12 (13/109)	5 (5/109)[*]
Pituitary gland tumors	29 (47/160)	21 (32/153)[*]	50 (55/109)	44 (48/109)
Mononuclear cell leukemia	52 (84/160)	46 (73/160)[a]	45 (49/110)	48 (53/110)
Mammary gland fibroadenoma	10 (16/160)	5 (8/160)	65 (72/110)	37 (41/110)[**]
Preputial/clitoral gland tumors	11 (17/160)	5 (8/160)[*]	18 (19/108)	7 (8/109)[*]
Pancreas acinar cell tumors	11 (18/160)	3 (5/157)[**]	0 (0/110)	0 (0/107)
Testis interstitial cell tumors	90 (144/160)	94 (150/160)	—	—
Uterine polyps	—	—	12 (13/110)	7 (8/110)
2-year survival (%)	57 (91/160)	64 (103/160)	55 (60/110)	72 (79/110)[**]

[a]Significantly (p < 0.05) reduced in 2-year studies (44/100 vs. 28/100), but not in lifetime study (40/60 vs. 45/60).
[*]Reduced (p < 0.05) relative to AL controls.
[**]Significant (p < 0.01) relative to AL controls.

Table 8. Comparison of survival and tumor rates (%) in ad libitum (AL) and food-restricted (FR) control B6C3F1 mice

Tumor type	Males		Females	
	AL	FR	AL	FR
Harderian gland tumors	6 (6/101)	4 (4/104)	6 (3/51)	6 (3/50)
Liver tumors	54 (55/101)	22 (23/104)[**]	45 (23/51)	8 (4/50)[**]
Lung tumors	28 (28/101)	17 (18/104)[*]	2 (1/51)	12 (6/50)
Malignant lymphoma	9 (9/101)	8 (8/104)	16 (8/51)	4 (2/50)[*]
Hemangioma/hemangiosarcoma	5 (5/101)	2 (2/104)	8 (4/51)	2 (1/50)
Survival (%)	80 (81/101)	88 (91/104)[a]	65 (33/51)	94 (47/50)[**]

[a]Significantly (p < 0.05) increased for chemical C (41/51 vs. 49/50), but not for chemical B (40/50 vs. 42/54).
[*]Reduced (p < 0.05) relative to AL controls.
[**]Significant (p < 0.01) relative to AL controls.

each study and then combined this information statistically across studies to obtain a single overall result. The most striking effects of food restriction were on mammary gland fibroadenoma in female rats. Other tumors showing reduced incidence in the food-restricted controls included clitoral/preputial gland neoplasms, pancreatic acinar cell tumors (males), leukemia (males), pituitary gland neoplasms (males), and thyroid c-cell tumors (females). All other tumors appeared to be unaffected by food restriction. Two-year survival was increased (p < 0.01) in food-restricted female rats and marginally increased in food-restricted males (Table 7).

Table 8 shows the association between body weight and tumor incidence for male and female B6C3F1 control mice. This table shows a striking correlation between reduced body weight and decreased liver tumor

Table 9. Incidence of liver tumors in male mice exposed to chemical B

Group	No. of survivors	Mean body weight[a]	Liver tumor rate Adenoma	Carcinoma	Combined
Ad libitum (AL) controls	40	—	23 (14/60)	22 (13/60)	43 (26/60)
Low dose	41	+1.1	55 (33/60)*	27 (16/60)	67 (40/60)*
Mid dose	41	−2.6	47 (28/60)*	42 (25/60)*	67 (40/60)*
High dose	46	−12.0	75 (45/60)*	13 (8/60)	75 (45/60)*
Weight-matched control	45	−10.8	10 (6/60)*	15 (9/60)	25 (15/60)*
Restricted control (2 yr)	42	−13.0	23 (14/62)	11 (7/62)	31 (19/62)
Restricted high dose (2 yr)	44	−29.4	13 (8/60)	3 (2/60)*	15 (9/60)*

[a]Expressed as a percentage increase or decrease relative to AL controls.
*$p < 0.05$ relative to AL controls.

incidence in both sexes. Other tumors showing weaker associations include lung tumors (males) and malignant lymphoma (females). The associations between decreased lung and liver tumor incidence and reduced body weight in male B6C3F1 mice have also been recently reported for control animals in the current NTP historical control database (Haseman et al. 1994). Note also the improved survival in food-restricted mice, particularly females (Table 8).

Perhaps the most unusual finding in the NTP dietary restriction study is the effect of chemical B on liver tumor incidence in male mice (Table 9). In the usual bioassay with AL feeding, all three doses of chemical B show a clear increase in liver tumor incidence. This increase occurs both for groups showing no body weight effects (low and mid dose) and for one group showing a 12% decrease in mean body weight (high dose). Comparison with a weight-matched control only increases the significance of the elevated liver tumor rate in the high-dose group. However, chemical B given in a food-restricted regimen (which resulted in animals with much lower body weights than those normally found in NTP studies) actually produces a decrease in liver tumor incidence (Table 9). This decrease is even more striking than the one noted above in the NTP o-nitroanisole study.

The implications of this finding on the interpretation of rodent bioassay results are currently under review. One possible interpretation is that chemicals whose carcinogenicity depends on the body weights of the animals being evaluated (such as chemical B) may not be as important from a public health point of view as those rodent carcinogens whose effects are evident regardless of body weight.

Although tumor rates in the NTP DR studies are preliminary and the data have not yet been fully evaluated, one can nevertheless estimate which increased tumor incidences will likely be regarded as chemically related

Table 10. Summary of carcinogenic effects detected by various protocols

Chemical	Sex/species	Tumor site	Estimated strength of carcinogenic effect[a]			
			Standard protocol	Weight matched	Food restricted	
					2 year	Lifetime
A	MR	Pancreas	+	+	−	−
A	MR	Leukemia	−	+	−	−
A	FR	Leukemia	−	E	−	−
A	FR	Urinary bladder	−	E	−	+
B	MR	Urinary bladder	+	+	−	−
B	MM	Liver	+	+	−	−
B	MM	Adrenal cortex	−	E	−	−
C	(No carcinogenic effects observed)					
D	MR	Preputial gland	−	+	NA	−
D	FR	Clitoral gland	−	+	NA	+
D	FR	Liver	E	E	NA	−

+ = carcinogenic, E = equivocal, − = not carcinogenic,
NA = not applicable (no food-restricted 2-year animals),
MR = male rats, FR = female rats, MM = male mice, FM = female mice.
[a]These studies have not yet been formally evaluated by the NTP; estimates may or may not reflect the final NTP judgment regarding a chemical's carcinogenic activity.

by the NTP for each protocol (i.e., AL feeding, comparison with weight-matched control, restricted feeding protocol for either 2 years or lifetime). These comparisons are summarized in Table 10 and suggest that the most sensitive protocol for detecting rodent carcinogenicity is the comparison with weight-matched controls. Use of food-restricted controls detects the fewest carcinogenic effects, regardless of whether the study duration is 2 years or extended to a lifetime study. One reason for this may be that the food restriction portion of the study was not designed to eliminate the body weight differences between high-dose and control groups. As a result, food-restricted high-dose animals remained lighter than food-restricted controls, and the protective effects of reduced body weight may well have masked some carcinogenic effects. Interestingly, none of the carcinogenic effects detected in the AL protocol was detected in the food-restricted protocol, and neither of the effects found in the food-restricted protocol was evident in the AL protocol (Table 10).

Conclusions

Examination of previous NTP studies has confirmed that some (but not all) site-specific tumors are correlated with body weight. However, most NTP long-term studies do not produce large body weight reductions in the top-dose group. Furthermore, the magnitude of body weight reduction in the top-dose group is not associated with the detection of rodent carcinogenicity. These findings suggest that there are relatively few if any rodent carcinogens that are undetected solely because of a correlation with body weight. The NTP data also suggest that a sizable reduction in body weight may be necessary before false-negative outcomes occur. False-positive outcomes associated with body weight differences do not appear to be a major problem, since only rarely does the chemical being evaluated increase body weight.

A preliminary statistical evaluation of the NTP DR studies confirmed that certain site-specific tumors (most notably mammary gland fibroadenoma in female F344 rats and liver neoplasms in B6C3F1 mice) are strongly correlated with body weight. In at least one case, the apparent carcinogenicity of the chemical under evaluation depended on whether the animals were food restricted, although the level of body weight reduction in the DR studies was much greater than that normally observed in NTP long-term studies.

Although no final evaluation of the DR studies has yet been carried out by the NTP, some preliminary conclusions can be made: 1) Weight-matched controls generally had mean body weights that were similar to those observed in the top-dose group. However, improvements may be possible in achieving this equivalency without the large weekly body weight fluctuations that were observed in some weight-matched control groups. 2) Using weight-matched controls may provide some improved study sensitivity relative to the use of AL controls. 3) The food-restricted protocol detected the fewest carcinogenic effects of the various methods studied.

References

Haseman JK (1992) Value of historical controls in the interpretation of rodent tumor data. Drug Inf J 26:191–200

Haseman JK (1993) The value and limitations of a large source of control animal pathology data. In McAuslane JAN, Parkinson C, Lumley CE (eds), Computerized control animal pathology databases: will they be used? Centre for Medicines Research, London, pp 11–16

Haseman JK, Bourbina J, Eustis SL (1994) Effect of individual housing and other environmental design factors on tumor incidence in B6C3F1 mice. Fundam Appl Toxicol 23:44–52

Haseman JK, Rao GN (1992) Effects of corn oil, time-related changes, and inter-laboratory variability on tumor occurrence in control Fischer 344 (F344/N) rats. Toxicol Pathol 20:52–60

Huff JE, McConnell EE, Haseman JK, et al. (1988) Carcinogenesis studies: results of 398 experiments on 104 chemicals from the U.S. National Toxicology Program. In Maltoni C, Seilkoff I (eds), Living in a chemical world—occupational and environmental significance of industrial carcinogens. Ann N Y Acad Sci 534:1–30

Rao GN, Haseman JK, Grumbein S, et al. (1990a) Growth, body weight, survival, and tumor trends in (C57BL/6 x C3H/HeN) F_1 (B6C3F1) mice during a nine-year period. Toxicol Pathol 18:71–77

Rao GN, Haseman JK, Grumbein S, et al. (1990b) Growth, body weight, survival, and tumor trends in F344/N rats during an eleven-year period. Toxicol Pathol 18:61–70

Rao GN, Piegorsch WW, Haseman JK (1987) Influence of body weight on the incidence of spontaneous tumors in rats and mice of long-term studies. Am J Clin Nutr 45:252–260

Fat, Calories, and Cancer

David Kritchevsky
The Wistar Institute

Watson and Mellanby (1930) were the first to show that the addition of fat in the form of 12.5–25.0% butter to a basal diet (3% fat) increased the incidence of experimental tumors. In their study the incidence of skin tumors in coal tar–treated mice increased from 34% to 57%. This observation was expanded by the research groups of Baumann at the University of Wisconsin and Tannenbaum at Michael Reese Hospital, in Chicago. Baumann et al. (1939) showed that increasing dietary fat from 5% to 25% increased the incidence of mouse tumors initiated by ultraviolet light or benzo(*a*)pyrene (BP) painting but had little effect on tumors resulting from injections of BP or methylcholanthrene (MCA). Their fat source was a commercially available saturated shortening (Crisco). Lavik and Baumann (1941) showed that the fatty acid portion of the fat and not the glycerol or unsaponifiable material was responsible for the increased incidence of MCA-induced skin tumors in mice. They also observed that the fat effect was most effective during the promotion phase of tumorigenesis.

The suggestion that the type of fat might influence carcinogenesis (Miller et al. 1944) led to a comparison of the effects of different fats on the carcinogenicity of *p*-dimethylaminoazobenzene (DAB). Rats were fed diets containing 5% fat. The fats were corn oil, hydrogenated coconut oil, coconut oil, trilaurin, and 4:1 and 1:4 mixtures of corn and hydrogenated coconut oils. The greatest incidence of hepatomas was seen in rats fed corn oil or corn oil–hydrogenated coconut oil 4:1. A later experiment (Kline et al. 1946) compared the effects of 5% corn oil, 5% olive oil, 20% corn oil, 20% Crisco, or 20% lard. The incidence of DAB-induced hepatomas seen after 4 months in rats fed these fats was 73%, 13%, 100%, 47%, and 60%, respectively. Rats fed a fat-free diet had a hepatoma incidence of 8% at 4 months but showed signs of essential fatty acid deficiency. Tannenbaum (1944) studied the effects of diets high (31%, 54.5 cal %) or low (2%, 4.8 cal %) in fat on BP-induced skin tumors in mice. He found that the high-fat diet enhanced tumorigenicity and showed that the fat effect was most evident during the promotion phase. In a later study (Silverstone and Tannenbaum 1950) it was demonstrated that a high-fat diet also led to greater incidence of spontaneous hepatomas in mice.

Carroll examined the effects of fat saturation on 7,12-dimethylbenz (a)anthracene (DMBA) on mammary tumorigenesis in rats (Gammal et al. 1967, Carroll and Khor 1971). They found that saturated fat was less cocarcinogenic than on saturated fat. Similar observations were made in the case of colon tumors induced by 1,2-dimethylhydrazine (DMH) (Broitman et al. 1977) and azaserine-induced pancreatic tumors (Roebuck et al. 1981). Ip et al. (1985) demonstrated a positive requirement for essential fatty acid for tumorigenesis. Adding linoleate to the saturated fat—rich diets of rats given DMBA showed a steady increase in tumorigenesis with increasing linoleate until 4.4%, after which incidence of tumors seemed to plateau. The requirement is for linoleic acid per se, which may explain why fish oil, which is very unsaturated but does not contain much linoleic acid, inhibits tumor growth (Karmali et al. 1984). Similarly, diets containing appreciable levels of *trans*-unsaturated fatty acid do not promote tumor growth compared with their *cis*-unsaturated counterparts (Selenskas et al. 1984, Sugano et al. 1989).

The level of dietary fat is important in determining the severity of induced tumors. Carroll and Khor (1971) fed DMBA-treated rats diets containing increasing levels of corn oil. A 10-fold increase from 0.5 to 5.0% of the diet had no effect on tumor incidence or multiplicity. Increasing corn oil from 5% to 10% of the diet led to a 22% increase in incidence and a 43% increase in multiplicity, but increasing corn oil concentration in the diet to 20% gave no further increase in either tumor incidence or multiplicity.

McCay et al. (1935) showed that male rats fed a severely restricted diet lived longer than ad libitum (AL) controls (820 ± 113 days vs. 483 ± 59 days) and showed fewer signs of degenerative disease, including tumors (McCay et al. 1939). The diet used by McCay contained 39.3 en % protein (casein), 27.5 en % carbohydrate (starch:sucrose 2.2:1), and 33.2 en % fat (lard:cod liver oil 2:1). Thirty years earlier Moreschi (1909) had demonstrated that underfeeding reduced the growth of transplanted tumors in mice. Rous (1914) showed that neither transplanted nor spontaneous tumors grew well in underfed rodents. For the next 30 years there was considerable interest in underfeeding and tumor growth. Many investigators observed decreased growth of various tumors in underfed mice and rats. Those data have been the subjects of several reviews (White 1961,

Kritchevsky and Klurfeld 1986). In the 1940s the laboratories of Tannenbaum and of Baumann were in the forefront of this research. Tannenbaum (1940) found that underfeeding reduced the incidence of both spontaneous and induced tumors in mice. Underfeeding of the same diet that is fed AL may lead to deficiencies in some micronutrients if we assume that the AL animals ingest limiting amounts of that substance. Tannenbaum (1942) recognized this and formulated diets of known energy content. Energy restriction studies using this diet showed that it reduced the incidence of spontaneous lung or mammary tumors or chemically induced skin tumors in several strains of mice (Tannenbaum 1942).

Energy restriction was most effective when imposed during the progression phase of tumorigenesis (Tannenbaum 1944). Tannenbaum (1945) showed that at the same level of caloric intake, increasing the fat content of the diet could lead to increased tumor incidence. Lavik and Baumann (1943) demonstrated that a low-energy, high-fat diet led to almost half the incidence (28%) of MCA-induced skin tumors in mice as did a high-energy, low-fat diet (54%). A diet high in both energy and fat led to a 66% incidence of tumors, and one low in both fat and energy yielded no tumors.

Caloric restriction (CR) was also shown to reduce the incidence of tumors induced in mice by ultraviolet irradiation (Rusch et al. 1945). In this situation there is no chance of diet affecting intermediate metabolism of the carcinogen. Tumor incidence in the high-calorie, low-fat group was 87%; that in the high-calorie, moderate-fat group was 63%; and that in the low-calorie, moderate- or low-fat groups was 24% and 7%, respectively.

After a 30-year hiatus, interest in caloric effects on carcinogenesis was rekindled. Kritchevsky et al. (1984) showed that 40% caloric restriction totally inhibited DMBA-induced mammary tumors in rats fed coconut oil despite the fact that the energy-restricted rats were fed twice as much fat as the AL controls. When the diet contained corn oil, 40% energy restriction did not completely inhibit tumorigenesis, but it was significantly lower than that seen in the AL rats. Boissonneault et al. (1986) administered DMBA to rats fed a high-fat diet (30% corn oil), a low-fat diet (5% corn oil) or the high-fat diet restricted by 18.5%. Tumor incidence in the three groups was 73%, 43%, and 7%, respectively. Klurfeld et al. (1987) found that 40% calorie restriction inhibited DMH-induced colon carcinogenesis in rats. The controls were fed half as much fat as were the test rats. Beth et al. (1987) treated rats with methylnitrosourea (MNU) to induce mammary tumors and fed them four diets of different fat content and composition. The diets were fed at levels of 35 kcal or 50 kcal/day. The lower-calorie diet resulted in a lower incidence of tumors in every case, and there was no effect of fat quantity or composition.

To determine the effective degree of energy restriction Klurfeld et al. (1989a) restricted energy in the diets of DMBA-treated rats by 10%, 20%, 30%, or 40%. When calories were restricted by 10%, tumor incidence was the same as that of the AL controls (60%), but tumor multiplicity and tumor burden fell by 32% and 47%, respectively. Restriction of calories by 20% reduced the incidence to 40%, but multiplicity and tumor burden were similar to those seen at 10% restriction. When energy was restricted by 30%, tumor incidence, multiplicity, and burden were reduced by 42%, 72%, and 91% compared with the controls. At 40% restriction, only one of 20 rats bore a tumor.

To determine whether a high dietary fat content can override the effects of caloric restriction, female rats were given DMBA by gavage and AL diets containing 5%, 15%, or 20% corn oil or given diets containing 20% or 26.7% corn oil but fed at an energy restriction of 25% (Klurfeld et al. 1989b). The AL rats yielded the expected result: going from 5% to 15% fat resulted in a 30% increase in tumor incidence, but there was no further increase when the dietary fat level was raised to 20%. The two 25% energy-restricted groups given 20% or 26.7% fat exhibited lower tumor incidence than the group fed 5% fat. Tumor multiplicity and tumor burden were lower as well. Welsch et al. (1990) fed DMBA-treated rats diets containing 5% or 20% fat fed AL or restricted by 29% in rats fed 5% fat and by 37% in those fed 20% fat. Energy restriction inhibited carcinogenesis in every case.

Virtually all studies of caloric restriction and tumorigenesis have been carried out in mice, rats, or hamsters. One study of eye cancer in cattle is worth noting (Anderson et al. 1970). Hereford heifers were fed either 1 pound of cottonseed cake/day, 2½ pounds of cottonseed cake/day, or 2½ pounds of cottonseed cake plus 3 pounds of oats. The incidence of ocular squamous cell carcinoma was 14% in cattle fed at the medium or high level and only 1.5% in those fed at the low level.

Energy restriction appears to be effective in reducing tumorigenesis even when instituted relatively late in the life span. Tannenbaum (1947) reported that the incidence of spontaneous breast tumors in 20-month-old DBA mice was zero when energy restriction began at weaning. When energy restriction began at 5 or 9 months of age, tumor incidence was reduced by 95% and 80%, respectively. Weindruch and Walford (1982) restricted by 44% the energy intake of 1-year-old cancer-prone mice. Incidence of lung tumors was reduced by 50%, lymphomas by 34%, and hepatomas by 7%. Life span was also increased. Kritchevsky et al. (1989) studied intermittent energy restriction in DMBA-treated rats and found tumor incidence to be related to feed efficiency. Ross and Bras (1971) found that when the caloric intake of rats was restricted by 60%, life span was increased by 40% and the incidence of spontaneous tumors fell by 90%. When the rats were energy restricted for 7 weeks postweaning and then returned to an AL regimen, survival did not increase but the incidence of spontaneous tumors fell by nearly 40%.

Exercise is another means of increasing caloric flux, and its effects on carcinogenesis have been investigated. Rusch and Kline (1944) subjected mice bearing a transplantable fibrosarcoma to forced exercise. In two experiments growth of the tumor was reduced by 25% and 40%. Vigorous treadmill exercise has been shown to reduce the incidence of DMH-induced colon tumors by about 50% (Klurfeld et al. 1988). Voluntary exercise (activity cage) has been reported to reduce mammary (Cohen et al. 1988) and pancreatic (Roebuck et al. 1990) tumors. Thompson et al. (1988, 1989) reported that treadmill exercise enhanced the growth of DMBA-induced mammary tumors in rats. Their exercised rats had a higher food intake and were heavier than the sedentary controls.

The mechanisms by which caloric restriction inhibits carcinogenesis are moot. Boutwell et al. (1949) suggested that energy-restricted female rats had been "pseudohypophysectomized," resulting in reduced sized of the uterus and ovaries. They also observed adrenal hypertrophy. Energy restriction has been shown to reduce levels of circulating mammotrophic hormones in rats (Sylvester et al. 1981) and mice (Sarkar et al. 1982). Oncogene expression in rats (Fernandes et al. 1987) and mice (Nakamura et al. 1989) is reduced by caloric restriction. Expression of c-fos and c-ki-ras mRNA is also reduced (Himeno et al. 1992). Energy restriction leads to enhanced DNA repair (Lipman et al. 1989).

Insulin deprivation inhibits tumor growth and cell division (Cohen and Hilf 1974). Tumor growth stops when tumor-bearing rats are rendered diabetic (Heuson and Legros 1972). Plasma insulin levels fall significantly when rats are subjected to CR (Klurfeld et al. 1989a,b). Insulin levels fall immediately upon institution of energy restriction and remain low throughout the period of restriction. Levels of IGF-I fall when energy restriction is introduced but return to normal levels within a few weeks; levels of IGF-II are unaffected (Ruggeri et al. 1989). Oxygen-derived free radicals have been implicated as possible factors in tumorigenesis, and there is a focus on the anticancer properties of antioxidant vitamins (A, C, and E) and carotenoids. Rao et al. (1990) reported that energy restriction increases the activities of superoxide dismutase, catalase and glutathione peroxidase in livers of aging rats. Zhu et al. (1991) measured levels of reduced growth-stimulating hormone (GSH) and oxidized glutathione (GSSG) in livers and tumor tissues of rats given MNU and subjected to 30% caloric restriction at two levels of dietary fat. Fat level had no effect on GSSG levels in livers or tumors. Hepatic GSH was reduced in calorie-restricted rats.

The ratio of large (greater than 100 mg) palpable to small (less than 100 mg) nonpalpable tumors changes with caloric restriction (Ruggeri et al. 1987). The ratio of large to small tumors is 4.88 in rats given DMBA and fed AL. At 10%, 20%, and 30% CR, the ratio is 3.76, 3.35, and 0.67, respectively. The data suggest that the tumors are established in all cases but fail to grow when nutrients are limited. Determination of the activities of various enzymes of carbohydrate metabolism in mammary tissue showed a 60% increase in hexokinase activity and a 74% decrease in phosphofructokinase activity in small, nonpalpaple versus large, palpable tumor tissue of rats subjected to 30% restriction. Other enzyme activities were not greatly affected. The conclusion from these studies is that the changes represent a compensatory mechanism for more efficient utilization of available substrate under conditions of reduced substrate activity. One might look on this as a competition for substrate between tumor and host in which the latter wins when available energy is reduced.

What is the relevance of these findings to human tumorigenesis? Hoffman (1927) suggested that energy excess was an important factor in cancer development. Berg (1975) proposed that cancers prevalent in the United States might be related to energy intake. Data compiled by the American Cancer Society show a relationship between overweight and cancer mortality in a cohort of over 1 million people (Garfinkel 1985). In a review of international data, Albanes (1987) found that increased body weight, high relative body weight, and high energy intake were associated with increased risk of cancer of the breast, colon, rectum, prostate, endometrium, kidney, cervix, ovary, thyroid, and gallbladder. An inverse correlation was seen in cancers of the bladder, stomach, and lung.

Jain et al. (1980) and Lyon et al. (1987) reported increasing risk of colon cancer with increasing caloric intake. Men in sedentary occupations are at greater risk of colon cancer than are those whose working life has been spent at hard physical work (Garabrant et al. 1984, Vena et al. 1985). The reverse may be true for prostate cancer (LeMarchand et al. 1991).

More than 400 hundred years ago Luigi Cornaro, who lived to be 102, suggested that "not to satiate oneself with food is the science of health"

(Cornaro 1918). Potter (1945), four centuries later, suggested that cancer risk could be reduced by eating less and exercising more. Almost all dietary guidelines, regardless of origin, suggest that we eat a variety of foods and maintain desirable weight. Nutritionally the easiest path to follow is variety, balance, and moderation (Kritchevsky 1993).

Acknowledgment

This review was supported, in part, by Research Career Award HL-00734 from the National Institutes of Health.

References

Albanes D (1987) Caloric intake, body weight and cancer: a review. Nutr Cancer 9:199–217

Anderson DE, Pope LS, Stephens D (1970) Nutrition and eye cancer in cattle. J Natl Cancer Inst 45:697–707

Baumann CA, Jacobi HP, Rusch HP (1939) The effect of diet on experimental tumor production. Am J Hygiene 30:1–6

Berg JW (1975) Can nutrition explain the pattern of international epidemiology of hormone-dependent cancer? Cancer Res 35:3345–3350

Beth M, Berger MR, Aksoy M, Schmahl D (1987) Comparison between the effects of dietary fat level and of calorie intake on methylnitrosourea-induced mammary carcinogenesis in female SD rats. Int J Cancer 39:737–744

Boissonneault GA, Elson CE, Pariza MW (1986) Net energy effects of dietary fat on chemically induced mammary carcinogenesis in F344 rats. J Natl Cancer Inst 76:335–338

Boutwell RK, Brush MK, Rusch HP (1949) Some physiological effects associated with chronic caloric restriction Am J Physiol 154:517–524

Broitman BA, Vitale JJ, Vavrousek-Jakuba E, Gottlieb LS (1977) Polyunsaturated fat, cholesterol and large bowel tumorigenesis Cancer 40:2453–2463

Carroll KK, Khor HT (1971) Effect of level and type of dietary fat on incidence of mammary tumors induced in female Sprague Dawley rats by 7.12 dimethylbenz (a) anthracene. Lipids 6:415–420

Cohen LA, Choi K, Wang CX (1988) Influence of dietary fat, caloric restriction and voluntary exercise on N-nitrosomethylurea-induced mammary tumorigenesis in rats. Cancer Res 48:4276–4283

Cohen ND, Hilf R (1974) Influence of insulin on growth and metabolism of 7.12 dimethylbenz(a)anthracene-induced mammary tumors. Cancer Res 34:3245–3252

Cornaro L (1918) The art of living long [english translation]. WF Butler, Milwaukee, WI

Fernandes G, Khare A, Langamere S, et al. (1987) Effect of food restriction and aging on immune cell fatty acids, functions and oncogene expression in SPF Fischer 344 rats. Fed Proc 46:567

Gammal EB, Carroll KK, Plunkett ER (1967) Effects of dietary fat on mammary carcinogenesis by 7.12-dimethylbenz(a)anthracene in rats. Cancer Res 27:1737–1742

Garabrant DH, Peters JM, Mack TM, Bernstein L (1984) Job activity and colon cancer risk. Am J Epidemiol 119:1005–1014

Garfinkel L (1985) Overweight and cancer. Ann Int Med 103:1034–1036

Heuson JC, Legros N (1972) Influence of insulin deprivation on growth of the 7,12 dimethylbenz(a)anthracene-induced mammary carcinoma in rats subjected to alloxan diabetes and food restriction. Cancer Res 32:226–232

Himeno Y, Engelman RW, Good RA (1992) Influence of caloric restriction on oncogene expression and DNA synthesis during liver regeneration. Proc Natl Acad Sci USA 89:5497–5501

Hoffman FL (1927) Cancer increase and overnutrition. Prudential Insurance Co, Newark, NJ

Ip C, Carter CA, Ip MM (1985) Requirement of essential fatty acid for mammary tumorigenesis in the rat. Cancer Res 45:1997–2001

Jain M, Cook GM, Davis EG, et al. (1980) A case-control study of diet and colorectal cancer. Int J Cancer 26:757–768

Karmali RA, Marsh J, Fuchs C (1984) Effect of omega-3 fatty acids on growth of a mammary rat tumor. J Natl Cancer Inst 73:457–461

Kline BE, Miller JA, Rusch HP, Baumann CA (1946) Certain effects of dietary fats on the production of liver tumors in rats fed p-dimethylamino-azobenzene. Cancer Res 6:5–7

Klurfeld DM, Weber MM, Kritchevsky D (1987) Inhibition of chemically-induced mammary and colon tumor promotion by caloric restriction in rats fed increased dietary fat. Cancer Res 47:2759–2762

Klurfeld DM, Welch CB, Davis MJ, Kritchevsky D (1989a) Determination of degree of energy restriction necessary to reduce DMBA-induced mammary tumorigenesis in rats during the promotion phase. J Nutr 119:286–291

Klurfeld DM, Welch CB, Einhorn E, Kritchevsky D (1988) Inhibition of colon tumor promotion by caloric restriction or exercise in rats. FASEB J 2:A433

Klurfeld DM, Welch CB, Lloyd LM, Kritchevsky, D (1989b) Inhibition of DMBA-induced mammary tumorigenesis by caloric restriction in rats fed high fat diets. Int J Cancer 43:922–925

Kritchevsky D (1993) Dietary guidelines: the rationale for intervention. Cancer 72:1011–1014

Kritchevsky D, Klurfeld DM (1986) Influence of caloric intake on experimental carcinogenesis: a review. Adv Exp Med Biol 206:55–68

Kritchevsky D, Weber MM, Klurfeld DM (1984) Dietary fat versus caloric content in initiation and promotion of 7.12-dimethylbenz(a)anthracene-induced mammary tumorigenesis in rats. Cancer Res 44:3174–3177

Kritchevsky D, Welch CB, Klurfeld DM (1989) Response of mammary tumors to caloric restriction for different time periods during the promotion phase. Nutr Cancer 12:259–269

Lavik PS, Baumann CA (1941) Dietary fat and tumor formation. Cancer Res 1:181–187

Lavik PS, Baumann CA (1943) Further studies on the tumor promoting action of fat. Cancer Res 3:749–756

LeMarchand L, Kolonel LN, Yoshizawa CN (1991) Lifetime occupational physical activity and prostate cancer risk. Am J Epidemiol 133:103–111

Lipman JM, Turturro A, Hart RW (1989) The influence of dietary restriction on DNA in rodents: a preliminary study. Mech Aging Dev 48:135–143

Lyon JL, Mahoney AW, West DW, et al. (1987) Energy intake: its relation to colon cancer. J Natl Cancer Inst 78:853–861

McCay CM, Crowell MF, Maynard LA (1935) The effect of retarded growth upon the length of life span and upon the ultimate body size. J Nutr 10:63–79

McCay CM, Ellis GH, Barnes LJ, et al. (1939) Chemical and pathological changes in aging and after retarded growth. J Nutr 18:15–25

Miller JA, Kline BE, Rusch HP, Baumann CA (1944) The effect of certain lipids on the carcinogenicity of p-dimethylaminoazobenzene. Cancer Res 4:756–761

Moreschi C (1909) Beziehungen Zwischen Ernahrung und Tumorwachstum. Z Immunitätsforsch 2:651–675

Nakamura KD, Duffy PH, Lu MS, et al. (1989) The effect of dietary restriction on myc protooncogene expression in mice: a preliminary study. Mech Ageing Dev 48:199–205

Potter VR (1945) The role of nutrition in cancer prevention. Science 101:105–109

Rao G, Xia E, Nadakavukaren MJ, Richardson A (1990) Effect of dietary restriction on the age-dependent changes in the expression of antioxidant enzymes in rat liver. J Nutr 120:602–609

Roebuck BD, McCaffrey J, Baumgartner KJ (1990) Protective effects of voluntary exercise during the postinitiation phase of pancreatic carcinogenesis in the rat. Cancer Res 50:6811–6816

Roebuck BD, Yager JD Jr, Longnecker DS, Wilpone SA (1981) Promotion by unsaturated fat of azaserine-induced pancreatic carcinogenesis in the rat. Cancer Res 41:3961–3966

Ross MH, Bras G (1971) Lasting influence of early caloric restriction on prevalence of neoplasms in the rat. J Natl Cancer Inst 47:1095–1113

Rous P (1914) The influence of diet on transplanted and spontaneous tumors. J Exp Med 20:433–451

Ruggeri BA, Klurfeld DM, Kritchevsky D (1987) Biochemical alterations in 7,12-dimethylbenz(a)anthracene–induced mammary tumors in rats subjected to caloric restriction. Biochem Biophys Acta 929:239–246

Ruggeri BA, Klurfeld DM, Kritchevsky D, Furlanetto RW (1989) Caloric restriction and 7.12-dimethylbenz(a)anthracene-induced mammary tumor growth in rats: alterations in circulating insulin, insulin-like growth factors I and II, and epidermal growth factor. Cancer Res 49:4130–4134

Rusch HP, Kline BE (1944) The effect of exercise on the growth of a mouse tumor. Cancer Res 4:116–118

Rusch HP, Kline BE, Baumann CA (1945) The influence of caloric restriction and of dietary fat on tumor formation with ultraviolet radiation. Cancer Res 5:431–435

Sarkar NH, Fernandes G, Telang NT, et al. (1982) Low calorie diet prevents the development of mammary tumors in C3H mice and reduces circulating prolactin level, mammary tumor virus expression and proliferation of mammary alveolar cells. Proc Natl Acad Sci USA 79:7758–7762

Selenskas SL, Ip MM, Ip C (1984) Similarity between trans fat and saturated fat in the modification of rat mammary carcinogenesis. Cancer Res 44:1321–1326

Silverstone H, Tannenbaum A (1950) The influence of dietary fat and riboflavin on the formation of spontaneous hepatomas in the mouse. Cancer Res 11:200–203

Sugano M, Watanabe M, Yoshida K, et al. (1989) Influence of dietary cis and trans fats on DMH-induced colon tumors, steroid excretion and eicosanoid production in rats prone to colon cancer. Nutr Cancer 12:177–187

Sylvester PW, Aylsworth CF, Meites J (1981) Relationship of hormones to inhibition of mammary tumor development by underfeeding during the "critical period" after carcinogen administration. Cancer Res 41:1383–1388

Tannenbaum A (1940) The initiation and growth of tumors: introduction: effects of undernutrition. Am J Cancer 38:335–350

Tannenbaum A (1942) The genesis and growth of tumors, II: effects of calorie restriction per se. Cancer Res 2:460–467

Tannenbaum A (1944) The dependence of the genesis of induced skin tumors on the caloric intake during different stages of carcinogenesis. Cancer Res 4:683–687

Tannenbaum A (1945) The dependence of tumor formation on the composition of the calorie-restricted diet as well as on the degree of restriction. Cancer Res 5:616–625

Tannenbaum A (1947) Effects of varying caloric intake upon tumor incidence and tumor growth. Ann N Y Acad Sci 49:5–17

Thompson HJ, Ronan AM, Ritacco KA, Tagliaferro AR (1989) Effect of type and amount of dietary fat on the enhancement of rat mammary tumorigenesis by exercise. Cancer Res 49:1904–1908

Thompson HJ, Ronan AM, Ritacco KA, et al. (1988) Effect of exercise on the induction of mammary carcinogenesis. Cancer Res 48:2720–2723

Vena JE, Graham S, Zielezny M, et al. (1985) Lifetime occupational exercise and colon cancer. Am J Epidemiol 122:357–365

Watson AF, Mellanby E (1930) Tar cancer in mice, II: the condition of the skin when modified by external treatment or diet as a factor in influencing the cancerous reaction. Br J Exp Pathol 11:311–322

Weindruch R, Walford RL (1982) Dietary restriction in mice beginning at 1 year of age: effect on life span and spontaneous cancer incidence . Science 215:1415–1418

Welsch CW, House JL, Herr BL, et al. (1990) Enhancement of mammary carcinogenesis by high levels of dietary fat: a phenomenon dependent on ad libitum feeding. J Natl Cancer Inst 82:1615–1620

White FR (1961) The relationship between underfeeding and tumor formation, transplantation and growth in rats and mice. Cancer Res 21:281–290

Zhu P, Frei E, Bunk B, et al. (1991) Effect of dietary calorie and fat restriction on mammary tumor growth and hepatic as well as tumor glutathione in rats. Cancer Lett 57:145–152

Influence of Caloric Intake on Drug-Metabolizing Enzyme Expression: Relevance to Tumorigenesis and Toxicity Testing

Julian E.A. Leakey, John Seng, Mikhail Manjgaladze, Nina Kozlovskaya, Shijun Xia, Min-Young Lee, Lynn T. Frame, Shu Chen, Crissy L. Rhodes, Peter H. Duffy, and Ronald W. Hart
National Center for Toxicological Research

Introduction

As noted in other reports in this volume, it is becoming increasingly apparent that caloric intake is a major uncontrolled variable in rodent bioassays. Caloric intake appears to markedly influence both the mortality and chronic morbidity of the rodents used in these assays through its associated effects on body growth and obesity (Maeda et al. 1985, Masoro 1988, Weindruch and Walford 1988, Hart et al. 1992, Turturro et al. 1993). These effects have made the control of caloric intake by dietary restriction (DR) an attractive means for controlling variability, decreasing background tumor incidence, and increasing survival in bioassay test animals (Hart 1993, Roe 1993). However, DR has also been shown in several cases to decrease the incidence and proliferative rate of chemically induced neoplasia (Allaben et al. 1991, Hart et al. 1992, Turturro et al. 1993). Thus, this potentially desensitizing effect of DR on chemical carcinogenesis must also be considered when designing optimized diets or idealized weight curves for rodents used in bioassays.

Caloric restriction differs from other forms of DR in that optimum ratios of protein, carbohydrate, and fat are maintained and the restricted diet is fortified with micronutrients (Yu et al. 1982, Masoro 1988, Witt et al. 1991). As such, caloric restriction has proved to be by far the most efficient method of nutritional intervention with respect to life span extension and reduction in the incidence of neoplastic and other degenerative diseases (Maeda et al. 1985, Yu et al. 1985). The National Center for Toxicological Research (NCTR)/National Institute on Aging (NIA) caloric restriction paradigm was used in much of the work quoted in this report. Under this paradigm, caloric restriction is initiated at 14 weeks of age and reaches 40% by 16 weeks. The calorically restricted rats receive vitamin-fortified NIH-31 diet at 60% of the amount consumed by age-matched ad libitum (AL) controls throughout the remainder of their lives.

Although the biochemical mechanisms by which caloric restriction affects chemical carcinogenesis are not completely understood, it is well established that caloric restriction can alter the pharmacokinetics and metabolism of many xenobiotics as well as rates of cell proliferation (Manjgaladze et al. 1993). This report reviews the effects of caloric restriction on drug pharmacokinetics, on drug metabolism and the expression of drug-metabolizing enzymes, and on hormonal factors that potentially influence drug toxicity.

Effects on Drug Disposition and Pharmacokinetics

In general, longevity and mechanistic studies investigating the effects of caloric restriction have used restriction levels ranging from 50–70% of AL consumption and vitamin-fortified diets that maintain vitamin consumption at AL levels on a per-animal basis. However, calorically restricted rodents partially compensate for their reduced energy intake by reducing body growth such that once they have adapted to their restricted dietary regime there is little difference in food consumption per lean body mass between the restricted rats and their AL counterparts (Duffy et al. 1989, Manjgaladze et al. 1993). Thus, under these conditions, the restricted animals are effectively consuming a vitamin-supplemented diet.

Calorically restricted rodents exhibit both reduced body fat and lean body mass as compared to their AL counterparts. Liver weight is also reduced, usually to a greater extent than is body weight. For example, liver weight as a percentage of total body weight was 2.78% in calorically restricted, 18-week-old male Fisher 344 (F344) rats and 3.52% in the corresponding AL controls (Table 1). Similar decreases were seen in 9-month-old rats (Manjgaladze et al. 1993), suggesting that liver weight is reduced as a fraction of body weight through much of the calorically restricted rat's lifetime. In contrast, kidney weight appeared to decrease proportionally with body weight (Table 1). Calorically restricted animals also tend to consume more water than their AL counterparts (Duffy et al. 1989, Turturro

Table 1. Effects of 40% caloric restriction on body, liver, and kidney weights in 18-week-old male Fischer 344 (F344) rats[a]

Parameter	Ad libitum (AL)	40% Caloric restriction	% of control
Body weight (g)	346.7 ± 10.0	262.2 ± 7.4[*]	76
Liver weight (g)	12.23 ± 0.52	7.29 ± 0.32[*]	60
% liver	3.52 ± 0.07	2.78 ± 0.09[*]	79
Kidney weight (g)	2.16 ± 0.03	1.73 ± 0.04[*]	80
% kidney	0.62 ± 0.01	0.65 ± 0.02	105

[a]Male 18-week-old F344 rats were fed vitamin-fortified NIH-31 at 60% of AL consumption from 14 weeks post partum as described previously (Duffy et al. 1989, Manjgaladze et al. 1993). At least 6 animals were used for each experimental group.
[b]Combined weight of both kidneys. Taken in part, with permission, from Manjgaladze et al. (1993).
[*]Significantly different from AL group ($p < 0.05$, by SAS LSD-GLM test).

et al. 1993). Taken together, these effects suggest that renal clearance of drugs may be selectively increased over biliary clearance in calorically restricted rodents.

The preferential loss of body fat under conditions of caloric restriction will also influence the pharmacokinetics and disposition of drugs and carcinogens. When hydrophilic compounds are administered to calorically restricted and AL rodents at equal doses per gram body weight, higher plasma concentrations would be expected in the AL animals, because their water soluble compartment will represent a smaller percentage of total body weight than that of the calorically restricted animal. In contrast, when lipophilic drugs are administered at equal doses per gram body weight, one would expect higher plasma concentrations in the calorically restricted animals because they have a proportionally smaller fat compartment than the AL animals. However, it has been shown for several lipophilic compounds that an equilibrium exists between fat and plasma and that the equilibrium constant depends on plasma triglyceride levels (Emmett 1985). Since plasma triglyceride levels are consistently reduced in calorically restricted animals (Turturro et al. 1993), plasma concentrations of lipophilic compounds may actually be reduced by caloric restriction. The exact effects of caloric restriction on lipophilic compounds would therefore be difficult to predict and would depend on the individual compound administered and the time after dosing that the plasma levels are measured.

Effects of Caloric Restriction on Drug-Metabolizing Enzyme Expression

Caloric restriction has been shown to alter the expression of several isoforms of the enzymes collectively known as drug-metabolizing enzymes

(Manjgaladze et al. 1993). While these drug-metabolizing enzymes are predominantly hepatic, they are also present in varying amounts in other tissues such as kidney, lung, intestine, and endocrine tissues. These enzymes are primarily responsible for activation and detoxication of carcinogens and mutagens, modulation of toxicity of many other xenobiotics, biotransformation and metabolic clearance of many endogenous compounds, and microsomal production of oxygen radicals. Thus, it is a reasonable assumption that drug-metabolizing enzymes may influence at least some of the effects of caloric restriction on chemical carcinogenesis, since changes in the relative activities of these enzymes can alter drug pharmacokinetics and biotransformation rates of chemicals under test (Manjgaladze et al. 1993).

Drug-metabolizing enzymes exist as families of isoforms that exhibit differential regulation by both xenobiotic inducers and hormonal factors (Manjgaladze et al. 1993). The major enzyme families are: the cytochrome P450-dependent monooxygenases (Nelson et al. 1993), the UDP-glucuronosyltransferases (UGT) (Burchell et al. 1991), the glutathione *S*-transferases (Mannervik and Danielson 1988), the epoxide hydrolases (Kato et al. 1986), and the sulfotransferases (Mulder and Jakoby 1990). Several other enzyme systems play a minor role in drug metabolism and can also be classified as drug-metabolizing enzymes. These include alcohol and aldehyde dehydrogenases and oxidoreductases, prostaglandin synthase, flavoprotein monooxygenases, esterases, and amino acid conjugating enzymes (Guengerich and Shimada 1991, Mulder and Jakoby 1990, Manjgaladze et al. 1993).

The effects of caloric restriction on hepatic drug-metabolizing enzyme expression can be divided into three types: (1) direct effects, i.e., those directly resulting from the organism's adaptation to reduced caloric intake; (2) aging-related effects, i.e., those resulting from a reduction of the physiological age of the organism relative to its chronological age; and (3) circadian effects, i.e., those resulting from the organism's altered feeding behavior. Individual isoforms that have been shown to be altered by caloric restriction are listed in Table 2.

Direct Effects

In rats, the most striking direct effect of caloric restriction is on sex-specific, growth-hormone-dependent liver enzymes. Forty percent caloric restriction in male and female F344 rats for as little as 4 weeks has been shown to dedifferentiate expression of these enzymes (Manjgaladze et al. 1993). Similar effects were observed in 9-month-old F344 rats (Leakey et

Table 2. Effects of 40% caloric restriction on isoforms of rat drug-metabolizing enzymes and their selective activities[a]

	Males	Females
Direct effects[b]		
	CYP2C11 ↓	Androgen 5α-reductase ↓
	(CYP2A2 ↓)[c]	Corticosterone sulfotransferase↓
	(CYP2C13 ↓)	(CYP2C12 ↓)
	(Arylsulfotransferase IV ↓)	CYP2D ↔
	Androgen 5α-reductase ↑	Phenol UGT[d] ↔
	Corticosterone sulfotransferase ↑	
	CYP3A2 ↔	
	CYP2D ↔	
	Phenol UGT[d] ↔	
Aging-related effects		
	CYP2C11 ↑	
	CYP3A2 ↑	—
	CYP2E1 ↑	
	Bilirubin UGT[d] ↑	
	Androgen 5α-reductase ↓	
	Corticosterone sulfotransferase ↓	
Circadian effects		
	CYP1A1 ↑	CYP1A1 ↑
	CYP2A1 ↑	CYP2A1 ↑
	CYP2B1 ↑	CYP2B1 ↑
	CYP2E1 ↑	CYP2E1 ↑
	CYP2A1 ↓ (testis)	

[a]Data compiled from Leakey et al. 1989a, 1989b, 1991, Manjgaladze et al. 1993, and unpublished observations. These studies used 40% caloric restriction in Fischer 344 rats initiated at 14 weeks of age (Manjgaladze et al. 1993). Listed isoforms are hepatic unless stated otherwise.
[b]↑, ↓, and ↔ = increased, decreased, and unchanged activities, respectively.
[c]Enzyme effects shown in parentheses are predicted on the basis of known regulatory mechanisms for the indicated isoform.
[d]UGT = UDP-glucuronosyltransferase.

al. 1991) and in 7 and 26–30-month-old Lobund-Wistar rats (30% dietary restriction from 6 weeks of age; Schmucker et al. 1991). For example, male-specific CYP2C11-dependent testosterone 16α-hydroxylase activity is decreased to 30–50% of AL values in liver from calorically restricted 18-week-old male F344 rats (Manjgaladze et al. 1993). Furthermore, Northern blot analysis of specific CYP2C11 mRNA suggests that this decrease is due to decreased CYP2C11 expression at the transcriptional level (Manjgaladze et al. 1993). In contrast, female-specific hepatic testosterone 5α-reductase and corticosterone sulfotransferase activities are both increased by caloric restriction in male rats but are decreased by caloric restriction in females (Manjgaladze et al. 1993). Not all sex-specific activities are altered by caloric restriction. For example, CYP3A-dependent testosterone 6β-hydroxylase activity, which is regulated differently from

CYP2C11, was not consistently decreased by caloric restriction in 18-week-old male rats, nor is there any significant decrease in CYP3A immunoreactive protein (Manjgaladze et al. 1993). Furthermore, caloric restriction does not appear to evoke expression of CYP2C11 or CYP3A2 in 18-week-old female rat liver.

Aging-Related Effects

The most striking aging-related effect of caloric restriction also involves the sex-specific, growth-hormone-dependent rat liver enzymes. In male rats, sex-specific CYP2C11 expression decreases in old age (Kamataki et al. 1985). We have recently found evidence that this decrease may be due to the secretion of abnormal amounts of growth hormone by pituitary adenomas, which are common in old rats (Witt et al. 1991). It has been known for a number of years that growth-hormone-secreting pituitary adenomas repress male-specific drug metabolism (Wilson 1969). When we examined 34-month-old male Brown Norway × F344 hybrid rats, we found that those exhibiting pituitary tumors exhibited decreased hepatic CYP2C11-dependent testosterone metabolism, whereas those that showed no tumors expressed normal CYP2C11 levels (Leakey et al. 1993). Forty percent caloric restriction significantly reduces the incidence and delays the onset of pituitary tumors (Witt et al. 1991) as well the decline in hepatic CYP2C11-dependent testosterone metabolism (Leakey et al. 1989a, Manjgaladze et al. 1993). Interestingly, hepatic CYP3A-dependent 6β-hydroxylase activity is also induced by caloric restriction in old rats (Leakey et al. 1989a).

Other aging-related effects of caloric restriction may involve age-related changes in serum insulin. AL rodents appear to exhibit progressive hyperinsulinemia as they age (Feuers et al. 1989, Lagopoulos et al. 1991, Leakey et al. 1991, Reed et al. 1993), and hyperinsulinemia appears to decrease the expression of hepatic CYP2E1 and the UGT isoform that conjugates bilirubin (Manjgaladze et al. 1993). Both CYP2E1-dependent monooxygenase and bilirubin UGT activities are increased by caloric restriction to a much greater extent in old rats than in young rats (Leakey et al. 1991, Manjgaladze et al. 1993).

Circadian Effects

Caloric restriction appears to cause minor changes in expression of several hepatic drug-metabolizing enzymes that are only observed at specific circadian time points. Such changes may be due to altered feeding behavior that is observed in restricted rodents (Duffy et al. 1989). Calorically restricted rats tend to eat their food allocation rapidly, as soon as they receive it, rather than consuming it more gradually during the dark period of their circadian cycle as is normal for AL rodents (Duffy et al. 1989). These changes in feeding behavior alter the circadian cycles of serum hormone levels such as corticosterone, insulin, or thyroid hormones (Leakey et al. 1991, 1994). Although the circadian profiles of these hormones can be somewhat synchronized by adjusting the time of restricted food allocation to the period within the dark phase when the AL animals are most active in eating (Duffy et al. 1989), the rapid food consumption in the calorically restricted animals may still produce amplitude differences in circadian profiles of certain hormones (Sabatino et al. 1991).

The circadian effects on drug-metabolizing enzymes are relatively small in calorically restricted rodents and can differ as a function of age, sex, and the feeding schedule used. For example, in female F344 rats, hepatic CYP1A-selective 7-ethoxyresorufin O-deethylase and CYP2A1-dependent testosterone 7α-hydroxylase did not show significant circadian variations in AL control animals but did show a circadian peak coincident with the onset of the light cycle in calorically restricted animals (Manjgaladze et al. 1993). In contrast, hepatic CYP2E1-selective chlorzoxazole 6-hydroxylase and CYP2B-selective 7-pentoxyresorufin O-dealkylase activities in the same animals exhibited circadian peaks at 8 hours into the light cycle and 4 hours into the dark cycle respectively, (M.-Y. Lee, S. Xia, and N. Kozlovskaya, unpublished observations). The latter peak was associated with increased expression of CYP2B1 immunoreactive protein. Neither the CYP2E1- or CYP2B-selective activity exhibited significant circadian rhythms in the AL controls.

In contrast to its effects on hepatic CYP2A1 expression, caloric restriction decreases the circadian expression of testicular CYP2A1. Leydig cells of sexually mature male rats express CYP2A1 and its dependent testosterone 7α-hydroxylase activity (Seng et al. 1991). This hydroxylase activity exhibits a circadian variation in AL rats with peak levels occurring from the mid-light phase to the mid-dark phase, which correlates with expression of CYP2A1 immunoreactive protein. These circadian peaks of both testosterone 7α-hydroxylase activity and CYP2A1 immunoreactive protein were completely suppressed by 40% caloric restriction (J. Seng, J. Gandy, and J. Leakey, unpublished observations).

Effects on Drug and Carcinogen Metabolism

Drug metabolism can be influenced by critical cofactor concentrations as well as by activities of drug-metabolizing enzymes. Although the effects of caloric restriction on such cofactor levels have not been systematically studied, it has been shown that caloric restriction will increase both hepatic glutathione (Manjgaladze et al. 1993) and NADPH levels (Feuers et al. 1993), but only at specific circadian time points.

Although the diet- and circadian-dependent changes in drug-metabolizing enzyme expression described above are readily apparent when isoform selective activities (i.e., activities predominantly catalyzed by just one isoform) are used, only small changes are observed with less specific substrates such as aminopyrine or ethylmorphine (Leakey et al. 1989a, 1989b). This is most probably due to more than one isoform catalyzing these reactions with changes in one isoform being compensated for by other isoforms. This would suggest that the effects of caloric restriction would be relatively minor for those drugs or carcinogens that are metabolized similarly by several isoforms. However, caloric restriction does appear to significantly alter the metabolism of several chemical carcinogens (Pegram et al. 1989, Chou et al. 1993, Manjgaladze et al. 1993). For example, the amount of genetic damage produced in isolated hepatocytes by exposure to either 2-acetylaminofluorene (2AAF) or aflatoxin B_1 (as measured by DNA repair activity) is altered by both the age and caloric intake of the source animals for the hepatocytes in a manner analogous to the expression of hepatic CYP2C11 (Manjgaladze et al. 1993, Shaddock et al. 1993). For example, both CYP2C11- and 2AAF-induced DNA repair are reduced by caloric restriction in young and middle-aged rats and are increased by caloric restriction in old rats (Manjgaladze et al. 1993, Shaddock et al. 1993). In addition, we have recently found that the ability of F344 rat testicular microsomes to metabolically activate aflatoxin B_1 is decreased by caloric restriction proportionally to the decrease in testicular CYP2A1 expression described in the previous section. However, the ability of the same microsomes to activate dimethylbezanthracene is not significantly altered by caloric restriction (J. Seng, J. Gandy, and J. Leakey, unpublished observations).

Effects on Drug Toxicity

Caloric restriction has been shown to markedly increase the survival of rodents exposed to several compounds, including isoproterenol (Duffy et al., this volume), and has also been shown to decrease the dermal and inflammatory responses of mice to phorbol esters (Pashko and Schwartz 1992) and carrageenan (Klebanov et al. 1993). The latter effects appear to be mediated by the hypercorticism that occurs in calorically restricted rodents (Klebanov et al. 1993, Leakey et al. 1994, Schwartz and Pashko 1994). It has recently been proposed that hypercorticism may mediate most, if not all, of the beneficial effects of caloric restriction on longevity and resistance to disease (Leakey et al. 1994). Glucocorticoids suppress the inflammatory response and are antagonistic to mitogenic hormones (Leakey et al. 1994). Thus, the elevated corticosterone levels observed in calorically restricted rodents (Sabatino et al. 1991, Leakey et al. 1994) would be expected to increase the maximum tolerated dose for many chemicals due to reduced inflammation. Likewise, such changes in serum corticosterone levels would be expected to reduce the potency of tumor-promoting chemicals by decreasing rates of cell proliferation. Glucocorticoid response elements have been found to be associated with several cytochrome P450 genes (Mathis et al. 1989, Jaiswal et al. 1990) and synergistically potentiate the effects of chemical inducers on the transcription of these genes (Mathis et al. 1986). This would suggest that caloric restriction may also potentiate cytochrome P450 induction, and it has recently been found that phenobarbital appears to induce more CYP2B-selective activity in calorically restricted than in AL rats (M. Alterman, D. Busbee, and J. Leakey, unpublished observations). Increased induction of drug-metabolizing enzymes would be expected to decrease the half-life and potential toxicity of the inducing chemical. However, in certain cases such induction may potentiate toxicity due to increased rates for formation of toxic intermediates.

Conclusions

From the data accumulated thus far it is a reasonable assumption that in rodents, restriction of caloric intake by 40% *may* result in significant alterations in a test chemical's pharmacokinetics, metabolism, toxicity, and ability to induce its own metabolism. However, it is also possible that for many chemicals caloric restriction will only result in small modifications in metabolism and pharmacokinetics since, from a metabolic standpoint, caloric restriction only produces large changes in the expression of a few selective isoforms of drug-metabolizing enzymes. Furthermore, even when large changes do occur (such as for CYP2C11), the resultant changes are no greater than differences in cytochrome P450 isoform expression that occur between test strains or are due to induction by the test chemical. For example, adult male Sprague-Dawley rats can express hepatic microsomal CYP2C13 at levels that are more than 60-fold greater than those expressed in hepatic microsomes from adult male F344 rats (McClellan-Green et al. 1987). Chronic feeding of mice with phenobarbital can induce CYP2B-selective monooxygenase activity by more than sevenfold (Wolff et al. 1991). Notwithstanding this, it is also probable that caloric restriction will

significantly reduce the carcinogenicity of chemicals that are solely activated by isoforms that are critically reduced by caloric restriction (e.g.,
CYP2C11) or are metabolically inactivated by isoforms that are increased
by caloric restriction. For such cases, caloric restriction would be expected
to significantly reduce the ability of the rodent bioassay to demonstrate
chemical carcinogenicity. Whether this will decrease the relevance of a
bioassay would depend on whether a similar activation pathway occurs in
human tissues. This can fortunately be readily tested in vitro by using
human tissue preparations or cell lines that express single isoforms of drug-
metabolizing enzymes (Crespi et al. 1990), which are now commercially
available.

Forty percent caloric restriction reduces total body weight by 25–45%
in male F344 rats (Manjgaladze et al. 1993). As shown by Turturro et al.
(this volume), greater variations in body weight occur between AL rodents of the same strain raised in different laboratories. Furthermore, increased body weight gain is positively correlated with spontaneous tumor
incidence in these rodents (Turturro et al. 1993, this volume). This suggests that any effect of caloric restriction that correlates with body weight
will show a greater variation between AL rats used in different laboratories than between the AL and 40% calorically restricted rats used in our
studies. It is not yet known whether body weight gain also influences drug-
metabolizing enzyme expression in an analogous way to caloric intake. It
is unlikely that the circadian effects of caloric restriction on drug-metabolizing enzyme expression will correlate with body weight in AL rodent
populations, because these effects appear to be a function of feeding behavior. However, it is probable that body weight does influence expression of sex-specific drug-metabolizing enzymes such as CYP2C11, since
expression of these enzymes is controlled by growth hormone (Manjgaladze
et al. 1993). It is also likely that the aging-related effects of caloric restriction also correlate with body weight, since the heavier AL rats are more
susceptible to degenerative disease (Turturro et al. 1993). If this proves to
be so, it is likely that the capability of rodents to metabolize many test
chemicals will vary between testing laboratories in a manner similar to
interlaboratory variability in body weight and spontaneous tumor incidence.
This will result in further decreases in bioassay reproducibility between
studies and between individual animals within studies.

Acknowledgments

This work was funded by the National Institute of Aging-National Center for Toxicological Research Biomarkers of Aging Program and by the

Food and Drug Administration (FDA). Postdoctoral support to M-Y.L., N.K., M.M., S.X., and S.C. was provided through an interagency agreement between the FDA and the Department of Energy that was administered by the Oak Ridge Institute for Science and Education. The conclusions and recommendations expressed in this paper are those of the authors and not necessarily those of the FDA.

References

Allaben WT, Chou MW, Pegram RA (1991) Dietary restriction and toxicological endpoints: An historical overview. In Fishbein L (ed), Biological effects of dietary restriction. Springer-Verlag, New York, pp 27–41

Burchell B, Nebert DW, Nelson DR, et al. (1991) The UDP-glucuronosyltransferase gene superfamily: suggested nomenclature based on evolutionary divergence. DNA Cell Biol 10:487

Chou MW, Kong J, Chung K-T, Hart RW (1993) Effect of caloric restriction on the metabolic activation of xenobiotics. Mutat Res 295:223

Crespi CL, Penman BW, Leakey JEA, et al. (1990) Human cytochrome P450IIA3: cDNA sequence, role of the enzyme in the metabolic activation of promutagens, comparison to nitrosamine activation by human cytochrome P450IIE1. Carcinogenesis 11:1293

Duffy PH, Feuers RJ, Leakey JEA, et al. (1989) Effect of chronic caloric restriction on the physiological variables related to energy metabolism in the Fischer 344 rat. Mech Ageing Devel 48:117

Emmett EA (1985) Polychlorinated biphenyl exposure and effects in transformer repair workers. Environ Health Perspect 60:185

Feuers RJ, Duffy PH, Leakey JA, et al. (1989) Effect of chronic caloric restriction on hepatic enzymes of intermediary metabolism in the male Fisher 344 rat. Mech Ageing Devel 48:179

Feuers RJ, Weindruch R, Hart RW (1993) Caloric restriction, aging and antioxidant enzymes. Mutat Res 295:191

Guengerich FP, Shimada T (1991) Oxidation of toxic and carcinogenic chemicals by human cytochrome P-450 enzymes. Chem Res Toxicol 4:391

Hart RW (1993) Metabolic and biochemical effects of reduced calorie intake in rodents. In The toxicology forum: 1993 annual winter meeting. Toxicology Forum Inc., Washington, DC, pp 243–250

Hart RW, Chou MW, Feuers RJ, et al. (1992) Caloric restriction and chemical toxicity/carcinogenesis. Quality Assurance Good Practice Regulation Law 1:120

Jaiswal AK, Haaparanta T, Luc P-V, et al. (1990) Glucocorticoid regulation of a phenobarbital-inducible cytochrome P-450 gene: the presence of a functional glucocorticoid response element in the 5'-flanking region of the CYP2B2 gene. Nucleic Acids Res 18:4237

Kamataki T, Maeda K, Shimada M, et al. (1985) Age-related alterations in the activities of drug-metabolizing enzymes and contents of sex-specific forms of cytochrome P-450 in liver microsomes from male and female rats. J Pharmacol Exp Ther 233:222

Kato R, Yamazoe Y, Shimada M, et al. (1986) Effect of growth hormone and ectopic transplantation of pituitary gland on sex-specific forms of cytochrome P-450 and testosterone and drug oxidations in rat liver. J Biochem 100:895

Klebanov S, Diais S, Stavinoha W, et al. (1993) Food restriction, elevated plasma corticosterone and an attenuated inflammatory response to carrageenan in male Balb/c mice. Gerontologist 33:96

Lagopoulos L, Sunahara GI, Wurzner H, et al. (1991) The effects of alternating dietary restriction and ad libitum feeding of mice on the development of diethylnitrosamine-induced liver tumors and its correlation to insulinemia. Carcinogenesis 12:311

Leakey JEA, Bazare Jr. J, Harmon JR, et al. (1991) Effects of long-term caloric restriction on hepatic drug metabolizing enzyme activities in the Fischer 344 rat. In Fishbein L (ed), Biological effects of dietary restriction. Springer-Verlag, New York

Leakey JEA, Chen S, Manjgaladze M, et al. (1994) Role of glucocorticoids and "caloric stress" in modulating the effects of caloric restriction in rodents. Ann N Y Acad Sci 719:171

Leakey JEA, Cunny HC, Bazare Jr. J, et al. (1989a) Effects of aging and caloric restriction on hepatic drug metabolizing enzymes in the Fischer 344 rat. I. The cytochrome P-450 dependent monooxygenase system. Mech Ageing Devel 48:145

Leakey JEA, Cunny HC, Bazare Jr. J, et al. (1989b) Effects of aging and caloric restriction on hepatic drug metabolizing enzymes in the Fischer 344 rat. II. Effects on conjugating enzymes. Mech Ageing Devel 48:157

Leakey JEA, Rhodes CL, Manjgaladze M, et al. (1993) Effect of Chronic caloric restriction on liver enzyme expression in aging rats. Gerontologist 33:77

Maeda H, Gleister CA, Masoro EJ, et al. (1985) Nutritional influences on aging of Fischer 344 rats II. Pathology. J Gerontol 40:671

Manjgaladze M, Chen S, Frame LT, et al. (1993) Effects of caloric restriction on rodent drug and carcinogen metabolizing enzymes: Implications for mutagenesis and cancer. Mutat Res 295:201

Mannervik B, Danielson UH (1988) Glutathione transferases—structure and catalytic activity. Crit Rev Biochem 23:283

Masoro EJ (1988) Food restriction in rodents: an evaluation of its role in the study of aging. J Gerontol 43:B59

Mathis JM, Houser WH, Bresnick E, et al. (1989) Glucocorticoid regulation of the rat cytochrome P450c (P450IA1) gene: receptor binding within intron 1. Arch Biochem Biophys 269:93

Mathis JM, Prough RA, Simpson ER (1986) Synergistic induction of monooxygenase activity by glucocorticoids and polycyclic aromatic hydrocarbons in human fetal hepatocytes in primary monolayer culture. Arch Biochem Biophys 244:650

McClellan-Green P, Waxman DJ, Caveness M, Goldstein JA (1987) Phenotypic differences in expression of cytochrome P-450g but not its MRNA in outbred male Sprague-Dawley rats. Arch Biochem Biophys 253:13

Mulder GJ, Jakoby WB (1990) Sulfation. In Mulder GJ (ed), Conjugation reactions in drug metabolism. Taylor and Francis, London, pp 107–161

Nelson DR, Kamataki T, Waxman DJ, et al. (1993) The P450 superfamily: update on new sequences, gene mapping, accession numbers, early trivial names of enzymes, and nomenclature. DNA Cell Biol 12:1

Pashko LL, Schwartz AG (1992) Reversal of food restriction-induced inhibition of mouse skin tumor promotion by adrenalectomy. Carcinogenesis 13:1925

Pegram RA, Allaben WT, Chou MW (1989) Effect of caloric restriction on aflatoxin B_1-DNA adduct formation and associated factors in Fischer 344 rats. Mech Ageing Devel 48:167

Reed MJ, Reaven GM, Mondon CE, Azhar S (1993) Why does insulin resistance develop during maturation. J Gerontol 48:B139

Roe FJC (1993) What does carcinogenicity mean and how should we test for it? Food Chem Toxicol 31:225

Sabatino F, Masoro EJ, McMahan CA, Kuhn RW (1991) Assessment of the role of the glucocorticoid system in aging processes and in the action of the food restriction. J Gerontol 46:B171

Schmucker DL, Wang RK, Snyder D, et al. (1991) Caloric restriction affects liver microsomal monooxygenases differentially in aging male rats. J Gerontol 46:B23

Schwartz AG, Pashko LL (1994) Role of adrenocortical steroids in mediating cancer-preventive and age-retarding effects of food restriction in laboratory rodents. J Gerontol 49:B37

Seng JE, Leakey JEA, Arlotto MP, et al. (1991) Cellular localization of cytochrome P450IIA1 in testes of mature Sprague-Dawley rats. Biol Reprod 45:876

Shaddock JG, Feuers RJ, Chou MW, et al. (1993) Effects of aging and caloric restriction on the genotoxicity of 4 carcinogens in the in vitro rat hepatocyte/DNA repair assay. Mutat Res 295:19

Turturro A, Duffy PH, Hart RW (1993) Modulation of toxicity by diet and dietary macronutrient restriction. Mutat Res 295:151

Weindruch R, Walford R (1988) Retardation of aging and disease by dietary restriction. Thomas Press, Springfield, IL

Wilson JT (1969) Identification of somatotropin as the hormone in a mixture of somatotropin, adrenocorticotropic hormone and prolactin which decreased liver drug metabolism in the rat. Biochem Pharmacol 18:2029

Witt WM, Sheldon WG,Thurman JD (1991) Pathological endpoints in dietary restricted rodents—Fischer 344 rats and B6C3F$_1$ mice. In Fishbein L (ed), Biological Effects of Dietary Restriction. Springer-Verlag, New York, pp 73–86

Wolff GL, Leakey JEA, Bazare JJ, et al. (1991) Susceptibility to phenobarbital promotion of hepatocarcinogenesis: correlation with differential expression and induction of hepatic drug metabolizing enzymes in heavy or light male (C3H × VY) F1 hybrid mice. Carcinogenesis 12:911

Yu BP, Masoro EJ, McMahan AC (1985) Nutritional influences on aging of Fisher 344 rats: I. Physical, metabolic, and longevity characteristics. J Gerontol 40:657

Yu BP, Masoro EJ, Murata I, et al. (1982) Life span study of SFR Fischer 344 male rats fed ad libitum or restricted diets: longevity, growth, lean body mass and disease. J Gerontol 37:130

Intermediary Metabolism and Antioxidant Systems

R.J. Feuers
National Center for Toxicological Research
University of Arkansas for Medical Sciences

P.H. Duffy, F. Chen, and V. Desai
National Center for Toxicological Research

E. Oriaku
Florida A&M University

J.G. Shaddock and J.W. Pipkin
National Center for Toxicological Research

R. Weindruch
University of Wisconsin

R.W. Hart
National Center for Toxicological Research

Introduction

Obesity complicates interpretation of results from long-term animal bioassays. The problems caused by obesity include increased incidence of spontaneous tumors, decreased health of animals, and reduced life span. These and other pathologies appear to be due to ad libitum (AL) feeding. Dietary control may delay or eliminate these conditions in experimental animals; however, it might also alter other biological events, which could confound interpretation of toxicity studies.

Dietary restriction (DR) extends life span, offsets the time to development of various degenerative diseases, and modulates the toxicity of a number of chemical, physical, and biological insults (Masoro 1988, Weindruch and Walford 1988). A reduction in caloric intake, by definition, imposes a state of reduced energy intake. Logically, it must follow that energy metabolism would be altered.

Based on evaluation of physiologic parameters such as body temperature, oxygen consumption, and CO_2 production, there has been considerable speculation and debate concerning the effect of DR on overall metabolic rates (Henson and Legros 1972, Kmiec and Mysliwske 1983), with the current consensus being that these rates do change and are of major importance. DR can induce a number of changes in energy metabolism, including an increase in the activities of hepatic hexokinase and decreases in glucokinase, malic enzyme, glucose 6-phosphate dehydrogenase, and fructose 1,6-biphosphatase (Appelbaun 1978, McCarter et al. 1985). Since a number of endocrine parameters, such as the level of circulating insulin, are altered by DR (Koizumi et al. 1987, Ruggeri et al. 1987), it is reasonable to assume that these factors might be responsible for modifying the endocrine-sensitive metabolic pathways.

One potential outcome of such changes might include a decrease in production of free radicals. It is widely speculated that free radicals play a primary role in drug toxicity and various pathologies including cancer and aging. DR may also decrease the production of free radicals and concurrently limit their persistence (and presumably macromolecular damage) by increasing the activities of enzymes such as catalase and superoxide dismutase (Semsei et al. 1989, Yu et al. 1989). This report summarizes our findings regarding the impact of DR on physiologic performance, energy metabolism, and the formation and elimination of free radicals.

Physiologic Changes with Chronic Caloric Restriction

A number of physiologic and behavioral variables were measured as a function of age in male and female mice and rats fed AL or DR diets. For these and all other studies described, DR was initiated at 14 weeks of age by providing the animals with 60% of the total amount of feed consumed by AL animals. The standard NIH-31 diet was vitamin-supplemented to avoid malnutrition, and all animals were singly housed. The diet and other aspects of animal care have been previously described by Duffy et al. (1989). In all cases DR was associated with a significant decrease in average and minimum body temperature; however, the magnitude of the decline was greater in mice than in rats and greater in females than in males of both species. As can be seen in Figure 1, maximums in motor activity were associated with the time of food intake, and changes in motor activity were closely correlated with body temperature changes. Therefore, in order to enhance the comparability of DR and AL groups, the DR rodents were fed at the beginning of the dark span in order to synchronize function. Food intake and the time of food presentation have dramatic effects on many physiologic and biological systems. Any time-of-day differences between DR and AL animals in resistance to drug toxicity would be minimized by phase shifting.

Alterations in the circadian rhythm of the respiratory quotient (RQ) suggests that significant changes have occurred in the metabolic output of the DR rodents. RQ varied little across time-of-day in AL animals, as illustrated in Figure 1. The circadian amplitudes of many physiologic and behavioral rhythms were significantly increased by a diet low in calories. These observations suggest that significant changes occur in the intermediary metabolism of DR animals. In AL animals, the RQ value was approximately 0.95 during the dark span and reached minimums of only 0.9 during the rest period (mid-light). These midrange RQ values suggest a mixed substrate metabolism; however, in DR animals the RQ rose rapidly to 1.0 whenever food was presented. A value of 1 indicates carbohydrate metabolism. During the inactive phase, which was during the dark for light-fed DR animals, the RQ was found to decrease to 0.83 (indicating lipid metabolism). When DR animals were presented food during the early dark span, the RQ would rise to its maximum value at a time that correlated to the maximum value seen for AL animals.

Pilot Studies Characterizing Enzymes of Intermediary Metabolism

The observation that feeding in DR rats and mice leads to rapid increases in RQ suggests that changes in basic metabolic pathways are being induced by DR. In DR animals, we found that hepatic insulin-dependent enzymes, such as the glycolytic and accessory enzymes as well as the enzymes of lipid metabolism, were decreased compared to AL animals. In contrast, hepatic glucagon-dependent enzymes, such as gluconeogenic and accessory enzymes as well as the enzymes of amino acid catabolism, were

increased in DR compared to AL animals (Feuers et al. 1989). These results suggested that carbohydrate catabolism is limited for the purpose of
increased capacity for glucose synthesis and export from the liver. Since
glycolysis is stimulated by insulin and gluconeogenesis is stimulated by

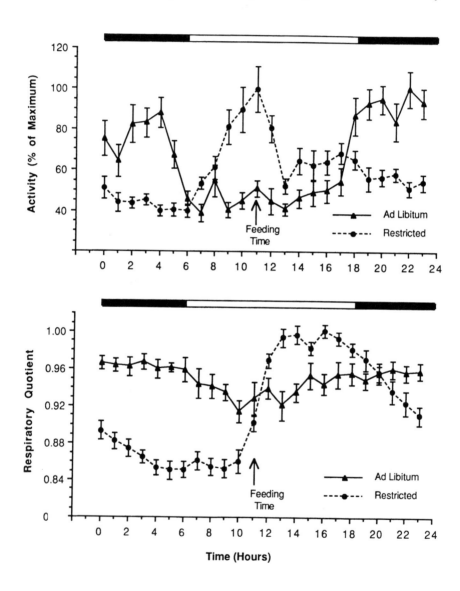

Figure 1. Motor activity and respiratory quotient from ad libitum (AL) and diet-
restricted (DR) female Fischer 344 rats.

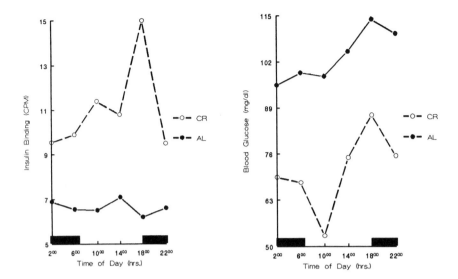

Figure 2. Blood glucose and hepatic insulin binding in 12-month-old ad libitum (AL) and diet-restricted (DR) B6C3F1 male mice. The dark bars represent the time of day when lights were out. DR animals were fed at the beginning of the dark.

glucagon, the insulin/glucagon ratio may be decreased by DR. Insulin levels were found to decline (data not shown); however, glucagon has not been adequately quantitated to date.

Glucose Regulation, Insulin, and Insulin and Glucagon Receptors

Blood glucose levels peak in all AL and DR animals at the time of feeding, but these levels are significantly higher in AL compared to DR rodents (Figure 2). The greatest glucose differences were found at mid-age, and the magnitude of difference was strain, sex, and species-dependent (data not shown, Feuers et al. 1990). If animals are not phase shifted, glucose levels in DR rodents would be at their lowest point when these levels reached their peak in AL animals (and vice versa). This could lead to errors of interpretation, since in non–phase-shifted rats it would falsely appear that DR blood glucose is higher in non–phase-shifted rats if blood sampling occurred during the mid-light span.

To investigate glucose regulation, ^{125}I-insulin (24 uCi) was administered, and levels of binding in liver were determined. Insulin binding was found to be significantly higher (approximately twofold) at all ages in the

DR rodents (Figure 2), while circulating insulin (Feuers et al. 1990) and blood glucose levels were lower in DR animals. There was a decline in total binding with age in both DR and AL animals, but the rate of decline was significantly greater in AL mice. The period of peak binding was associated with food intake but appeared at progressively later times in the dark span as the animals got older (Feuers et al. 1990). As can be seen in Figure 2, at mid-age, the phase-shifted DR animals have a pronounced peak at the beginning of the dark cycle, while slightly earlier the AL mice had a minor peak of binding. Changes across time-of-day were minimal in AL animals compared to DR animals. This indicates that there is a loss of efficiency for the up- or down-regulation of the insulin receptor with age in AL mice. An efficient regulation of receptor number or the affinity of the receptor for insulin was maintained in DR animals. Circadian regulation of insulin binding was lost in 24-month-old AL rodents but not in DR animals. These data suggest that the insulin receptor is upregulated by DR and/or that binding affinity was improved. Therefore, the insulin receptor-mediated regulation of glucose appears to be maintained more efficiently in DR as opposed to AL animals. These changes occur at times of day that coincide with switches in RQ from low to high values suggesting their involvement in the control of RQ regulation in DR rodents (Feuers 1991).

Glucagon binding was also found to be significantly higher at all ages in the DR rodents (data not shown). As with insulin, binding declined with age, but the decline was offset by DR. Again, the effect of DR was to up-regulate the glucagon receptor. As with the insulin receptor, it remains to be determined if this change is the result of increased receptor number or increased affinity for glucagon. Peak binding was "out of phase" with insulin binding, suggesting that the glucagon receptor was up-regulated when insulin binding was low. This coordination of regulation of flux through appropriate metabolic pathways was better maintained with age in the DR population.

Mechanisms for Changes in DR Enzyme Activities

Pyruvate kinase (PK) catalyzes an irreversible ATP-generating step of glycolysis and is activated by insulin-induced dephosphorylation. This results in a dramatic increase in the affinity of PK for its substrate, phosphoenol pyruvate (PEP), as indicated by a low Km (0.1 mM in DR rodents versus 0.4 mM in AL rodents). Thus, even low levels of carbohydrate can be metabolized through glycolysis, allowing relatively high levels of ATP generation under conditions of DR. PK is inactivated by glucagon-induced phosphorylation and results in a dramatic decrease in

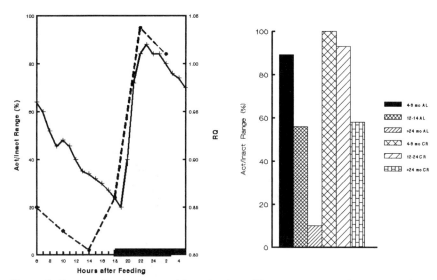

Figure 3. Range of activation and inactivation of hepatic pyruvate kinase (PK) from ad libitum (AL) and diet-restricted (DR) male Fischer 344 rats. Total phosphorylation of the enzyme molecules present yields 100% inactivation, while total dephosphorylation of the enzyme molecules present yields 100% activation. This was taken to occur in young (4-month-old) DR rats. The solid line in the left panel represents the respiratory quotient (RQ), while the broken line represents the range of activity/inactivity.

affinity (i.e., higher Km) for PEP. When the enzyme becomes maximally phosphorylated, metabolism through this pathway halts, and glucose synthesis is initiated through gluconeogenesis. We have found that in young animals, the range of phosphorylation/dephosphorylation approaches 100% in both AL and DR rodents. That is, in the presence of insulin, the dephosphorylation of PK can go essentially to completion. In the presence of glucagon (low blood glucose, low RQ, etc.), the phosphorylation and inactivation of PK goes to completion. With age, the AL animals lose the ability to activate and deactivate the enzyme, presumably through a loss of efficiency in the insulin and glucagon receptor-mediated mechanisms (Figure 3, left). As the loss of ability to dephosphorylate the enzyme occurs in the older AL rats, significantly more carbohydrate must be available for metabolism through glycolysis to produce any ATP.

PK is efficiently regulated in DR animals, even at older ages, and there is a strong correlation between changes in RQ and time of feeding (Figure 3, right). Indeed, it appears that this maintenance of regulation may be involved in the ability of DR to preserve the circadian rhythm for

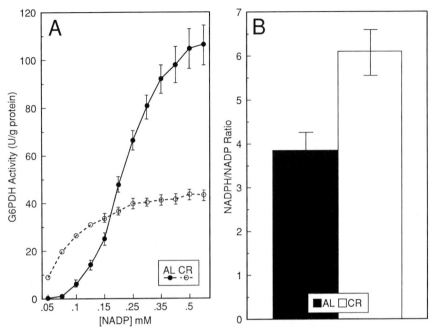

Figure 4. NADPH metabolism of ad libitum (AL) and calorie-restricted (CR) male (Fischer 344 × Brown Norway) F1 rats. A. Activity of glucose 6-phosphate dehydrogenase (G6PDH) as a function of NADP$^+$ concentration. B. The ratio of hepatic NADPH/NADP$^+$ in AL and CR rats.

activation/inactivation of PK across life span. The dephosphorylation of PK in DR rats correlates well to the increase in RQ seen just after feeding, and the phosphorylation of the enzyme increases as RQ decreases (Figure 3). Additionally, Vmax was significantly higher (twofold) and Km slightly higher in AL animals. This suggests that in AL rodents, compensation for the inactivity of the phosphorylated enzyme is attempted by "wasteful" synthesis, that is, there seems to be an increase in total enzyme that is less active, thus requiring expenditure of energy and excessive use of amino acid pools.

Glucose 6-phosphate dehydrogenase (G6PDH) provides another example of basic changes in DR regulation of enzymatic activity. G6PDH catalyzes the phosphorylation of glucose and the concomitant reduction of NADP$^+$ to NADPH. In addition, the NADPH pool is used to maintain those enzyme systems responsible for free radical detoxification. As was the case with PK, G6PDH is synthesized in excess in AL animals, but the enzyme is not active at low (endogenous) NADP$^+$ levels (Figure 4). DR

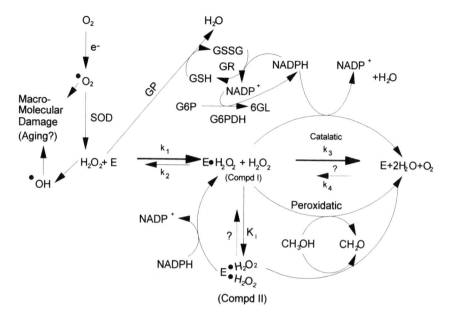

Figure 5. Antioxidant enzymes and their support systems. See text for definitions and discussion of mechanisms of action. E = catalase enzyme.

animals maintain far lower levels of total enzyme, but the enzyme activity is several-fold higher for DR animals at the levels of $NADP^+$ present in the cell (0.05–0.12 mM $NADP^+$). This may be why the $NADPH/NADP^+$ ratio is significantly higher in liver from DR animals (Figure 4, panel B).

Antioxidant Enzymes

As we have shown, basic optimal metabolic function declines with age, and much of the decline is offset by DR. This age-associated decline may be accompanied by or even produce increased levels of free radicals. This is a particularly appealing notion since accumulated free radical damage to macromolecules may be involved in the mechanism of aging. Induction of free radical production may also be involved in the mechanism of toxicity for a variety of drugs and xenobiotics.

During normal metabolic function, oxygen may receive an extra electron (Figure 5), which can happen as a result of xenobiotic and drug metabolism. Measurement of active oxygen species is difficult and not reliable; however, Leakey et al. (1989) have shown that cytochrome P4502C11 increases with age, and DR offsets this increase. This enzyme

is known to generate free radicals as a byproduct of the reaction it catalyzes. Superoxide dismutase (SOD) rapidly converts these superoxide radicals to hydrogen peroxide (H_2O_2). Glutathione peroxidase (GPX) and catalase (CAT) can then convert H_2O_2 to water and oxygen. In each case there
is a requirement for NADPH, which is supplied by G6PDH. Additionally, GPX must be supported by a recycling of oxidized to reduced glutathione. This is accomplished by glutathione reductase (GR) using
NADPH as a reductant.

The detoxification of H_2O_2 by CAT proceeds through two types of reactions. The enzyme combines with one molecule of H_2O_2 as an obligatory
step to form an intermediate known as compound I. Catalatic activity is
expressed when another molecule of H_2O_2 is used to yield free enzyme
plus $2H_2O$ and O_2. Alternatively, compound I may be oxidized by another
molecule of H_2O_2 to produce the inactive compound II (yielding abnormal
kinetics). This can be converted back to compound I and H_2O_2 through an
enzymatically catalyzed reduction using NADPH as substrate. The second type of reaction is in the presence of compound I, other appropriate
hydrogen donors (e.g., lower aliphatic alcohols) that protect against oxidation of CAT and stimulate peroxidatic activity that yields free enzyme,
H_2O, and an aldehyde as product (Figure 5).

Because CAT displays complex kinetics, quantifying its activity is complicated when an alcohol is not added to the reaction mixture. By calcu-
lating activity as µmol/min/g liver and multiplying this by the time span
of proportional conversion of substrate-to-product at maximal rates, the
effective amount of enzyme activity is defined. We have found that effective CAT activity is significantly increased in liver from DR rodents (Figure 6), which suggests that the accumulation of the inactive form of CAT
(compound II) occurred at a slower rate in cytosols from DR rats. There
are many potential nonmutually exclusive explanations for this action of
DR, including 1) an increased rate of catalysis of a second molecule of
H_2O_2 in the presence of compound I, 2) an altered enzyme structure having a higher affinity for H_2O_2 or an increased rate of dissociation of the
enzyme-substrate complex, 3) an increased concentration of CAT, and 4)
higher levels of NADPH to facilitate the NADPH-dependent reduction of
compound II. GPX and SOD activity responded to DR in a similar fashion. These results are similar to what others have reported in rodent liver
(Rao et al. 1990, Waggoner et al. 1990), although use of effective CAT
activity demonstrates that these DR effects occur at earlier ages than previously demonstrated.

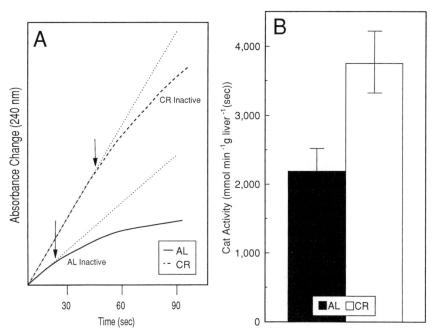

Figure 6. Effective catalase activity in ad libitum (AL) and calorie-restricted (CR) male (Fischer 344 × Brown Norway) F1 rats. The arrows indicate the time to deviation from nonlinearity. This time was multiplied by the initial slope to obtain effective activity.

As discussed in the preceding section, G6PDH activity was reduced by DR at Vmax, and NADPH was increased (Figure 4). Additionally, the NADPH/NADP$^+$ ratio was significantly increased, particularly just after feeding. GR activity was also increased. These changes may play a role in the increased activity noted for GPX and effective CAT in DR animals. Therefore, in liver, antioxidant enzymes and their support systems decrease with age, and this is opposed by DR. Improved efficiency of intermediary metabolism may be accompanied by relatively lower rates of free radical production.

Further investigation of these systems in a more metabolically active tissue (skeletal muscle) yielded a different result, thus revealing a more complicated situation. We found that activities of CAT and GPX increased with age and that these increases were opposed by DR. Furthermore, we found that most (if not all) of the age-related increases of CAT and GPX in muscle from AL rats were in the 12,000 × g fraction, which was enriched

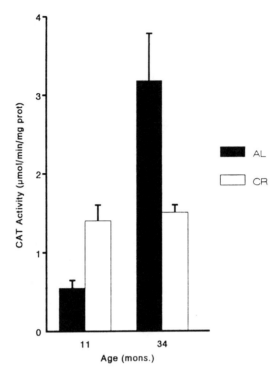

Figure 7. Muscle mitochondrial/peroxisomal catalase activity from ad libitum (AL) and diet-restricted (DR) male (Fischer 344 × Brown Norway) F1 rats. The enzyme appears to be induced in muscle from old (34-month-old) AL rats.

for peroxisomes and mitochondria (Figure 7; CAT as a representative example). This raises the possibility that these enzyme activities may be substrate-induced. In this case, free radical production in mitochondria would be increasing with age in animals on AL diets. Since free radicals may be produced at higher rates in mitochondria from AL rats, enzymes involved in mitochondrial electron transport would be likely candidates for this aberrant oxygen radical formation. If this is the case, there are profound implications for alterations in drug metabolism leading to toxicity through free radical production with differing levels of caloric intake. DR appears to oppose the deleterious changes in free radical production and detoxification associated with age.

Summary

Numerous physiological parameters are altered by caloric restriction. One of these is the pattern of food consumption. It is well-established that many variables are synchronized to food intake. Because DR animals eat their food when it is presented, if meal timing is not taken into consideration, misinterpretation of data can occur. DR induces significant changes in the circadian rhythms for RQ, indicating a rapid switch in metabolic pathways and output. Productive use of any available substrate in DR animals appears to increase metabolic efficiency. Periods of high carbohydrate metabolism correlate with increased body temperature, while periods of high fat metabolism trigger low body temperature. Enzyme activities correlate with these observations, since glycolytic and fat metabolism appear to decrease while gluconeogenesis and amino acid catabolism increase in response to DR, especially as RQ goes down. The insulin receptor is greatly up-regulated by DR, and consequently insulin sensitivity is increased in comparison to AL rodents.

Enzymes whose activities are regulated by insulin and glucagon also respond to DR. The range of phosphorylation (glucagon-dependent inactivation) and dephosphorylation (insulin-dependent activation) for PK is maximized and maintained across age in DR rodents. In DR animals there is a high degree of correlation between activation (glycolysis), inactivation (gluconeogenesis), and changes in RQ. This improved efficiency is a general phenomenon for enzymes and receptor systems from animals on DR. Rates of metabolic activity and efficiency seem to correlate with levels of free radical production in specific tissues and organelles. In AL animals, the antioxidant enzymes of the liver decline with age but are opposed by DR. On the other hand, in metabolically active muscle tissue where antioxidant enzyme activities have been shown to increase with age, the activities are kept low in DR animals. Most, if not all, of these effects seen in muscle were associated with the mitochondrial fraction, suggesting that the increases associated with age and AL feeding were probably due to increasing rates of free radical production in mitochondria. A likely source of such free radical production would be enzyme-catalyzed reactions of electron transport.

A primary responsibility of the Food and Drug Administration (FDA) is to regulate the food and pharmaceutical industry and to develop a comprehensive database that can be used for risk assessment. Clearly, from the standpoint of potential risk to humans, it is important to evaluate the response of whole animals under various nutritional states and at different ages to a variety of stressors. To make these types of studies relevant, appropriate biomarkers must be established and used to develop baseline data on metabolism for the DR model. The results of the physiologic and metabolic studies presented here provide crucial information that relates to the nutritional modulation of toxicity and may directly impact the implementation of any bioassay that uses DR.

References

Appelbaun A (1978) Adaptation to changes in caloric intake. Proc Food Nutr Sci 2:543–559

Duffy PH, Feuers RJ, Leakey JA, et al. (1989) Effect of chronic caloric restriction on physiological variables related to energy metabolism in male Fischer 344 rats. Mech Ageing Dev 48:117–133

Feuers RJ (1991) The relationship of dietary restriction to circadian variation in physiologic parameters and regulation of metabolism. Aging 3:399–401

Feuers RJ, Duffy PH, Leakey JEA, et al. (1989) Effect of chronic caloric restriction of hepatic enzymes of intermediary metabolism in the male Fischer 344 rat. Mech Ageing Dev 48:179–189

Feuers RJ, Hunter JD, Duffy PH, et al. (1990) ^{125}I-insulin binding in liver and the influence of insulin on blood glucose in calorically restricted B6C3F1 male mice. Annu Rev Chronopharmacol 7:189–192

Heuson JC, Legros N (1972) Influence of insulin deprivation on growth of the 7,12-dimethylbenz[a]anthracene-induced mammary carcinoma in rats subjected to alloxan diabetes and food restriction. Cancer Res 32:226–232

Kmiec Z, Mysliwske A (1983) Age-dependency change of hormone-stimulated gluconeogenesis in isolated rat hepatocytes. Exp Gerontol 18:173–183

Koizumi A, Weindruch R, Walford RC (1987) Influences of dietary restriction and age on liver enzyme activities and lipid peroxidation in mice. J Nutr 117:361–367

Leakey JEA, Cunny HC, Bazare J, et al. (1989) Effects of aging and caloric restriction on hepatic drug metabolizing enzymes in Fischer 344 rat. I. The cytochrome P-450 dependent monooxygenase system. Mech Ageing Dev 48:145

Masoro EJ (1988) Food restriction in rodents: an evaluation of its role in the study of aging. J Gerontol Biol Sci 43:B59–64

McCarter R, Masoro EJ, Byung PY (1985) Does food restriction retard aging by reducing the metabolic rate? Ann J Physiol 248:488–489

Rao G, Xia E, Nadakavukaren MJ, Richardson A (1990) Effect of dietary restriction on the age-dependent changes in the expression of antioxidant enzymes in rat liver. J Nutr 120:602–609

Ruggeri BA, Klurfeld DM, Kritchevesky D (1987) Biochemical alterations in 7,12-dimethylbenz[a]anthracene induced mammary tumors from rats subjected to caloric restriction. Biochim Biophys Acta 929:239–246

Semsei I, Rao G, Richardson A (1989) Changes in the expression of superoxide dismutase and catalase as a function of age and dietary restriction. Biochem Biophys Res Commun 164:620–629

Waggoner SM, Gu MZ, Chiang WH, Richardson A (1990) The effect of dietary restriction on the expression of a variety of genes. In Harrision DE (ed), Genetic effects on aging II. Telfore Press, Inc., Caldwell, NJ, pp 255–272

Weindruch R, Walford RL (1988) The retardation of aging and disease by dietary restriction. Charles C. Thomas, Springfield, IL

Yu BP, Laganier S, Kim JW (1989) Influence of life-prolonging food restriction on membrane lipoperoxidation and antioxidant states. In Simic MG (ed), Oxygen radicals in biology and medicine. Plenum Press, New York, pp 1067–1077

Effects of Dietary Calories and Fat on Levels of Oxidative DNA Damage

Z. Djuric
Wayne State University

M.H. Lu and S.M. Lewis
National Center for Toxicological Research

D.A. Luongo
Wayne State University

X.W. Chen
National Center for Toxicological Research

L.K. Heilbrun and B.A. Reading
Wayne State University

P.H. Duffy and R.W. Hart
National Center for Toxicological Research

Introduction

Calories, Fat, and Mammary Tumorigenesis

Breast cancer continues to be a major cause of death and disability among women in the United States, with a lifetime incidence of one in nine. Treatment modalities have improved over the years, but mortality from breast cancer has changed little since 1930 (U.S. Department of Health and Human Services 1991). There is a need for active efforts towards disease prevention. Breast cancer etiology is complex and multifactorial, with contributions from both genetic and epigenetic factors. Breast cancer prevention depends on alterations of the factors that can be controlled. One important prevention strategy to reduce breast cancer incidence may be dietary change.

Many epidemiological studies have suggested a relationship between fat in the diet and breast cancer risk (reviewed in Greenwald 1988, Howe et al. 1990, Schatzkin et al. 1989), although cohort studies generally have not indicated such a relationship (Howe et al. 1991, Van den Brandt et al. 1993, Willett et al. 1992). One difficulty in determining the influence of fat intake on breast cancer risk in epidemiological studies is that within a cultural group, the range of fat intakes is often narrow (Wynder and Stellman 1992). In most studies done in the United States, it is difficult to find sufficient numbers of individuals who have a fat intake as low as 15–20% of calories, the level used in most low-fat intervention studies (Chlebowski et al. 1991).

Both human and animal studies have pointed to the importance of including diets very low in fat in studies that aim to examine the impact of dietary fat on breast cancer and mammary gland carcinogenesis. There may be a threshold effect for the promoting effects of dietary fat between 20–30% of calories from fat, above which there is no further effect of dietary fat on tumor promotion (reviewed in Cohen et al. 1993). For example, if a threshold exists at 25% of calories from fat, then it would be difficult to detect differences in the tumor-promoting effects of diets that provide 45%–25% of calories from fat. Thus, cohort studies that do not include women who consume diets very low in fat may not be able to detect dietary fat as a significant risk factor for breast cancer incidence.

Studies comparing breast cancer incidence between the United States and Japan (with intakes of 37% versus 11% of calories from fat, respectively) have shown more consistent associations of fat intake and breast cancer risk than cohort studies (reviewed in Chlebowski et al. 1991, Schatzkin et al. 1989, Carroll 1992). Similarly, in an Italian study where the diet varied considerably due to regional differences in culinary practices, increased fat intake was associated with increased breast cancer risk (Toniolo et al. 1989). Intervention clinical trials, which involve a more dramatic reduction in fat intake, may show clearer results, and the initial cancer incidence results from the Women's Health Trial that had a target of 20% of calories from fat are promising (Henderson et al. 1991).

Caloric content of the diet may also be a key factor. Increased body weight, which may reflect caloric intake, has been associated with increased cancer risk (Albanes 1987, Kritchevsky 1990). The independent and interactive effects of fat and caloric intake are, however, difficult to separate in human epidemiological studies. Numerous animal studies have shown a protective effect of reduced dietary fat and calories on the development of mammary gland tumors (reviewed in Freedman et al. 1990, Schatzkin et al. 1989, Welsch 1992).

The importance of both caloric intake and fat content of the diet on mammary gland tumorigenesis was first demonstrated by Tannenbaum (1945). More recent studies indicate that caloric content of the diet may influence mammary gland tumor incidence in rats to a greater extent than fat content. A review of 100 animal studies indicated that both higher caloric intake and higher fat intake independently increase mammary tumor incidence in rats and mice, with dietary fat intake having an effect that is two-thirds that of dietary calories intake (Freedman et al. 1990). Most studies on dietary fat and calories, however, have generally failed to include a low-fat, calorie-restricted group, and the effect of dietary fat is typically evaluated by comparing a high-fat diet to a control diet. To fully evaluate the independent and interactive effects of fat and calories, four diets need to be used: control, low-fat, low-calorie, and a diet that is both low-fat and low-calorie. Type of fat also may be a key issue (Carroll 1992, Eaton 1992). The relative importance of dietary fat and calories on breast cancer incidence in humans remains to be elucidated.

Oxidative DNA Damage As a Marker of Cancer Risk

Oxygen radicals are products of normal cell metabolism, and DNA damage by oxygen radicals has been suggested to play a role in endogenous carcinogenesis (Loeb 1989, Lutz 1990). There is strong evidence that oxidative DNA damage can play a role in first-stage promotion, initiation, and progression of tumors (reviewed in Cerutti 1987, Cochrane 1991, Frenkel 1992). Many tumor promoters and complete chemical carcinogens cause increased formation of reactive oxygen species in cells, resulting in oxidative DNA damage (Frenkel 1992). Oxidative DNA damage includes both strand breakage and formation of hydroxylated DNA bases (Cochrane 1991). Strand breakage, however, can be induced by numerous types of DNA damage (e.g., unstable DNA adducts) and may not be useful as a specific marker of oxidative stress.

Numerous studies have examined the role of oxidative DNA damage in carcinogenesis. Hydroxylated DNA bases have been shown to be mutagenic when incorporated into the genome (McBride et al. 1991, Reid and Loeb 1992, Shibutani et al. 1991, Shirname-More et al 1987). Cellular oxidation can induce changes in gene expression (Allen and Venkatraj 1992).

Oxidative DNA damage can also induce aberrant oncogene expression and the malignant transformation of cells (Cerutti 1987, Cochrane 1991, Frenkel 1992). Later in the carcinogenic process, oxidant-induced DNA damage may be part of the inherent genetic changes associated with tumor cell growth. Many types of tumor cells have been shown to produce large amounts of hydrogen peroxide (Szatrowski and Nathan 1991). Accordingly, in cultured human mammary gland epithelial cells, the levels of 5-hydroxymethyluracil, a hydroxylated thymine residue, were higher in DNA from MCF-7 human breast epithelial tumor cells than from MCF-10A human breast epithelial normal-like cells (Djuric et al. 1993). Endogenous production of hydrogen peroxide by tumor cells has been associated with tumor cell proliferation and may confer a growth advantage to tumor cell populations via mutation that favors heterogeneity, invasion, and metastasis (Burdon et al. 1990, Galeotti et al. 1991, Szatrowski and Nathan 1991).

Of the various roles oxidative DNA damage can have in carcinogenesis, promotion is perhaps the role that is most closely tied to dietary fat and calories. Promotion is thought to be reversible and may be influenced by intervention strategies such as diet (Frenkel 1992, Pitot et al. 1988). Furthermore, the promotion phase of carcinogenesis is less likely to involve factors exclusive to a particular target tissue; therefore, its prevention does not depend on elucidation of specific and as yet unknown factors that are unique for a given type of cancer. This means that the effects of a promoter may be evident in many tissues other than the target organ, including peripheral blood, thus making human biomarker studies feasible.

In humans, endogenous DNA damage and oxidative stress are reported to be increased in leukocytes of breast cancer patients and high-risk individuals. DNA repair in response to hydrogen peroxide exposure is compromised in leukocytes of patients with cancer of various sites (Pero et al. 1989b), as is DNA repair in leukocytes of individuals with a family history of cancer (Pero et al. 1989a). DNA repair ability in lymphocytes from breast cancer patients and from women at high risk for breast cancer is also lower relative to control women (Kovacs and Almendral 1987, Kovacs et al. 1986). Lipid peroxidation is increased in women with mammographic dysplasia relative to control women, as determined by urinary malonaldehyde levels (Boyd and McGuire 1990). High levels of oxidative DNA damage are present in surgically obtained human breast cancer tissues (Malins et al. 1993). Thus, individuals at increased cancer risk may have increased levels of oxidative stress and/or compromised DNA repair. These biological changes should be amenable to modification by diet.

Experimental Methods

The effects of changes in dietary fat and calories on oxidative DNA damage levels were examined in the rat model. Female Fischer 344 (F344) rats were placed on the control diet (5% fat) at 38 days of age. Food intake was monitored from 38–42 days of age, such that the amount of food necessary to maintain rats under 40% dietary restriction (DR) could be determined. The food was presented in jars, and uneaten food was weighed. In order to calculate food consumption accurately, the animals were housed singly in hanging, mesh-bottom cages fitted with trays to facilitate collection of spilled food, which was then weighed. At 45 days of age the animals were weighed and divided arbitrarily into four diet groups, with 12 animals per group. The animals were maintained on the experimental diets for 2 weeks with daily food changes.

The diets consisted of ingredients obtained from Teklad Premier Laboratory Diets (a division of Harlan Sprague Dawley, Inc., Madison, WI). The diets prepared were 1) 3% corn oil fed ad libitum (AL) (low-fat), 2) 5% corn oil fed AL (control), 3) 20% corn oil fed AL (high-fat) and 4) 5% corn oil fed at 40% DR relative to the AL control diet (40% less food offered than the observed AL intake). The diets were formulated as shown in Table 1. The dextrose content of the diets was modified to accommodate the desired percentage of corn oil, but the other ingredients were kept constant. The percentages of all ingredients, except dextrose, were increased by 66% in the formulation of the restricted diet, such that when fed at 40% restriction, the rats maintained under DR would consume the same amount of nutrients as the control animals (Djuric et al. 1992). The diets were stored refrigerated.

The total energy values of the diets were calculated using the standard Atwater value estimates for physiological fuel values. Carbohydrates, protein, and cellulose were calculated at 4 kcal/g, and fat at 9 kcal/g. In our calculation of caloric intake, 12% digestibility of cellulose was used (Table 1).

After 2 weeks of feeding, the animals were sacrificed by carbon dioxide inhalation and decapitation. The entire mammary gland tissue and livers were removed. Whole livers were frozen. The mammary gland epithelium was isolated from fresh mammary gland tissue. The mammary gland epithelium and liver tissue were homogenized in 1% sodium dodecyl sulfate and 1 mM EDTA, and the DNA was extracted using phenol and chloroform. The purified DNA samples were hydrolyzed enzymatically to nucleosides, and trimethylsilyl derivatives of the nucleosides

Table 1. Formulation of diets

Ingredient	Diet[a]			
	3% Fat	5% Fat	20% Fat	DR
Corn oil	3	5	20	8.33
Casein	20	20	20	33.3
Dextrose	67	65	50	41.8
Cellulose	5	5	5	8.33
AIN-76A vitamin mix	1	1	1	1.67
AIN-76A mineral mix	3.7	3.7	3.7	6.16
Choline bitartrate	0.2	0.2	0.2	0.2
Dl-methionine	0.3	0.3	0.3	0.5
Kcal/100 g diet	377	387	462	379

[a]The percentage composition of all ingredients is given by weight. The diet fed under dietary restriction (DR) was formulated as shown and fed at 40% restriction relative to the ad libitum animals. Data is from Djuric et al. 1992.

were prepared for analysis by gas chromatography–mass spectroscopy. This procedure has been described previously (Djuric et al. 1991b, 1992).

Results and Discussion

Effects of Low-Fat and Restricted Diets on Oxidative DNA Damage Levels

Fat in the diet is one factor that may contribute to increased levels of oxidants in cells, and this may provide a mechanistic explanation for the cancer-promoting effects of dietary fat. Unsaturated fats can be readily oxidized, leading to production of reactive oxygen species. These reactive oxygen species, and perhaps lipid radicals, can oxidize DNA bases (Ames and Saul 1988). For example, autoxidized linolenic or linoleic acids can form 8-hydroxy-2'-deoxyguanosine in DNA (Kasai and Nishimura 1988). Indirectly, high fat intake may influence the level of cellular oxidative stress by changing metabolic routes from glycolysis to gluconeogenesis, with a concomitant decrease in available reducing equivalents within the cell (Hietanen et al. 1991). In addition, increased fat intake in humans is associated with increased levels of serum estrogens (Rose et al. 1987). The increased availability of estrogens would be expected to lead to oxidative damage, since estrogens can redox cycle, which can result in the formation of free radicals, including reactive oxygen species (Liehr and Roy 1990, Nutter et al. 1994). One might then expect that the levels of oxidized DNA bases can be used as an index of the carcinogenic effect of fat in the diet.

Women who consume a low-fat diet have been shown to have relatively low levels of oxidative DNA damage in their peripheral nucleated blood

Table 2. Mean dietary intakes and endogenous oxidative DNA damage levels by diet group

Variable	Diet group[a]				p-value[b]
	3% Fat	5% Fat	20% Fat	DR	
Food intake (g/day/animal)	11.84 ± 0.54 c e	12.10 ± 1.07 c	10.67 ± 0.79 e	7.14 ± 0.07 c e	<0.0001
Calories (kcal/day/animal)	44.62 ± 0.27 e	46.81 ± 3.64 c	49.27 ± 2.04 c	26.77 ± 0.27 c e	<0.0001
Fat intake (g/day/animal)	0.36 ± 0.0 c e d	0.61 ± 0.05 e d	2.13 ± 0.16 c d e	0.60 ± 0.01 e d	<0.0001
Liver DNA damage[f]	2.30 ± 0.68	3.28 ± 1.32 d	2.59 ± 0.66 e	1.86 ± 0.36 e d	0.002
Mammary gland DNA damage[c]	1.19 ± 0.40 e	1.76 ± 0.49 e e d	1.12 ± 0.26 e	1.10 ± 0.26 d	0.001

DR = diet restriction.
[a]Table entries are the mean for 12 rats ± SD. For dietary variables, the mean over the 14 days of the study was first calculated for each rat, and the mean of those 12 animal-specific means was then calculated for each diet group shown. Data are from Djuric et al. 1992.
[b]p Values shown for the simultaneous comparison of the four diet group means by the Kruskal-Wallis rank sum test.
[c-e]Pairs of means significantly different by Miller's multiple comparisons procedure with: c = $p < 0.001$; d = $p < 0.01$; e = $p < 0.05$. Pairs of values indicated by the same letter on the same line are significantly different at the level noted.
[f]DNA damage is expressed as 5-hydroxymethyluracil/10^4 thymine residues.

cells (Djuric, et al. 1991a). In rats, reduced dietary fat intake also decreases oxidative stress and oxidative DNA damage. Oxidative stress, as measured by ethane exhalation, correlates very closely with fat intake and tumor incidence in rats (Hietanen et al. 1990, 1987). With regard to oxidative DNA damage, the effects of reducing dietary fat on 5-hydroxymethyluracil levels in DNA from both liver and mammary gland epithelium were evident in rats after just 2 weeks of dietary change (Table 2).

Diets reduced in calories can also decrease the levels of oxidative DNA damage. As shown in Table 2, 40% DR significantly reduced 5-hydroxymethyluracil levels in rat liver and mammary gland, and the effect was greater than that of a 40% reduction in fat intake, i.e., from 5–3% corn oil (Djuric et al. 1992). In another study, 8-hydroxy-2'-deoxyguanosine levels in both nuclear and mitochondrial rat liver DNA were decreased by

40% DR (Chung et al. 1992). DR in one human subject has also been shown to decrease urinary oxidized DNA bases (Simic and Bergtold 1991). These studies indicate that diets restricted in calories, similar to low-fat diets, may decrease cancer risk through at least one common mechanism: a decrease in oxidative DNA damage levels.

There are several possible biochemical mechanisms by which DR can protect from tumor formation. Many of the biochemical changes elicited by DR could account for the observed decrease in oxidative DNA damage levels. Oxidative DNA damage levels will reflect the net result of oxidant levels, detoxification ability, and DNA repair. In rats and mice restricted in caloric intake by 40%, lipid metabolism is diminished, suggesting a decreased potential for generation of reactive species from fats (Feuers et al. 1989, 1990). Insulin levels are decreased during DR, which would be expected to decrease lipolysis and gluconeogenesis (Klurfeld et al. 1989a). Consistent with these observations, the levels of thiobarbituric acid-reactive substances and membrane lipid peroxidation are decreased in rats maintained under DR (Albrecht et al. 1992, Chipalkatti et al. 1983, Koizumi et al. 1987, Yu et al. 1988). Several laboratories have shown that DR increases the activity of antioxidant enzymes such as catalase and glutathione peroxidase (Koizumi et al. 1987, Langaniere and Yu 1989, Lang et al. 1989, Mote et al. 1991, Pieri et al. 1990, Rao et al. 1990, Yu 1991, Yu et al. 1988). Finally, DR also leads to enhanced DNA repair and reverses, to a degree, the age-related loss of specific activity and fidelity of DNA polymerases (Hart et al. 1990, Lipman et al. 1989, Srivastava et al. 1991, Srivastava and Busbee 1992, Tilley et al. 1992, Weindruch and Walford 1988, Weraarchakul et al. 1989).

Effects of High-Fat Diets on Oxidative DNA Damage Levels

The effects of a high-fat diet on oxidative DNA damage levels are not as straightforward as that of low-fat or restricted diets. It was somewhat surprising to find relatively decreased 5-hydroxymethyluracil levels in both mammary gland and liver of rats fed a high-fat diet for 2 weeks (Table 2). This could, however, be due to compensatory mechanisms that can be induced initially in response to the oxidative stress that results from a high-fat diet. In Sprague-Dawley rats, a 20% corn oil diet has been shown to increase glutathione (GSH) levels in liver and kidney relative to a 5% corn oil diet after 3 weeks of feeding (Nonavinakere et al. 1992). In another study, the activities of GSH peroxidase, GSH transferase, and catalase in colonic mucosa were all increased in Sprague-Dawley rats fed a 20% corn oil diet relative to rats fed 5% corn oil after 1 month. After 3 or more months of feeding these diets, however, the detoxification enzyme activities were lower in the rats fed a 20% corn oil diet relative to those fed a 5% corn oil diet (Kuratko and Pence 1991).

Diets high in fat thus may induce certain detoxification systems initially, and this can reduce the levels of oxidizing species within cells, but this increased detoxification ability apparently cannot be sustained. Using ethane exhalation as an index of oxidative stress in rats fed diets containing sunflower seed oil, rats fed either low-fat (2% oil) or high-fat (25% oil) diets exhibited decreased ethane exhalation relative to control rats (12.5% oil) after 10 weeks. Increased ethane exhalation in the high-fat group was observed only after increased time (20–27 weeks) on the diets (Hietanen et al. 1987, 1990). These data indicate that there may be a time-dependent factor for the promoting effects of fat on mammary gland tumorigenesis.

Another difficulty in interpreting the influence of dietary fat on mammary gland tumorigenesis is that there may be a threshold for the tumor-promoting effects of dietary fat, i.e., the promoting effects of fat may not be evident above a certain level of fat intake. Klurfeld et al. (1989b) found little difference in rat mammary gland tumor incidence using diets containing 15% and 20% corn oil. Similarly, Cohen et al. (1986) found no difference in the promoting effect of fat on mammary gland tumor incidence in rats using diets containing 16–23% corn oil (32–45% of the calories from fat). Zevenbergen et al. (1992) obtained essentially the same result using a mouse model, with no increase in mammary gland tumor incidence using 11–20% fat diets (22–40% of the caloric intake as fat). The studies of both Zevenbergen et al. (1992) and Cohen et al. (1986) also were not able to distinguish a significant difference in tumor incidence among animals fed diets that were low in fat (5–10% fat), although there was clearly a significant difference in incidence between the lower and higher-fat diets. These results indicate that either a nonlinear relationship exists between mammary tumor promotion and fat intake or that a linear dose response relationship exists over a narrow range of fat intakes.

Summary

The data shown here indicate that a 40% reduction in fat intake from a control diet (5% corn oil) decreased oxidative DNA damage in rat liver and mammary gland to a somewhat lesser extent than did a 40% reduction in caloric intake after 2 weeks of feeding. To fully evaluate the independent and interactive effects of dietary fat and calories, however, four diets should be used: control, low-fat, low-calorie, and a combination low-fat and low-calorie. The results could vary, however, depending on the type of control diet selected.

The effects of increases in dietary fat on oxidative stress are more difficult to interpret since there is a time-dependent change in the oxidant/antioxidant balance after dietary fat is increased. With a high-fat diet, there appears to be an initial induction of antioxidant defenses. This could be responsible for the relatively low DNA damage levels we found in rats fed diets containing 20% corn oil for 2 weeks. Several studies have indicated that antioxidant defenses are induced initially in rats by feeding of high-fat diets (20% oil), with oxidative modification of cellular macromolecules occurring only after prolonged feeding of the high-fat diet (Nonavinakere et al. 1992, Kuratko and Pence 1991, Hietanen et al. 1987, Hietanen et al. 1990). To fully assess the effects of changes in fat intake on oxidative damage, dietary studies of longer duration, such as 3 months or more, should be used.

A final consideration is how to best design rat diets to model human exposures to fat. A 20% corn oil diet provides 39% of calories from fat, which is similar to the dietary intake of fat in the U.S. population (Subcommittee on the 10th Edition of the RDAs 1989). The effects of dietary fat on tumor promotion have largely been determined by comparing control rats fed 5% corn oil diets with rats fed 5% corn oil initially and then 20% corn oil later (Freedman et al. 1990). Different results may be obtained, however, by maintaining animals on a 20% fat diet and then reducing fat intake, since that is what would occur in human intervention studies that aim to lower dietary fat.

Acknowledgment

This work was supported in part by a grant from the American Institute for Cancer Research.

References

Albanes D (1987) Caloric intake, body weight and cancer: a review. Nutr Cancer, 9:199

Albrecht R, Pelissier MA, Atteba S, Smaili M (1992) Dietary restriction decreases thiobarbituric acid-reactive substances generation in the small intestine and in the liver of young rats. Toxicol Lett 63(1):91

Allen RG, Venkatraj VS (1992) Oxidants and antioxidants in development and differentiation. J Nutr 122:631

Ames BN, Saul RL (1988) Cancer, ageing, and oxidative damage. In Iverson OH (ed), Theories of carcinogenesis. Hemisphere, Cambridge, pp 203–270

Boyd NF, McGuire V (1990) Evidence of lipid peroxidation in premenopausal women with mammographic dysplasia. Cancer Lett 50:31

Burdon RH, Gill V, Rice-Evans C (1990) Oxidative stress and tumour cell proliferation. Free Radic Res Commun 11:65

Carroll KK (1992) Dietary fat and breast cancer. Lipids 27:793

Cerutti PA (1987) Prooxidant states and tumor promotion. Science 227:375

Chipalkatti S, De AK, Aiyar AS (1983) Effect of diet restriction on some biochemical parameters related to aging in mice. J Nutr 113:944

Chlebowski RT, Rose D, Buzzard IM, et al. (1991) Adjuvant dietary fat intake reduction in postmenopausal breast cancer patient management. Breast Cancer Res Treat 20:73

Chung MH, Kasai H, Nishimura S, Yu BP (1992) Protection of DNA damage by dietary restriction. Free Radic Biol Med 12:523

Cochrane CG (1991) Cellular injury by oxidants. Amer J Med 91(suppl 3C):23s

Cohen LA, Choi K, Weisburger JH, Rose, DP (1986) Effect of varying proportions of dietary fat on the development of N-nitrosomethylurea-induced rat mammary tumors. Anticancer Res 6:215

Cohen LA, Rose DP, Wynder EL (1993) A rationale for dietary intervention in postmenopausal breast cancer patients: an update. Nutr Cancer 19:1

Djuric Z, Everett CK, Luongo DA (1993) Toxicity, single strand breaks, and 5-hydroxymethyluracil formation in human breast epithelial cells treated with hydrogen peroxide. Free Radic Biol Med 14:541

Djuric Z, Heilbrun LK, Reading BA, et al. (1991a) Effects of a low-fat diet on levels of oxidative damage to DNA in human peripheral nucleated blood cells. J Natl Cancer Inst 83:766

Djuric Z, Lu MH, Lewis S, et al. (1992) Oxidative DNA damage in rats fed low-fat, high fat or calorie-restricted diets. Toxicol Appl Pharmacol 115:156

Djuric Z, Luongo DA, Harper DA (1991b) Quantitation of 5-(hydroxymethyl)uracil in DNA by gas chromatography with mass spectral detection. Chem Res Toxicol 4:687

Eaton SB (1992) Humans, lipids and evolution. Lipids 27:814

Feuers RJ, Duffy PH, Leakey JA, et al. (1989) Effect of chronic caloric restriction on hepatic enzymes of intermediary metabolism in the male Fischer 344 rat. Mech Ageing Devel 48:179

Feuers RJ, Leakey JE, Duffy PH, et al. (1990) Effect of chronic caloric restriction on hepatic enzymes of intermediary metabolism in aged B6C3F1 female mice. Prog Clin Biol Res 341B:177

Freedman LS, Clifford C, Messina M (1990) Analysis of dietary fat, calories and body weight, and the development of mammary tumors in rats and mice: a review. Cancer Res 50:5710

Frenkel K (1992) Carcinogen-mediated oxidant formation and oxidative DNA damage. Pharmacol Ther 53:127

Galeotti T, Masotti L, Borrello S, Casali E (1991) Oxy-radical metabolism and control of tumour growth. Xenobiotica 21:1041

Greenwald P (1988) Issues raised by the Women's Health Trial. J Natl Cancer Inst 80:788

Hart RW, Turturro A, Pegram RA, Chou MW (1990) Effects of caloric restriction on the maintenance of genetic fidelity. Basic Life Sci 53:351

Henderson MH, White E, Thompson RS (1991) Cancer incidence in Seattle Women's Health Trial participants by group and time since randomization. J Natl Cancer Inst 83:1260

Hietanen E, Ahotupa M, Bereziat J-C, et al. (1987) Elevated lipid peroxidation in rats induced by dietary lipids and N-nitrosodimethylamine and its inhibition by indomethacin monitored via ethane exhalation. Toxicol Pathol 15:93

Hietanen E, Bartsch H, Ahotupa M, et al. (1991) Mechanisms of fat-related modulation of N-nitrosodiethylamine-induced tumors in rats: organ distribution, blood lipids, enzymes and pro-oxidant state. Carcinogenesis 12:591

Hietanen E, Bartsch H, Bereziat J-C, et al. (1990) Quantity and saturation degree of dietary fats as modulators of oxidative stress and chemically-induced liver tumors in rats. Int J Cancer 46:640

Howe GR, Friedenreich CM, Jain M, Miller AB (1991) A cohort study of fat intake and risk of breast cancer. J Natl Cancer Inst 83:336

Howe GR, Hirohata T, Hislop TG, et al. (1990) Dietary factors and risk of breast cancer: combined analysis of 12 case-control studies. J Natl Cancer Inst 82:561

Kasai H, Nishimura S (1988) Formation of 8-hydroxydeoxyguanosine in DNA by auto-oxidized unsaturated fatty acids. In Hayaishi O, et al. (eds), Medical, biochemical and chemical aspects of free radicals. Elsevier, New York, pp 1021–1023

Klurfeld DM, Welch CB, Davis MJ, Kritchevsky D (1989a) Determination of the degree of dietary restriction necessary to reduce DMBA-induced mammary tumorigenesis in rats during the promotion phase. J Nutr 119:286

Klurfeld DM, Welch CB, Lloyd LM, Kritchevsky D (1989b) Inhibition of DMBA-induced mammary tumorigenesis by caloric restriction in rats fed high-fat diets. Int J Cancer 43:922

Koizumi A, Weindruch R, Walford RL (1987) Influence of dietary restriction and age on liver enzyme activities and lipid peroxidation in mice. J Nutr 117:361

Kovacs E, Almendral A (1987) Reduced DNA repair synthesis in healthy women having first degree relatives with breast cancer. Eur J Cancer Clin Oncol 23:1051

Kovacs E, Stucki D, Weber W, Muller HJ (1986) Impaired DNA-repair synthesis in lymphocytes of breast cancer patients. Eur J Cancer Clin Oncol 22:863

Kritchevsky D (1990) Nutrition and breast cancer. Cancer 66:1321
Kuratko C, Pence B (1991) Changes in colonic antioxidant status in rats
    during long-term feeding of different high fat diets. Lipids 121:1562
Lang CA, Wu WK, Chen T, Mills BJ (1989) Blood glutathione: a biochemical
    index of life span enhancement in the diet restricted Lobund-Wistar rat.
    Prog Clin Biol Res 287:241
Langaniere S, Yu BP (1989) Effect of chronic food restriction in aging rats II.
    Liver cytosolic antioxidants and related enzymes. Mech Aging Devel
    48:221
Liehr JG, Roy D (1990) Free radical generation by redox cycling of estro-
    gens. Free Radic Biol Med 8:415
Lipman JM, Turturro A, Hart RW (1989) The influence of dietary restriction
    on DNA repair in rodents: a preliminary study. Mech Ageing Devel
    48(2):135
Loeb LA (1989) Endogenous carcinogenesis: molecular oncology into the
    twenty-first century—presidential address. Cancer Res 49:5489
Lutz WK (1990) Endogenous genotoxic agents and processes as a basis of
    spontaneous carcinogenesis. Mutat Res 238:287
Malins DC, Holmes EH, Polissar NL, Gunselman SJ (1993) The etiology of
    breast cancer: characteristic alterations in hydroxyl radical-induced DNA
    base lesions during oncogenesis with potential for evaluating incidence
    risk. Cancer 71:3036
McBride TJ, Preston BD, Loeb LA (1991) Mutagenic spectrum resulting from
    DNA damage by oxygen radicals. Biochemistry 30:207
Mote PL, Grizzle JM, Walford RL, Spindler SR (1991) Influence of age and
    caloric restriction on expression of hepatic genes for xenobiotic and oxy-
    gen metabolizing enzymes in the mouse. J Gerontol 46(3):B95
Nonavinakere VK, Black CL, Curtis S, et al. (1992) Effects of dietary fat and
    selenium (Se) on weight gain; and liver, kidney, colon glutathione (GSH)
    of 1,2-dimethylhydrazine dihydrochloride (DMH) treated rats. The Toxi-
    cologist 12:409
Nutter LM, Wu Y-Y, Ngo EO, et al. (1994) An o-quinone form of estrogen
    produces free radicals in human breast cancer cells: correlation with DNA
    damage. Chem Res Toxicol 7:23.
Pero RW, Johnson DB, Miller DG, et al. (1989b) Adenosine diphosphate
    ribosyltransferase responses to a standardized dose of hydrogen peroxide
    in the mononuclear leukocytes of patients with a diagnosis of cancer.
    Carcinogenesis 10:1657
Pero RW, Johnson DB, Markowitz M, et al. (1989a) DNA repair synthesis in
    individuals with and without a family history of cancer. Carcinogenesis
    10:693
Pieri C, Falasca M, Moroni F, et al. (1990) Antioxidant enzymes in erythro-
    cytes from old and diet restricted old rats. Boll Soc Ital Biol Speri
    66(10):909

Pitot HC, Beer D, Hendrich S (1988) Multistage carcinogenesis: the phenomenon underlying the theories. In Iverson OH (ed), Theories of carcinogenesis. Hemisphere, Cambridge, pp 159–177

Rao G, Xia E, Nadakavukaren MJ, Richardson A (1990) Effect of dietary restriction on the age-dependent changes in the expression of antioxidant enzymes in rat liver. J Nutr 120:602

Reid TM, Loeb LA (1992) Mutagenic specificity of oxygen radicals produced by human leukemia cells. Cancer Res 52:1082

Rose DP, Boyar AP, Cohen C, Strong LE (1987) Effect of a low-fat diet on hormone levels in women with cystic breast disease I. Serum steroids and gonadotropins. J Natl Cancer Inst 78:623

Schatzkin A, Greenwald P, Byar DP, Clifford CK (1989) The dietary fat-breast cancer hypothesis is alive. JAMA 261:3284

Shibutani S, Takeshita M, Grollman AP (1991) Insertion of specific bases during DNA synthesis past the oxidation damaged base 8-oxodG. Nature 349:431

Shirname-More L, Rossman TG, Troll W, et al. (1987) Genetic effects of 5-hydroxymethyl-2'-deoxyuridine, a product of ionizing radiation. Mutat Res 178:177

Simic MG, Bergtold DS (1991) Dietary modulation of DNA damage in human. Mutat Res 250:17

Srivastava VK, Busbee DL (1992) Decreased fidelity of DNA polymerases and decreased DNA excision repair in aging mice: effects of caloric restriction. Biochem Biophys Res Commun 182(2):712

Srivastava VK, Tilley RD, Hart RW, Busbee DL (1991) Effect of dietary restriction on the fidelity of DNA polymerases in aging mice. Exp Gerontol 26(5):453

Subcommittee on the Tenth Edition of the RDAs (1989) Food and Nutrition Board, Commission on Life Sciences, National Research Council, Recommended Dietary Allowances, 10th edition. National Academy Press, Washington, DC

Szatrowski TP, Nathan CF (1991) Production of large amounts of hydrogen peroxide by human tumor cells. Cancer Res 51:794

Tannenbaum A (1945) The dependence of tumor formation on the composition of the calorie-restricted diet as well as on the degree of restriction. Cancer Res 5:616

Tilley R, Miller S, Srivastava V, Busbee D (1992) Enhanced unscheduled DNA synthesis by secondary cultures of lung cells established from calorically restricted aged rats. Mech Ageing Devel 63(2):165

Toniolo P, Riboli E, Protta F, et al. (1989) Calorie-providing nutrients and risk of breast cancer. J Natl Cancer Inst 81:278

U.S. Department of Health and Human Services (1991) National Cancer Institute: 1990 Annual Cancer Statistics Review. NIH Publication No. 91-2789, National Cancer Institute, Bethesda, MD

Van den Brandt PA, Van't Veer P, Goldbohm RA, et al. (1993) A prospective cohort study on dietary fat and the risk of postmenopausal breast cancer. Cancer Res 53:75

Weindruch R, Walford RL (1988) The retardation of ageing and disease by dietary restriction. Charles C. Thomas, Springfield, IL

Welsch CW (1992) Relationships between dietary fat and experimental mammary tumorigenesis: a review and critique. Cancer Res 52:2040s

Weraarchakul N, Strong R, Wood WG, Richardson A (1989) The effect of aging and dietary restriction on DNA repair. Exp Cell Res 181:197

Willett WC, Hunter DJ, Stampfer MJ, et al. (1992) Dietary fat and fiber in relation to risk of breast cancer: an 8-year follow-up. JAMA 268:2037

Wynder EL, Stellman SD (1992) The "over-exposed" control group. Am J Epidemiol 135:459

Yu BP (1991) Free radicals and modulation by dietary restriction. Age Nutr 2:84

Yu BP, Langaniere S, Kim J-W (1988) Influence of life-prolonging food restriction on membrane lipoperoxidation and antioxidant status. Basic Life Sci 49:1067

Zevenbergen JL, Verschuren PM, Zalberg J (1992) Effect of the amount of dietary fat on the development of mammary tumors in BALB/c-MTV mice. Nutr Cancer 17:9

Effect of Dietary Restriction on the Genome Function of Cells: Alterations in the Transcriptional Apparatus of Cells

Ahmad R. Heydari, Astrid Gutsmann, Shenghong You, Ryoya Takahashi, and Arlan Richardson

Audie L. Murphy Memorial VA Hospital
University of Texas Health Science Center

Introduction

The molecular basis for gene control in eukaryotic cells is currently one of the most actively studied areas in biology. Regulation of expression of specific genes occurs at four levels: 1) transcription of specific messenger RNA (mRNA), i.e., initiation, elongation, and termination of mRNA, 2) processing of the primary transcripts within the nucleus, 3) stability of the mRNA in the cell cytoplasm, and 4) translation of the mRNA transcript. While several biological systems have been shown to regulate expression of specific genes at the levels stated above, the primary site for regulation of gene expression is the initiation stage of transcription (Conaway and Conaway 1993, Kollmar and Farnham 1993). The selective and accurate initiation of mRNA synthesis by RNA polymerase II is governed primarily by cooperative interaction of two distinct regulatory components: consensus elements of specific DNA sequences in the genes and regulatory proteins.

The DNA consensus elements comprise three major functional classes: core promoter elements, upstream regulatory elements, and enhancers (Conaway and Conaway 1993). The transcription of specific genes is regulated by the interaction of specialized regulatory proteins and these consensus elements. The core promoter elements are common to all genes transcribed by RNA polymerase II. These elements contain the binding site for RNA polymerase II and control the location of the site-specific initiation by interacting with basal transcription factors, which are general transcription factors essential for initiation and are sufficient to direct a basal level of transcription from core promoters in an orderly fashion (Kollmar and Farnham 1993, Conaway and Conaway 1993). Thus, the specificity of the start site is governed through specific interactions among basal transcription proteins and the core promoter elements. Regulatory transcription factors, which bind to upstream regulatory elements, control the rate of initiation of specific genes in specific tissues and/or at specific stages of cell growth and development. The regulatory transcription factors are believed to activate or repress transcription either by interacting with RNA polymerase II and the basal factors or by changing chromatin structure (Kollmar and Farnham 1993). Therefore, the basal and regulatory transcription factors act in concert to regulate the transcription of specific genes during cellular development, homeostasis, and perhaps senescence.

Although it is well demonstrated that dietary restriction (DR) increases the mean and maximum life span of laboratory rodents (Masoro 1988, Masoro 1990, Heydari and Richardson 1992, Masoro 1992a,b), the molecular mechanism by which DR enhances survival and retards senescence remains unknown. It has been proposed that changes in the expression of specific genes play a major role in the biological mechanism by which DR affects senescence and survival (Heydari and Richardson 1992). This hypothesis was initially advanced by Young (1979) and later by Richardson and Cheung (1982). Young proposed that diet or nutrition might influence aging and senescence by interacting at the structural and functional levels of the gene, in particular by influencing the translational and/or posttranslational processes associated with protein synthesis. Subsequently, our laboratory (Richardson 1985, Heydari and Richardson 1992) argued that changes in gene expression were a potentially important element in the action of DR on senescence for the following reasons: 1) changes in the expression of genes represent one of the primary sites of cellular regulation for all living organisms, 2) changes in the expression of even one gene can have a profound effect on a cell or organism, and 3) various steps of gene expression are altered by senescence (Heydari and Richardson 1992). Studies from our laboratory provided the first evidence that DR altered the transcription of specific genes using a DR regimen in which animals were fed 60% of the diet consumed by ad libitum (AL) rats (Richardson et al. 1987). As shown in Figure 1A, the level of the mRNA transcript for $\alpha_2\mu$-globulin is 1.8-fold higher in liver tissue from rats maintained under conditions of DR than in liver tissue from AL animals (Richardson et al. 1987). Subsequently, our laboratory showed a significant increase (40–50%) in the expression of the mRNA transcripts for the

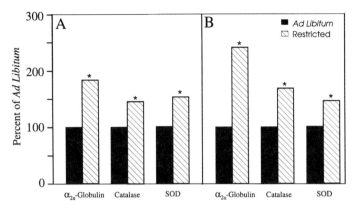

Figure 1. Effect of DR on the expression of α_{2u}-globulin, catalase, and superoxide dismutase (SOD) in the liver of Fischer 344 rats. The mRNA levels (panel A) and nuclear transcription (Panel B) of α_{2u}-globulin, catalase, and SOD from 18-month-old AL rats or 18-month-old rats maintained under conditions of DR are shown. Data for α_{2u}-globulin were taken from Richardson et al. (1987) and for catalase and SOD were taken from Semsei et al. (1989). The rats were maintained on two dietary regimens: AL rats were given free access to food, and the DR group received 60% of the diet consumed by the AL animals. DR was initiated when rats were 6 weeks old. The values are expressed as the percentage of values obtained for 18-month-old AL-fed rats. These values (*) are significantly different from the values for AL rats at the $p \leq 0.01$ level.

antioxidant genes of superoxide dismutase (SOD) and catalase in liver of rats maintained under conditions of DR (Semsei et al. 1989).

Since the initial study by Richardson et al. (1987), numerous investigators have measured the mRNA levels of a variety of genes in various tissues of AL rodents or animals maintained under conditions of DR. Many of these studies have focused on genes that play a potentially important role in the survival of the tissue/organism. Genes coding for proteins that play strategic roles in protecting cells from toxic factors found in the environment or produced as by-products of cellular metabolism are likely to be logical candidates for genes that would have a significant impact on senescence and survival. Table 1 compares the expression of specific genes in various tissues of rodents under AL and calorically restricted feeding regimens. It is clear from these data that DR alters the expression of a variety of mRNA transcripts in liver, adrenal gland, lymphocytes, thyroid, and trigeminal ganglion. More importantly, the data clearly demonstrate that changes in gene transcription associated with DR are not due simply to an overall alteration in the transcriptional apparatus of the cell, because not all genes are affected in the same way. While DR enhances the expression of many genes, the expression of other genes is reduced or does not change with DR (Table 1).

Table 1. Effect of dietary restriction (DR) on expression of specific genes in rats and mice

Species/strain	Tissue	Gene	Ages studied (months)	Alteration with caloric restriction	References
Rat (male)					
F344	Liver	α₂μ-globulin	18	Increase[a]	Richardson et al. (1987)[b]
F344	Liver	α₂μ-globulin	27	Increase	Chatterjee et al. (1989)[b]
F344	Liver	Senescence marker protein (SMP-2)	27	Decrease	Chatterjee et al. (1989)[1]
F344	Liver	Superoxide dismutase (Cu/Zn)	18	Increase[a]	Semsei et al. (1989)[b]
F344	Liver	Catalase	18	Increase[a]	Semsei et al. (1989)[b]
F344	Liver	α₂μ-globulin	21	Increase	Waggoner et al. (1990)[b]
F344	Liver	Apolipoprotein A1	18	Decrease	Waggoner et al. (1990)[b]
F344	Liver	Apolipoprotein B	18	No change	Waggoner et al. (1990)[b]
F344	Liver	c-myc	18	No change	Waggoner et al. (1990)[b]
F344	Liver	Catalase	28	Increase	Rao et al. (1990)[b]
F344	Liver	Cytochrome P450 (P450IIB1/IIB2)	20	Increase	Horbach et al. (1990)[b]
F344	Liver	Superoxide dismutase (Cu/Zn)	28	Increase	Rao et al. (1990)[b]
F344	Adrenal gland	Tyrosine hydroxylase	21	Increase	Strong et al. (1990)[b]
F344	Spleen lymphocytes	Interleukin II (stimulated with conA)	28	Increase	Pahlavani et al. (1990)[b]
F344	Liver	Glutathione peroxidase	28	Increase	Rao et al. (1990)[b]
F344	Liver	Androgen receptor	27	Increase	Song et al. (1991)[b]
SD	Liver	T-kininogen	24	Decrease	Sierra et al. (1992)[b]
F344	Liver	Heat shock protein 70	28	Increase[a]	Heydari et al. (1993)[b]
F344	Thyroid	Calcitonin	18	Decrease[a]	Salih et al. (1993)[b]
F344	Thyroid	Calcitonin gene-related peptide	24	Decrease[a]	Salih et al. (1993)[b]
F344	Thyroid	Somatostatin	24	Decrease[a]	Salih et al. (1993)[b]
F344	Trigeminal ganglion	Calcitonin gene-related peptide	24	No change	Salih et al. (1993)[b]
Mouse (female)					
C3B10RF1	Liver	Glucose-regulated protein 78	31	Decrease	Spindler et al. (1990)[b]
C3B10RF1	Liver	Glucose-regulated protein 94	31	Decrease	Spindler et al. (1990)[b]
C3B10RF1	Liver	Insulin-like growth factor 1	31	Increase	Spindler et al. (1990)[b]
C3B10RF1	Liver	RNA polymerase II elongation factor SII	31	No change	Spindler et al. (1991)[b]
C3B10RF1	Liver	Transcription factor Sp1	31	No change	Spindler et al. (1991)[b]
C3B10RF1	Liver	CAAT enhancer binding Protein (C/EBP)	31	No change	Spindler et al. (1991)[b]
C3B10RF1	Liver	c-jun	31	No change	Spindler et al. (1991)[b]
C3B10RF1	Liver	Glucocorticoid receptor	31	No change	Spindler et al. (1991)[b]
C3B10RF1	Liver	Cytochrome P_3-450 (cyp 1A2)	31	Increase	Mote et al. (1991)[c]
C3B10RF1	Liver	Epoxide hydrolase	31	No change	Mote et al. (1991)[c]

F344 = Fischer 344, SD = Sprague-Dawley.
[a]Nuclear transcription was also measured in this study.
[b]These animals were maintained under conditions of DR and were fed 60% of diet consumed by ad libitum (AL) animals.
[c]These animals were maintained under conditions of DR and were fed 48% of diet consumed by AL animals.

Because aging is generally characterized by a reduced ability of an organism to maintain homeostasis in response to stress, we are interested in how DR affects the ability of cells to respond to one form of environmental stress, heat shock. Thus, our laboratory has studied extensively the effect of age and DR on induction of heat shock protein 70 (hsp70) gene in response to heat shock (Heydari et al. 1993). Here, we will review how aging and DR alter the transcription of hsp70 and then present preliminary data supporting our hypothesis that DR triggers changes in the expression of hsp70 by altering the levels/activities of transcription factors that regulate hsp70 transcription (Heydari and Richardson 1992).

Heat Shock Protein 70

A variety of physiological stresses, including hyperthermia, induce the synthesis of a group of proteins known as heat shock proteins (hsps) or stress-induced proteins (Lindquist 1986, Lindquist and Craig 1988). The proteins in the heat shock protein 70 (HSP70) family are the most prominent and evolutionarily conserved of the heat shock proteins. The HSP70 family is made up of at least four groups of genes: hsp70, hsc70, grp78, and grp75 (Lindquist and Craig 1988). Hsp70 (with molecular weights of 68 kDa and 72 kDa in rodent and human cells, respectively) is expressed only slightly in nonstressed cells and is highly heat-inducible. Hsc70 (also known as p72 or hsc73, with molecular weights of 70 and 73 in rodent and human cells, respectively) is constitutively expressed and only slightly induced by hyperthermia (Sorger et al. 1987). Grp78 has a molecular weight of 78 kDa in both rodent and human cells and is induced significantly by glucose starvation, calcium ionophores, inhibitors of glycosylation, and amino acid analogs. However, the synthesis of grp78 is not significantly affected by hyperthermia (Cairo et al. 1989). Grp75 (with a molecular weight of 75 kDa) is also a constitutively expressed protein and is presumed to function in a manner similar to grp78 (Mizzen et al. 1989). The hsps that make up the HSP70 family are found in specific cellular compartments. For example, hsp70 and hsc70 are found in the cytoplasm and nucleus and become associated with nucleoli after heat shock. Grp78 is found in the endoplasmic reticulum (Lindquist and Craig 1988, Cairo et al. 1989), whereas grp75 is localized within the matrix of mitochondria (Mizzen et al. 1989). In contrast to hsp70, the other members of the HSP70 family are expressed in relatively high amounts in unstressed cells (Lindquist and Craig 1988).

Hsp70 is the most prominent heat shock protein expressed after heat shock and is the most conserved hsp found among both prokaryotes and eukaryotes. Hsp70 appears to play a critical role in protecting cells against the adverse effects of hyperthermia, because the thermosensitivity of cells is increased if the expression of hsp70 is reduced and is decreased with the enhanced expression of hsp70 gene (Riabowol et al. 1988, Johnson and Kucey 1988, Angelidis et al. 1991, Li et al. 1991). The gene coding for hsp70 has been cloned from a variety of eukaryotes, and its regulation by hyperthermia has been studied extensively (Heydari et al. 1993). Hsp70 expression is regulated primarily at the level of transcription (Kingston et al. 1987, Mosser et al. 1990, Sorger et al. 1987, Zimarino and Wu 1987), and this regulation is an excellent example of a cellular mechanism that has evolved to protect all living organisms from hyperthermia and other types of stress. Therefore, changes in this system could seriously compromise the capacity of an organism to respond to changes in its environment.

The induction of hsp70 expression (mRNA levels) decreases with increasing age in various tissues, e.g., lung, skin, brain, liver, and lymphocytes, of both rats and humans (Liu et al. 1989, Fassen et al. 1989, Blake et al. 1991, Fargnoli et al. 1990, Campanini et al. 1992). Our laboratory has compared the ability of hepatocytes isolated from young and old rats to express hsp70 in response to increased temperature (Wu et al. 1993, Heydari et al. 1993). The induction of hsp70 synthesis by heat shock was significantly reduced in hepatocytes isolated from old rats compared to hepatocytes isolated from young adult rats (Wu et al. 1993, Heydari et al. 1993). Because an age-related decline in hsp70 induction by stress is a universal phenomenon in mammals, it was of interest to study the mechanism of the age-related decrease in the expression of hsp70 and also to analyze the effect of DR on the age-related alteration in expression of this gene.

Our laboratory has shown that the age-related decrease in the induction of hsp70 synthesis and mRNA occurred at the level of transcription; nuclear transcription of hsp70 was significantly lower in hepatocytes isolated from old rats than in those isolated from young adult rats. Although the initial rate of induction of hsp70 transcription was similar for hepatocytes isolated from young adult and from old rats, the maximum level of hsp70 transcription achieved with heat shock was significantly lower for hepatocytes isolated from old rats (Heydari et al. 1993). In addition, using in situ hybridization, we have shown that the decrease in the induction of hsp70 expression with age is not due to an inability of subpopulations of cells from old rats to respond to hyperthermia but, rather, because all hepatocytes from old rats show a decrease in hsp70 expression (Heydari et al. 1993). This observation is important from an aging perspective because previous studies have suggested that age-related deficits in the response of

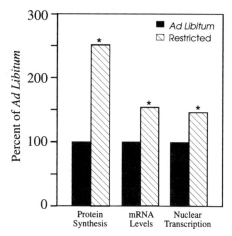

Figure 2. Effect of DR on the induction of hsp70 synthesis, mRNA levels, and nuclear transcription in Fischer 344 rat hepatocytes. The induction of hsp70 synthesis, mRNA levels, and nuclear transcription in heat-shocked (42.5° C, 30 minutes) hepatocytes isolated from 28-month-old AL rats or 28-month-old rats maintained under conditions of DR. Data were taken from Heydari et al. (1993). The AL rats were given free access to food, whereas the DR group received 60% of the diet consumed by the AL animals. Dietary restriction was initiated when rats were 6 weeks old. The values are expressed as the percentage of values obtained for 28-month-old AL rats. These values (*) are significantly different from the values for AL rats at the $p \leq 0.01$ level.

cells to signals may be due to an accumulation of cells that cannot respond to the stimuli.

Figure 2 shows the effect of DR on the ability of hepatocytes to express hsp70 after a mild heat shock. The induction of hsp70 synthesis was significantly higher in hepatocytes isolated from old rats fed a restricted diet than in old AL rats. It is apparent from Figure 2 that DR enhanced the induction of hsp70 mRNA and nuclear transcription of the hsp70 gene. We also have shown previously that the degradation of hsp70 mRNA was reduced with age in hepatocytes; i.e., the age-related decline in expression of hsp70 gene is not due to a decline in the half-life of hsp70 mRNA with age (Heydari et al. 1993). Interestingly, the half-life of hsp70 mRNA from hepatocytes isolated from old DR rats was similar to the half-life of hsp70 mRNA of hepatocytes isolated from young AL rats and shorter than the half-life of hsp70 mRNA of hepatocytes isolated from old AL rats (Heydari et al. 1993). Thus, our study shows that DR alters the expression of hsp70 mRNA, and this alteration is due to changes in the hsp70 transcript (Heydari et al. 1993).

At present there are only three other studies in the which the effect of DR on nuclear transcription of specific genes was analyzed. As shown in

Figure 1, our laboratory reported that the increase in the levels of $\alpha_2\mu$-globulin (Richardson et al. 1987), SOD, and catalase (Semsei et al. 1989) mRNA transcripts is paralleled by an increase in the level of the nuclear transcription of these genes. More recently, Salih et al. (1993) reported that DR altered the transcription of two genes by the thyroid. Dietary restriction resulted in a decrease in the nuclear transcription of the genes coding for the calcitonin gene-related peptide and for the somatostatin peptide. The decrease in the transcription of these genes was paralleled by a similar decrease in the mRNA of these two genes. Thus, DR appears to alter the expression of specific genes at the level of transcription.

Effect of DR on Levels/Activities of Transcription Factors

It is clear that the effect of DR on transcription varies considerably from gene to gene. Based on these observations, we previously proposed that DR alters the transcription of specific genes through changes in the levels and/or activities of specific transcription factors that regulate the expression of specific genes (Heydari and Richardson 1992). Transcription factors represent one of the largest and most diverse classes of DNA-binding proteins. Because a transcription factor can potentially regulate the expression of more than one gene, changes in the level/activity of a transcription factor or a few transcription factors have the potential of altering the expression of a variety of genes simultaneously. Over the past decade, it has become clear that these proteins play a critical role in development, differentiation, and cell growth. Thus, changes in the activities or levels of transcription factors could be a mechanism whereby DR alters the transcription of genes in a specific manner.

The induction of hsp70 expression in response to heat shock (Figure 3) is mediated by the binding of a transcriptional activator, the heat shock transcription factor (HSF), to the highly conserved DNA sequence known as the heat shock element (HSE) (Goldenberg et al. 1988, Sorger 1991, Wu et al. 1987). An increase in temperature in mammalian cells results in the conversion of HSF from an inactive monomer that does not bind DNA to an active trimer that binds the HSE (Kingston et al. 1987, Mosser et al. 1990, Sorger et al. 1987, Zimarino and Wu 1987). Using a gel shift assay, we measured the levels of HSF binding activity to the HSE in nuclear extracts from heat-shocked hepatocytes isolated from young and old AL rats and old DR rats (Heydari et al. 1993). Cell extracts from hepatocytes isolated from AL rats or DR rats (Figure 4) showed similar patterns of

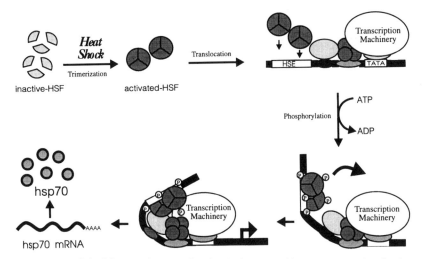

Figure 3. A model of the mechanism for the induction of hsp70 expression by heat shock.

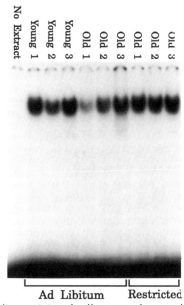

Figure 4. The HSF binding activity of cell extracts from rat hepatocytes. Hepatocytes isolated from young (4–6 months) and old (26–28 months) AL rats or old (26-28 month) rats maintained under conditions of DR were heat shocked for 30 minutes at 42.5° C. Pooled nuclear extracts (20 μg protein/assay) from the hepatocytes of two rats per group were analyzed by the gel shift assay using radiolabeled HSE as described by Heydari et al. (1993).

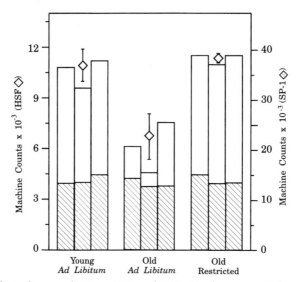

Figure 5. Effect of age and DR on HSF and SP-1. Nuclear extracts from young (4–6 months) and old (26–28 months) AL rats or old (26–28 months) rats maintained under conditions of DR were analyzed by the gel shift assay using radiolabeled HSE or SP-1 oligonucleotides. The dried gels were scanned and the relative amounts of radiolabeled HSE or SP-1 oligonucleotides bound to HSF or to transcription factor SP1, respectively, were quantified using a Molecular Dynamics (Sunnyvale, CA) 400B PhosphorImager™. The data are expressed as the amount of radioactivity bound (machine counts) per 20 µg of nuclear extract and were taken from Heydari et al. (1993). The amount of HSF binding to HSE by cell extracts from old AL rats was significantly different from the amount bound by cell extracts from both young AL rats and old rats fed the calorically restricted diet ($p \leq 0.01$ for both comparisons).

binding to the HSE; i.e., the slowly migrating band corresponds to the HSE-HSF complex that has been described by Heydari et al. (1993). Similar HSE binding patterns have been observed with cell extracts from various mammalian cell lines (Mosser et al. 1990, Choi et al. 1990). The data in Figure 5 show that the level of HSF binding activity decreases with age and parallels the decrease in hsp70 transcription (Heydari et al. 1993). More importantly, DR resulted in increased HSF binding activity, and this was correlated with the increase in hsp70 transcription (Figure 4). Therefore, it appears that DR alters the transcription of hsp70 through increased activity of HSF. Interestingly, as shown in Figure 5, the level of transcription factor SP1 binding activity in the hepatocyte extracts, as measured by the gel shift assay, did not change with age. This is in agreement with the report by Spindler et al. (1991) in which the levels of SP1 mRNA in the livers of female mice did not change significantly with age or DR (Table 1).

Although our research suggests that DR can enhance the transcription of hsp70 through increased binding of HSF to the heat shock element, the molecular mechanism(s) responsible for this response remains unknown. HSF is found in an inactive (i.e., non-DNA-binding) form in nonstressed cells in mammals. Heat shock results in the activation of HSF to a form that binds the HSE through oligomerization and translocation of HSF into the nucleus of the heat-shocked cells (Baler et al. 1993, Sarge et al. 1993). The binding of HSF to the HSE on the promoter region of the hsp70 gene is followed by the phosphorylation of the HSF to create a complex with high transcriptional activity (Baler et al. 1993, Sarge et al. 1993). Based on our knowledge of the molecular mechanism of HSF activation of hsp70 transcription, we suggest that DR could alter the HSF binding activity through two separate mechanisms: 1) the increased HSF binding activity could arise from an increase in the expression of HSF, or 2) the increased HSF binding activity could arise from an increase in the posttranslational activation (oligomerization, translocation, and/or phosphorylation) of HSF by heat shock; i.e., similar levels of HSF are present in nonstressed cells of AL and DR rats; however, in response to heat shock, the conversion of inactive HSF to its HSE binding form is enhanced in cells from calorically restricted animals. Using a polyclonal antiserum raised against mouse HSF (Sarge et al. 1993), we have preliminary data that indicate that the level of HSF protein does not decrease with increasing age for AL animals. Thus, these data suggest that the age-related decrease in HSF binding activity and its reversal by DR might arise from a decline in the ability of cells to convert HSF from its inactive form to its HSE-binding form.

It is very inviting to suggest that one possible mechanism for the age-related decline in the binding activity of HSF is due to an accumulation of abnormal HSF in hepatocytes isolated from old rats and that DR reverses the age-related decrease in hsp70 expression by preventing the accumulation of abnormal HSF. Research over the past two decades has shown that altered enzymes accumulate with age and that these alterations arise from conformational changes rather than from errors in translation (Rothstein 1979, Sharma and Rothstein 1978, Sharma et al. 1980, Sharma and Rothstein 1980). These conformational changes are believed to be brought about by postsynthetic modifications of enzymes (proteins) in older organisms that are often subtle and have minor effects on enzymatic activity (Sharma et al. 1980, Hiremath and Rothstein 1982, Rothstein 1977, Sharma and Rothstein 1980). Furthermore, this suggestion is supported by a recent experiment in which we found no decrease in the level of HSF protein in cells isolated from old AL rats compared with cells isolated from young AL rats and old rats maintained under conditions of DR. In addition, it has been shown that protein turnover in the liver declines with age in AL rats but not in calorically restricted rats, whereas the protein levels remain unchanged with age and DR (Ward and Richardson 1991, Richardson and Ward 1994). Furthermore, HSF has been shown to have a relatively slow turnover rate (Lindquist 1986, Lindquist and Craig 1988). Thus, the age-related decline in protein turnover in hepatocytes coupled with the relatively slow HSF turnover would appear to predispose HSF to age-related alteration.

Conclusion

Although DR is the only experimental manipulation that has been consistently shown to prolong the longevity of mammals, the biological mechanism by which DR enhances the survival of laboratory rodents is unknown. It has been proposed that changes in gene expression are potentially important elements in the action of DR. Because transcription factors represent one of the largest and most diverse classes of DNA-binding proteins and regulate the transcription of genes, we previously proposed that DR alters the transcription of specific genes through changes in the levels and/ or activities of specific transcription factors (Heydari and Richardson 1992). Experiments conducted in our laboratory provided the first evidence that DR alters gene expression at the level of transcription factors. We found that the increase in transcription of hsp70 associated with DR was correlated with a change in the transcription factor, HSF: DR enhanced binding activity of the HSF to the HSE. Preliminary data indicate that the effect of age and DR on HSF binding activity is not due to a change in HSF protein level. We propose that DR alters HSF binding activity through one of the following mechanisms:

1) by altering the signaling mechanism of the cells, i.e., cells from old AL rats do not sense the stress signal and thus do not activate HSF, or

2) by preventing the accumulation of abnormal HSF in cells, i.e., abnormal HSF, which is unable to oligomerize and acquire HSE binding ability, does not accumulate in rats maintained under conditions of DR.

Acknowledgments

This research was supported by grants AG01548 and AG01188 from the National Institute on Aging and by the Office of Research and Development, Department of Veterans Affairs.

References

Angelidis CE, Lazaridis I, Pagoulatos GN (1991) Constitutive expression of heat-shock protein 70 in mammalian cells confers thermoresistance. Eur J Biochem 199:35

Baler R, Dahl G, Voellmy R (1993) Activation of human heat shock genes is accompanied by oligomerization, modification, and rapid translocation of heat shock transcription factor HSF1. Mol Cell Biol 13:2486

Blake MJ, Fargnoli J, Gershon D, Holbrook NJ (1991) Concomitant decline in heat-induced hyperthermia and HSP70 mRNA expression in aged rats. Am J Physiol 260:663

Cairo G, Schiaffonati L, Rappocciolo E, et al. (1989) Expression of different members of heat shock protein 70 gene family in liver and hepatomas. Hepatology 9:740

Campanini C, Petronini PG, Alfieri R, Borghetti AF (1992) Decreased expression of heat shock protein 70 mRNA and protein in WI-38 human fibroblasts aging in vitro. Ann N Y Acad Sci 663:442

Chatterjee B, Fernandes G, Yu BP, et al. (1989) Calorie restriction delays age-dependent loss in androgen responsiveness of the rat liver. FASEB J 3:169

Choi H, Lin Z, Li BS, Liu AY (1990) Age-dependent decrease in the heat-inducible DNA sequence-specific binding activity in human diploid fibroblasts. J Biol Chem 265:18,005

Conaway RC, Conaway JW (1993) General initiation factors for RNA polymerase II. Annu Rev Biochem 62:161

Fargnoli J, Kunisada T, Fornace AJ, et al. (1990) Decreased expression of heat shock protein 70 mRNA and protein after heat treatment in cells of aged rats. Proc Natl Acad Sci USA 87:846

Fassen AE, O'Leary JJ, Rodysill KJ, et al. (1989) Diminished heat-shock protein synthesis following mitogen stimulation of lymphocytes from aged donors. Exp Cell Res 183:326

Goldenberg CJ, Luo Y, Fenna M, et al. (1988) Purified human factor activates heat shock promoter in a HeLa cell-free transcription system. J Biol Chem 263:19,734

Heydari AR, Richardson A (1992) Does gene expression play any role in the mechanism of the antiaging effect of dietary restriction? Ann N Y Acad Sci 663:384

Heydari AR, Wu B, Takahashi R, et al. (1993) Expression of heat shock protein 70 is altered by age and diet at the level of transcription. Mol Cell Biol 13:2909

Hiremath LS, Rothstein M (1982) Regenerating liver in aged rats produces unaltered phosphoglycerate kinase. J Gerontol 37:680

Horbach GJ, Venkatraman JT, Fernandes G (1990) Food restriction prevents the loss of isosafrole inducible cytochrome P-450 mRNA and enzyme levels in aging rats. Biochem Int 20:725

Johnson RN, Kucey BL (1988) Competitive inhibition of hsp70 gene expression causes thermosensitivity. Science 242:1551

Kingston RE, Schuetz TJ, Larin Z (1987) Heat inducible human factor that binds to a human hsp70 promoter. Mol Biol Cell 7:1530

Kollmar R, Farnham PJ (1993) Site-specific initiation of transcription by RNA polymerase II. Proc Soc Exp Biol Med 203:127

Li GC, Li L, Liu YK, et al. (1991) Thermal response of rat fibroblasts stably transfected with the human 70-kDa heat shock protein-encoding gene. Proc Natl Acad Sci USA 88:1681

Lindquist S (1986) The heat-shock response. Annu Rev Biochem 55:1151

Lindquist S, Craig EA (1988) The heat-shock proteins. Annu Rev Genet 22:631

Liu AY, Lin Z, Choi H, et al. (1989) Attenuated induction of heat shock gene expression in aging diploid fibroblasts. J Biol Chem 264:12,037

Masoro EJ (1988) Food restriction in rodents: an evaluation of its role in the study of aging. J Gerontol 43:B59

Masoro EJ (1990) Assessment of nutritional components in prolongation of life and health by diet. Proc Soc Exp Biol Med 193:31

Masoro EJ (1992a) Retardation of aging processes by nutritional means. Ann N Y Acad Sci 673:29

Masoro EJ (1992b) Potential role of the modulation of fuel use in the antiaging action of dietary restriction. Ann N Y Acad Sci 663:403

Mizzen LA, Chang C, Garrels JI, Welch WJ (1989) Identification, characterization, and purification of two mammalian stress proteins present in mitochondria, grp75, a member of hsp70 family and hsp58, a homologue of the bacterial groEL protein. J Biol Chem 264:20,664

Mosser DD, Kotzbauer PT, Sarge KD, Morimoto RJ (1990) In vitro activation of heat shock transcription factor DNA-binding by calcium and biochemical conditions that affect protein conformation. Proc Natl Acad Sci USA 87:3748

Mote PL, Grizzle JM, Walford RL, Spindler SR (1991) Influence of age and caloric restriction on expression of hepatic genes for xenobiotic and oxygen metabolizing enzymes in the mouse. J Gerontol 46:B95

Pahlavani MA, Cheung HT, Cai NS, Richardson A (1990) Influence of dietary restriction and aging and gene expression in the immune system of rats. In Goldstein AL (ed), Biomedical advances in aging. Plenum Publishing Corp., New York, pp 259

Rao G, Xia E, Nadakavukaren MJ, Richardson A (1990) Effect of dietary restriction on the age-dependent changes in the expression of antioxidant enzymes in rat liver. J Nutr 120:602

Riabowol KT, Mizzen LA, Welch WJ (1988) Heat shock is lethal to fibroblasts microinjected with antibodies against hsp70. Science 242:433

Richardson A (1985) The effect of age and nutrition on protein synthesis by cells and tissues from mammals. In Watson RR (ed), Handbook of nutrition and aging. CRC Press, Boca Raton, FL, p 31

Richardson A, Butler JA, Rutherford MS, et al. (1987) Effect of age and dietary restriction on the expression of $\alpha_{2\mu}$-globulin. J Biol Chem 262:12,821

Richardson A, Cheung HT (1982) The relationship between age-related changes in gene expression, protein turnover, and the responsiveness of an organism to stimuli. Life Sci 31:605

Richardson A, Ward WF (1994) Handbook of nutrition in the aged. In Watson RR, (ed), Changes in protein turnover as a function of age and nutritional status. CRC Press, Boca Raton,Florida, pp. 309

Rothstein M (1977) Recent developments in the age-related alteration of enzymes: a review. Mech Ageing Dev 6:241

Rothstein M (1979) The formation of altered enzymes in ageing animals. Mech Ageing Dev 9:197

Salih MA, Herbert DC, Kalu DN (1993) Evaluation of the molecular and cellular basis for modulation of thyroid c-cell hormones by aging and food restriction. Mech Ageing Dev 70:1

Sarge KD, Murphy SP, Morimoto RI (1993) Activation of heat shock gene transcription by heat shock factor 1 involves oligomerization, acquisition of DNA-binding activity, and nuclear localization and can occur in the absence of stress. Mol Cell Biol 13:1392

Semsei I, Rao G, Richardson A (1989) Changes in the expression of superoxide dismutase and catalase as a function of age and dietary restriction. Biochem Biophys Res Commun 164:620

Sharma HK, Prasanna HR, Rothstein M (1980) Altered phosphoglycerate kinase in aging rats. J Biol Chem 255:5043

Sharma HK, Rothstein M (1978) Age-related changes in the properties of enolase from *Turbatrix aceti.* Biochemistry 17:2869

Sharma HK, Rothstein M (1980) Altered enolase in aged *Turbatrix aceti* results from conformational changes in the enzyme. Proc Natl Acad Sci USA 77:5865

Sierra F, Coeytaux S, Juillerat M, et al. (1992) Serum T-kininogen levels increase two to four months before death. J Biol Chem 267:10,665

Song CS, Rao TR, Demyan WF, et al. (1991) Androgen receptor messenger ribonucleic acid (mRNA) in the rat liver: changes in mRNA levels during maturation, aging, and calorie restriction. Endocrinology 128:349

Sorger PK (1991) Heat shock factor and the heat shock response. Cell 65:363

Sorger PK, Lewis MJ, Pelham HRB (1987) Heat shock factor is regulated differently in yeast and HeLa cells. Nature 329:81

Spindler SR, Crew MD, Mote PL, et al. (1990) Dietary energy restriction in mice reduces hepatic expression of glucose-regulated protein 78 (BiP) and 94 mRNA. J Nutr 120:1412

Spindler SR, Grizzle JM, Walford RL, Mote PL (1991) Aging and restriction of dietary calories increases insulin receptor mRNA, and aging increases glucocorticoid receptor mRNA in the liver of female C3B10RF1 mice. J Gerontol 46:B233

Strong R, Moore MA, Hale C, Burke WJ, et al. (1990) Age-related changes in adrenal catecholamine content and tyrosine hydroxylase gene expression: effects of dietary restriction. In Armbrecht JA, Coe R, Wongsurawat N (eds), Endocrine function and aging. Springer-Verlag, Berlin, p 218

Waggoner SM, Gu MZ, Chiang WH, Richardson A (1990) The effect of dietary restriction on the expression of a variety of genes. In Harrison DE (ed), Genetic effects on aging II. Telford Press, Caldwell, NJ, p 255

Ward W, Richardson A (1991) Effect of age on liver protein synthesis and degradation. Hepatology 14:935

Wu B, Gu MJ, Heydari AR, Richardson A (1993) The effect of age on the synthesis of two heat shock proteins in the HSP70 family. J Gerontol 48:B50

Wu BJ, Williams GT, Morimoto RI (1987) Detection of three protein binding sites in the serum-regulated promoter of the human gene encoding the 70-kDa heat shock protein. Proc Natl Acad Sci USA 84:2203

Young VR (1979) Diet as a modulator of aging and longevity. Fed Proc 38:1994

Zimarino V, Wu C (1987) Induction of sequence-specific binding of *Drosophila* heat shock activator protein without protein synthesis. Nature 327:727

Age-Dependent Indigenous Covalent DNA Modifications (I-Compounds)

E. Randerath
Baylor College of Medicine

G.-D. Zhou
Baylor College of Medicine

R.W. Hart
National Center for Toxicological Research

K. Randerath
Baylor College of Medicine

Analysis of I-Compounds

I-compounds were first detected (K. Randerath et al. 1986) in tissues of untreated animals by the nuclease P1-enhanced bisphosphate version of ^{32}P-postlabeling assay (Reddy and Randerath 1986); this method has played a major role in exploring the properties and possible biological roles of I-compounds. The final products generated from DNA are 5'-^{32}P-labeled modified deoxyribonucleoside 3',5'-bisphosphates (pXp), which are separated by multidirectional anion-exchange thin-layer chromatography (TLC) on polyethyleneimine (PEI)-cellulose (K. Randerath et al. 1988, 1990b, 1992). This separation involves a one-dimensional prefractionation and purification step and subsequent two-dimensional separation of three compound groups with increasing polarity originating from lower (L), central (C), and upper (U) cuts of the initial chromatogram. Resolution of these compound groups is achieved with aqueous solvents containing electrolytes and urea. Labeled I-compounds are detected by autoradiography and are quantified by counting excised spots directly in the absence of scintillation fluid. I-compound levels are expressed as relative adduct labeling (RAL) values, which are calculated according to

$$RAL = \text{I-compound(s) [cpm] / (DNA-P [pmol]} \times \text{Spec. Act.}_{ATP} \text{ [cpm/pmol]}),$$

where DNA-P represents the amount of DNA assayed (expressed as mononucleotide units in pmol) and Spec. Act.$_{ATP}$ is the specific activity of ATP (Reddy and Randerath 1986).

Major Properties, Occurrence, and Origins of I-Compounds

I-compounds are modified deoxyribonucleotides as inferred from their detection by ^{32}P-postlabeling. The covalent nature of these modifications is further supported by their stability under strongly denaturing conditions. The wide range of polarities and the complexity of the chromatographic profiles suggest that I-modifications represent diverse molecular structures and, hence, may be formed from different precursors.

The presence of I-compounds in untreated animals suggests that they are not derived from exogenous carcinogens but from endogenous DNA-reactive metabolites, and therefore they have been termed I (indigenous)-compounds. A characteristic property of most I-compounds is that their levels increase with age. I-compounds have been detected in all rodent tissues studied, including liver, kidney, lung, skin, heart, colon, brain, spleen, mammary gland, uterus, and ovary, and white blood cells, as well as in human brain (K. Randerath et al. 1986, 1990a, 1992, 1993c). The number and levels of I-compounds are greatest in metabolically active organs, i.e., liver and kidney, where their total amounts are substantial (about one modification in 10^7 DNA nucleotides). I-compounds also occur in mitochondrial DNA (Gupta et al. 1990).

Total cellular DNA I-compound profiles exhibit marked species, tissue, and gender differences, and smaller strain differences are also observed (K. Randerath et al. 1990a, 1992). These findings suggest that I-compound formation is genetically controlled. However, exogenous factors such as hormonal manipulations, diet, and carcinogen exposure also modulate I-compound profiles and levels as detailed below.

Fundamental differences exist between I-compounds and carcinogen-DNA adducts. A comparison of both types of DNA modifications in Sprague-Dawley (SD) rats, ICR mice, and Syrian hamsters, which were all raised concurrently on the same natural ingredient diet and under the same environmental conditions, showed that each of three carcinogens

administered (dibenz[a,j]acridine, safrole, and 7,12-dimethylbenz[a]anth-
racene) produced qualitatively identical adduct patterns across the tissues
(liver, kidney, and skin) of the three species examined, whereas I-compounds exhibited distinct species- and tissue-dependent profiles (Li et al.
1990b). Moreover, a number of I-compounds in rodent liver display circadian variations, which are not observed for carcinogen-DNA adducts (Nath
et al. 1992).

The origins of the majority of I-compounds need to be explored, a difficult task in view of the insufficient sensitivity of available methodology
for structural characterization. Therefore, our present knowledge on the
origins of I-compounds comes largely from indirect experimental evidence.
In view of the increases in I-compound levels (K. Randerath et al. 1986,
1990a, 1992, 1993c) and oxidative stress (Sohal and Orr 1992) with age in
mammals, it has been of great interest to determine whether and to what
extent I-compound formation is due to endogenous oxidative stress. Recent work from our laboratory has demonstrated that several I-compounds
represent direct DNA oxidation products (K. Randerath et al. 1991; Chang
et al. 1993). On the other hand, most I-compounds do not appear to be
derived from lipid peroxides produced by oxidative stress, as discussed
below. The majority of I-compounds do not represent advanced
glycosylation end products (K. Randerath and E. Randerath, unpublished
experiments), i.e., derivatives resulting from the nonenzymatic reaction
of glucose with DNA, which have been postulated to increase with age
(Cerami 1985). Furthermore, it is unlikely that I-compounds contain protein or common carbohydrate moieties, because the chromatographic properties of I-compounds are not changed by the presence or the absence of
proteinase K during DNA isolation or by treating I-compounds with various lectins (E. Randerath and K. Randerath, unpublished experiments).
However, at least some I-compounds may be derived from lipid metabolism, as certain specific I-compounds in rat liver are induced by oat lipids
(Li et al. 1992a).

Two classes of I-compounds, Type I and Type II, have recently been
defined (K. Randerath et al. 1993d). Type I I-compounds form as a consequence of normal metabolism and display positive linear correlations with
median life span, while Type II I-compounds represent bulky DNA lesions, which result from oxidative stress (K. Randerath et al. 1991, 1993d;
Chang et al. 1993) and also increase with age (K. Randerath, M. Gaeeni,
G.-D. Zhou, and E. Randerath, unpublished observations). These two types
of I-compounds differ in their chromatographic properties. Under the standard chromatographic mapping conditions (K. Randerath et al. 1988, 1990b,

1992, 1993d), two polarity classes of prefractionated I-compounds are further resolved on L (low polarity) and C (higher polarity) maps by solvents containing urea and electrolytes, i.e., lithium formate for the first dimension and sodium phosphate and Tris or sodium phosphate for the second dimension. Resolution of Type I I-compounds on L and C maps is achieved by using different electrolyte and urea concentrations, the nonpolar L fraction requiring higher concentrations than the more polar C fraction. However, under the conditions for optimal resolution of Type I I-compounds on L and C maps, Type II I-compounds move with the solvent fronts on both maps. Thus, to resolve Type II I-compounds, electrolyte and urea concentrations are employed that are lower than those needed for Type I I-compounds, resulting in decreased mobilities of both Type II and I I-compounds (Chang et al., 1993; E. Randerath et al., 1995). The two types of I-compounds will be discussed in more detail later.

Modulation of I-Compound Profiles and Levels by Exogenous Factors

Modulation by Hormonal Manipulation

Steroid hormones, especially estrogens, are involved in the regulation of I-compound profiles and levels. Administration of sex hormones affects I-spot patterns in rodent liver and kidney (K. Randerath et al. 1989). While hepatic I-compound profiles of untreated rats exhibit pronounced sexual dimorphism, castration feminizes and ovariectomy masculinizes I-compound profiles and levels (E. Randerath et al. 1991b). The dependence of I-compound formation on both age and sex hormones suggests that the levels of these DNA modifications are developmentally controlled.

Modulation by Diet

I-compound profiles and levels are strongly diet-dependent. Two types of studies on the effects of diet on I-compound profiles and levels have been conducted, namely 1) investigations concerned with different diet compositions and 2) investigations using caloric restriction. The former studies will be reported in this section, while the effects of the latter will be described below.

Liver and kidney DNA of female SD rats fed natural ingredient diets show a greater complexity and 2.5- to 5.4-fold higher levels of I-compounds than rats fed a purified diet (AIN-76A) (Li and Randerath 1990a). Three classes of I-compounds were recognized in this study: those that were common to both kinds of diet and those that were either chow- or

AIN-76A-specific. Furthermore, variation of macro- and micronutrient diet composition produces specific quantitative effects on I-compounds. A few I-compounds observed predominantly in female rat liver could be traced to the presence of oats (oat lipids) in the diet (Li et al. 1992a,b). These findings strongly support the original hypothesis that I-compounds are produced through endogenous mechanisms during normal nutrient metabolism.

Modulation by Carcinogens/Tumor Promoters

The levels of most hepatic I-compounds, when assayed under standard chromatographic conditions (K. Randerath et al. 1988, 1990b, 1992, 1993d), are reduced by a number of carcinogens/tumor promoters (reviewed in K. Randerath et al. 1993b); these include 1) the nonmutagenic carcinogens 2,3,7,8-tetrachlorodibenzo(p)dioxin (TCDD), choline-devoid diet, peroxisome proliferators, and carbon tetrachloride; 2) the mutagenic carcinogens 2-acetylaminofluorene and 3-methylcholanthrene; and 3) tumor promoters, such as certain polychlorinated dibenzofurans, Aroclor 1254, and phenobarbital. As found for TCDD (K. Randerath et al. 1988, 1990b) and choline-devoid diet (Li et al. 1990a), these reductions are target organ-specific. Levels of individual I-compounds respond differentially to carcinogens, indicating that reduction of hepatic I-compound levels is not merely caused by dilution of modified DNA with newly synthesized unmodified DNA as a consequence of cell proliferation but involves altered I-compound formation/removal. It was further shown that changes in cytochrome P450 activities contribute to the reduction of the levels of many I-compounds by carcinogens/tumor promoters (Moorthy et al. 1993).

I-Compounds in Preneoplastic and Neoplastic Tissues

I-compound levels are not only reduced during chemical carcinogenesis, but smaller numbers and lesser amounts of I-compounds are also found in target organs of spontaneous carcinogenesis before tumor appearance (Li and Randerath 1990b). I-compound levels are very low in spontaneous hepatic adenomas of susceptible (C3H) mice (Li et al. 1991a) and in chemically induced rat hepatic tumors (K. Randerath et al. 1990c). Notably, in Morris hepatomas there is little or no correlation between tumor growth rates and the extent of I-compound depletion, again implicating impaired I-compound synthesis rather than cell proliferation in the loss of these DNA modifications. There appears to be a continuum from the partial reduction of I-compound levels during carcinogenesis/tumor promotion to the almost complete loss of these modifications in neoplasms. Thus, a close relationship between permanent impairment of I-compound formation and malignant transformation is apparent.

Age Dependence of I-Compound Profiles and Levels in Ad Libitum–Fed and Calorically Restricted Rats

Rats were raised at the National Center for Toxicological Research (NCTR, Jefferson, AR) in a specific pathogen-free environment at 23°C as described in detail by Duffy et al. (1989). Rats were singly housed and maintained on a light-dark cycle with the lights on from 0600 to 1800 hours. All rats were fed a conventional NIH-31 open formula diet (about 4.0 kcal/g) at 1100 hours daily and were divided into an ad libitum (AL)– fed control group and a calorically restricted (CR) group that received 60% of the AL consumption beginning at 3.5 months of age and continuing for the duration of the experiment. The CR diet was vitamin-supplemented so that both animal groups received the same amounts of vitamins. Thirty days before sacrifice, the CR rats were phase-shifted by 12 hours so that their time of feeding was synchronized to the time within the dark phase when the control rats were most active in feeding (Duffy et al. 1989).

The profiles of nonpolar (lower) and polar (central) I-compounds from liver DNA of 1-, 30-, and 41-month-old male AL and CR F344xBN (progeny of Fischer 344 and Brown Norway) rats obtained under standard chromatographic conditions are shown in Figures 1 and 2. While these DNA modifications are barely detectable in the 1-month-old animals, they display strong autoradiographic intensities in animals   30 months of age. As is evident from the figures, intensities further increase from 30 to 41 months in both AL and CR animals and are greater in the CR animals. On the other hand, AL and CR animals display qualitatively similar spot patterns. No new I-compounds are formed between 30 and 41 months. Most I-compounds are detected by 4 months of age, but their intensity increases continuously with age (E. Randerath et al. 1991a, K. Randerath et al. 1993a,d).

I-compound levels in animals ≥ 8 months of age were remarkably high, resembling those of persistent carcinogen-DNA adducts in rodent organs during carcinogenesis (K. Randerath et al. 1993a,d). For example, total I-compound levels in liver and kidney DNA of 24-month-old male AL F344 rats amounted to 112 and 156 modifications in 10^9 DNA nucleotides, respectively, while the corresponding values for CR animals were 2.0 and 1.5 times higher. At 24 months of age, total RAL values for liver and kidney DNA in AL animals exceeded the 1-month values by 2.3 and 5.2

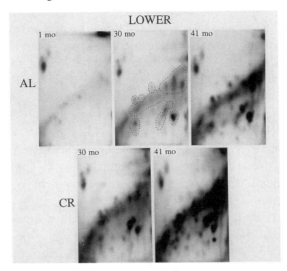

Figure 1. Profiles of nonpolar (LOWER) I-compounds of liver DNA from 1-, 30-, and 41-month-old male ad libitum–fed (AL) and calorie-restricted (CR) F344 x BN rats, as analyzed by the nuclease P1/bisphosphate version of the ^{32}P-postlabeling assay. Screen-enhanced autoradiography employed Kodak XAR-5 X-ray film for 16 hours at –80°C. Dotted lines indicate how individual spots or clusters of spots are excised for quantitative determination of I-compound levels.

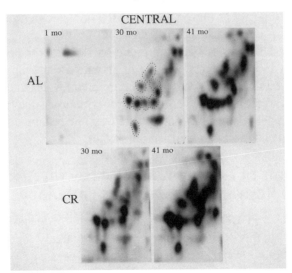

Figure 2. Profiles of polar (CENTRAL) I-compounds of liver DNA from 1-, 30-, and 41-month-old male ad libitum–fed (AL) and calorie-restricted (CR) F344 x BN rats, as analyzed by the nuclease P1/bisphosphate version of the ^{32}P-postlabeling assay. Screen-enhanced autoradiography was for 16 hours at –80°C. Dotted lines indicate how individual spots or clusters of spots are excised for quantitative determination of I-compound levels.

times, and in CR animals by 4.5 and 8.0 times, respectively. Polar I-compounds (C fractions) were more prevalent in kidney than in liver. In AL animals, polar kidney I-compounds represented 76% of total modifications, and polar liver I-compounds represented 47%. In CR animals, the corresponding values were 82% and 48%.

Age-dependent increases of I-compounds were also observed in white blood cells (WBC) (E. Randerath et al. 1991a) and skin (K. Randerath, G.-D. Zhou, and E. Randerath, unpublished experiments). Again, CR significantly increased I-compound levels compared to AL animals, but total I-compound levels were about 50 and 9 times lower in WBC and skin, respectively, compared with liver.

As shown by these results, CR, the most effective measure correctly known to extend the median and maximum life spans of experimental animals, improve resistance to carcinogenesis, and retard the rate of age-associated degenerative processes, increased rather than decreased I-compound levels when age-matched AL and CR animals were compared. Thus, I-compounds, in spite of bearing physical similarities to bulky carcinogen adducts, do not resemble chemical DNA lesions, whose formation would be reduced by CR (Pashko and Schwartz 1983, Chou et al. 1991). Similarly, the increased I-compound levels in CR animals argue against the idea that the majority of these DNA modifications are derived from lipid peroxidation products, as CR is known to reduce lipid peroxidation (Laganiere and Yu 1987). This interpretation is further supported by the finding that a diet deficient in vitamin E, an antioxidant known to inhibit lipid peroxidation, does not decrease I-compound levels (Li et al. 1991b). Also, hepatocarcinogens that cause lipid peroxidation, such as carbon tetrachloride and peroxisome proliferators, reduce rather than increase I-compound levels (K. Randerath et al. 1992, 1993b).

Correlations of I-Compound Levels with Age: Effects of Calorie Restriction and Organ Pathology

Both F344 and BN rats are susceptible to progressive chronic nephropathy, but this disorder is more severe and largely accounts for earlier mortality in F344 rats. Thus, the median life span of AL F344 rats is about 20% shorter than that of AL BN rats; CR causes a 20% increase in median life span in both strains, according to data from the NCTR (K. Randerath et al. 1993d). Plots of I-compound levels in liver and kidney DNA of F344 and BN rats as a function of animal age were examined for linearity, i.e., a desired property of biomarkers of aging, and were also compared for strain differences (K. Randerath et al. 1993a).

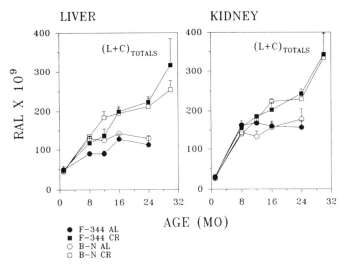

Figure 3. Effects of calorie restriction (CR) and age on total I-compound levels, expressed as RAL x 10^9 values, in combined lower + central (L + C) fractions from liver and kidney DNA of male F344 and BN rats. Mean RAL values (± standard deviations) are based on data from three or four animals. Simple linear regression analysis (Zar 1984) gave the following values for r (correlation coefficient), b (regression coefficient or slope of regression line), and p (probability that b is different from 0 at the 0.05 level of significance) in CR animals: 0.98, 8.7, and < 0.001 in F344 liver; 0.94, 6.4, and 0.006 in BN liver; 0.96, 9.3, and 0.002 in F344 kidney; and 0.95, 9.2, and 0.003 in BN kidney. No significant linear correlations were found for the corresponding tissues in AL animals.

As shown in Figure 3, total I-compound levels in liver and kidney of CR animals, when analyzed under standard chromatographic conditions, display steeper increases with age than do corresponding levels of AL animals, which tend to plateau between 12 and 24 months of age. Linear regression analysis of liver and kidney data for CR F344 and BN rats (K. Randerath et al. 1993a) reveals that levels of total I-compounds (Figure 3) and of many individual and summed polar (C fraction) I-compounds (data not shown) increase linearly with age in both tissues of the two strains. However, summed nonpolar (L fraction) I-compounds of kidney DNA increase linearly in BN but not F344 rats, i.e., in the strain that is less susceptible to renal disease (data not shown). In AL BN and F344 rats, the majority of individual I-compounds tend to level off at higher ages, as reflected in the plots for total I-compounds (Figure 3). Several individual I-compounds, especially from the polar groups, exhibit linear increases with age in AL animals; this is the case in liver of both strains, but interestingly, not a single kidney I-spot of AL F344 rats increases linearly with age (data not shown). All results taken together indicate that better health of the animals is paralleled by linear increases of I-compound levels with age. Although I-compound data display overall similarities for the two strains, they reflect the more severe kidney pathology in F344 rats.

Linear Correlations of I-Compound Levels with Median Life Span in Calorie-Restricted and Ad Libitum–Fed Rats and Mice

A comprehensive study (K. Randerath et al. 1993d) was conducted in AL and CR male animals from three strains of rats (BN, F344, and F344 × BN) and two strains of mice (C57BL/6N and B6D2F1). The study addressed whether correlations exist between I-compound levels in middle-aged animals, i.e., 24- and 20-month-old rats and mice, respectively, and their median life span (animal age at 50% survival). All animals were raised as described for rats in the preceding section. The following major results were obtained:

1) In both AL and CR animals, the levels of a number of individual liver and kidney I-compounds tend to increase in the order F344 < BN < F344 × BN and C57BL/6N < B6D2F1, paralleling strain-specific median life span increases. Thus, the longer-lived hybrid F344 × BN rats and B6D2F1 mice exhibit higher I-compound levels than the parent strains.

2) CR causes an overall elevation of I-compound levels in both rats and mice.

3) The levels of many individual I-compound fractions in tissues of both rats and mice exhibit significant linear correlations with life span, as illustrated by the examples in Figure 4. Significant linear correlations as in Figure 4 have been found for 53% and 73% of the individual I-compound fractions analyzed in rat and mouse liver, respectively (K. Randerath et al. 1993d). The corresponding values for rat and mouse kidney are 23% and 37%. These numbers represent minimum estimates, since they depend on the analytical conditions used. Linear correlations with life span appear to be more common among the polar than nonpolar I-compounds, particularly in rats, where such correlations were observed for six and two C fractions from liver and kidney DNA, respectively, but only for two L fractions from liver DNA.

4) The statistically significant correlations between the levels of specific I-compounds and median life span make it possible to predict effects of both strain and diet on survival rates.

As indicated by these linear relationships, many I-compounds represent molecular biomarkers of aging. The positive linear correlations between levels of many I-compounds with life span suggest that these

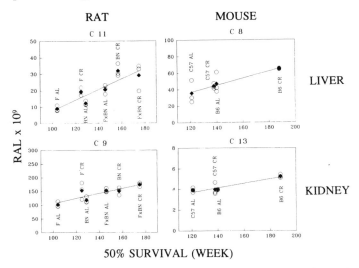

Figure 4. Correlations between the levels (expressed as RAL x 10^9 values) of selected individual I-compounds in liver and kidney DNA of ad libitum–fed (AL) and calorie-restricted (CR) male rats or mice and median life spans, expressed as animal age in weeks at 50% survival (K. Randerath et al. 1993d). Rat strains are F344 (F), BN, and F344 x BN at 24 months of age; mouse strains are C57BL/6N (C57) and B6D2F1 (B6) at 20 months of age. DNA was isolated from three individual animals per diet group for the three strains of rats and two strains of mice. All DNA preparations from rats or mice, respectively, were analyzed in parallel. Note that the selected I-compounds are all different due to species and tissue specificities of I-compound profiles. Open symbols = RAL x 10^9 values for three individual animals; closed symbols = means from the three animals. Linear regression lines (Zar 1984) are based on data for three individual animals of each diet group from three rat strains or two mouse strains, respectively. r and p values are 0.83 and < 0.001 for rat liver fraction C11, 0.76 and < 0.001 for rat kidney fraction C9, 0.80 and < 0.002 for mouse liver fraction C8, and 0.87 and < 0.001 for mouse kidney fraction C13.

modifications may be functionally important and thus do not represent endogenous DNA lesions, consistent with the circadian rhythms of certain I-compounds and the results discussed earlier with respect to neoplasms. DNA modifications displaying such positive linear correlations with median life span have been termed Type I I-compounds. On the other hand, endogenously produced DNA lesions, such as bulky DNA oxidation products (see the following section), which have been classified as Type II I-compounds, are expected to display negative rather than positive relationships with median life span according to the DNA damage hypothesis of aging (Hart and Turturro 1981, Ames and Gold 1991, Holmes et al. 1992, Kirkland 1992). While preliminary experiments have shown that Type II I-compounds increase with age (K. Randerath, M. Gaeeni, G.-D. Zhou, and E. Randerath, unpublished observations), relationships between their levels and life spans have not yet been investigated.

I-Compounds and Oxidative Stress

In vitro oxidation of DNA by the Fenton reaction (i.e., in the presence of $FeSO_4$ and H_2O_2) results in a number of novel DNA oxidation products that appear to represent intrastrand crosslinks (K. Randerath et al. 1991). Several of these products were identical to I-compounds in rodent liver, kidney, and lung DNA (Chang et al. 1993). In particular, two major kidney I-compounds are enhanced several-fold by the renal carcinogen and pro-oxidant Ni(II) acetate. Most recently, it was found that ferric nitrilotriacetate (Fe-NTA), which also acts as a renal carcinogen and pro-oxidant, enhances the same two nonpolar I-compounds in rat kidney DNA, but not in DNA from other organs (E. Randerath et al. 1995). An additional polar I-compound increased by Fe-NTA treatment was also identical to a DNA modification induced by the Fenton reaction, while two other products appear distinct from products generated by the Fenton reaction and may be the result of DNA-protein crosslinking or binding of lipid peroxides to DNA.

These findings show that not all I-compounds are innocuous or fit the concept of potentially "beneficial" DNA modifications outlined in the preceding section; rather, some I-compounds represent DNA lesions. The preliminary findings concerning target organ specificity implicate these lesions in mutagenesis/carcinogenesis.

Conclusions

I-compounds are indigenous covalent DNA modifications that accumulate with age in tissue DNA of mammals in the absence of exposure to mutagens/carcinogens. Their origins appear to be in DNA-reactive intermediates of nutrient metabolism, with their profiles being determined by genetic and environmental factors. The levels of most or all I-compounds, when assayed under standard chromatographic conditions, are substantially reduced by carcinogens/tumor promoters, depending on dose and type of agent. In spontaneous and chemically induced hepatomas, such levels are further decreased. On the other hand, CR consistently increases I-compound levels in comparison with age-matched AL animals.

Two types of endogenous DNA modifications have been recognized: Type I I-compounds, when measured at intermediate age, exhibit positive linear correlations with median life span (K. Randerath et al. 1993d) and apparently reflect innocuous or potentially functional changes in DNA. Type II I-compounds represent DNA lesions resulting from oxidative stress. Other mechanisms leading to the formation of endogenous DNA adducts have been reviewed by Marnett and Burcham (1993). While occurrence of age-dependent Type II I-compounds is consistent with the DNA damage hypothesis of aging, roles of Type I I-compounds remain to be defined. One potential explanation for the paradoxical properties of Type I I-compounds—namely their age-dependent increases, higher levels in the healthier CR animals, and positive linear correlations with median life span—is that these DNA modifications may be influenced by age-dependent metabolic and hormonal changes and thus may be associated with genetically programmed aging (K. Randerath et al. 1993d). Higher I-compound levels in the healthier CR animals would then appear to reflect programmed normal aging, while lower I-compound levels in AL animals would indicate a disturbance of the program resulting in early development of tissue lesions (Bronson and Lipman 1991; K. Randerath et al. 1993a). The idea that certain I-compounds are involved in programmed aging is attractive, since cellular transformation leading to immortality (Hayflick 1991) appears to be accompanied by Type I I-compound losses.

Acknowledgments

Work from the authors' laboratory reviewed in this article has been supported by USPHS grants R37 CA32157, RO1 AG07750, and P42 ES04917, awarded by the National Cancer Institute, the National Institute on Aging, and the National Institute of Environmental Health Sciences, respectively, and by a research grant from Shell Research Limited.

References

Ames BN, Gold LS (1991) Endogenous mutagens and the causes of aging and cancer. Mutat Res 250:3

Bronson RT, Lipman RD (1991) Reduction in rate of occurrence of age-related lesions in dietary restricted laboratory mice. Growth Dev Aging 55:169

Cerami A (1985) Hypothesis: Glucose as a mediator of aging. J Am Geriatr Soc 33:626

Chang J, Watson WP, Randerath E, et al. (1993) Bulky DNA-adduct formation induced by Ni(II) in vitro and in vivo as assayed by ^{32}P-postlabeling. Mutat Res 291:147

Chou MW, Pegram RA, Gao P, et al. (1991) Effects of caloric restriction on aflatoxin B1 metabolism and DNA modification in Fischer 344 rats. In Fishbein L (ed), Biological Effects of Dietary Restriction. ILSI Monographs, Springer-Verlag, Berlin, New York, pp 42–54

Duffy PH, Feuers RJ, Leaky KD, et al. (1989) Effects of chronic caloric restriction on physiological variables related to energy metabolism in the
male Fischer 344 rat. Mech Ageing Devel 48:117

Gupta KP, van Golen KL, Randerath E, et al. (1990) Age-dependent covalent
DNA alterations (I-compounds) in rat liver mitochondrial DNA. Mutat Res
237:17

Hart RW, Turturro A (1981) Evolution and longevity-assurance processes.
Naturwissenschaften 68:552

Hayflick L (1991) Aging under glass. Mutat Res 256:69

Holmes GE, Bernstein C, Bernstein H (1992) Oxidative and other DNA damages as the basis of aging: a review. Mutat Res 275:305

Kirkland JL (1992) The biochemistry of mammalian senescence. Clin Biochem
25:61

Laganiere S, Yu BP (1987) Anti-lipoperoxidation action of food restriction.
Biochem Biophys Res Commun 145:1185

Li D, Chen S, Becker FF, et al. (1991a) Specific reduction of I-compound
levels in DNA from spontaneous hepatomas of 22–24 month old male
C3H mice. Carcinogenesis 12:2389

Li D, Chen S, Randerath E, et al. (1992a) Oat lipid-induced covalent DNA
modifications (I-compounds) in female Sprague-Dawley rats, as determined
by ^{32}P-postlabeling. Chem Biol Interact 84:229

Li D, Chen S, Randerath K (1992b) Natural dietary ingredients (oats and
alfalfa) induce covalent DNA modifications (I-compounds) in rat liver and
kidney. Nutr Cancer 17:205

Li D, Randerath K (1990a) Association between diet and age-related DNA
modifications (I-compounds) in rat liver and kidney. Cancer Res 50:3991

Li D, Randerath K (1990b) Strain differences of I-compounds in relation to
organ sites of spontaneous tumorigenesis and non-neoplastic renal disease in mice. Carcinogenesis 11:251

Li D, Wang Y-M, Nath RG, et al. (1991b) Modulation by dietary vitamin E of
I-compounds (putative indigenous DNA modifications) of rat liver and
kidney. J Nutr 121:65

Li D, Xu D, Chandar N, et al. (1990a) Persistent reduction of I-compound
levels in liver DNA from male Fischer rats fed choline-devoid diet and in
DNA of resulting neoplasms. Cancer Res 50:7577

Li D, Xu D, Randerath K (1990b) Species and tissue specificities of I-compounds as contrasted with carcinogen adducts in liver, kidney and skin
DNA of Sprague-Dawley rats, ICR mice and Syrian hamsters. Carcinogenesis 11:2227

Marnett LJ, Burcham PC (1993) Endogenous DNA adducts: potential and
paradox. Chem Res Toxicol 6:771

Moorthy B, Chen S, Li D, et al. (1993) 3-Methylcholanthrene-inducible liver
cytochrome P450 in female Sprague-Dawley rats: possible link between

P450 turnover and formation of DNA adducts and I-compounds. Carcinogenesis 14:879

Nath RG, Vulimiri SV, Randerath K (1992) Circadian rhythm of covalent modifications in liver DNA. Biochem Biophys Res Commun 189:545

Pashko LL, Schwartz AG (1983) Effect of food restriction, dehydroepiandrosterone, or obesity on the binding of [3H]7,12-dimethylbenz[a]anthracene to mouse skin DNA. J Gerontol 38:8

Randerath E, Hart RW, Turturro A (1991a) Effects of aging and caloric restriction on I-compounds in liver, kidney, and white blood cell DNA of male Brown-Norway rats. Mech Ageing Devel 58:279

Randerath E, Randerath K, Reddy R, et al. (1991b) Sexual dimorphism of the chromatographic profiles of I-compounds (endogenous deoxyribonucleic acid modifications) in rat liver. Endocrinology 129:3093

Randerath E, Watson WP, Zhou G-D, et al. (1995) Intensification and depletion of specific bulky renal DNA adducts (I-compounds) following exposure of male F344 rats to the renal carcinogen ferric nitrilotriacetate (Fe-NTA). Mutat Res 341:265

Randerath K, Hart RW, Zhou G-D, et al. (1993a) Enhancement of age-related increases in DNA I-compound levels by calorie restriction: comparison of male B-N and F-344 rats. Mutat Res 295:31

Randerath K, Li D, Moorthy B, et al. (1993b) I-compounds—Endogenous DNA markers of nutritional status, aging, tumor promotion and carcinogenesis. In Phillips DH, Castegnaro M, Bartsch H (eds), Postlabelling Methods for Detection of DNA adducts. IARC Scientific Publications No 124, International Agency for Research on Cancer, Lyon, France, pp 157–165

Randerath K, Li D, Nath R, et al. (1992) Exogenous and endogenous DNA modifications as monitored by ^{32}P-postlabeling: relationships to cancer and aging. Exp Gerontol 27:533

Randerath K, Li D, Randerath E (1990a) Age-related DNA modifications (I-compounds): modulation by physiological and pathological processes. Mutat Res 238:245

Randerath K, Liehr JG, Gladek A, et al. (1989) Age-dependent covalent DNA alterations (I-compounds) in rodent tissues: species, tissue and sex specificities. Mutat Res 219:121

Randerath K, Putman KL, Osterburg HH, et al. (1993c) Age-dependent increases of DNA adducts (I-compounds) in human and rat brain DNA. Mutat Res 295:11

Randerath K, Putman KL, Randerath E, et al. (1988) Organ-specific effects of long term feeding of 2,3,7,8-tetrachlorodibenzo-p-dioxin and 1,2,3,7,8-pentachlorodibenzo-p-dioxin on I-compounds in hepatic and renal DNA of female Sprague-Dawley rats. Carcinogenesis 9:2285

Randerath K, Putman KL, Randerath E, et al. (1990b) Effects of 2,3,7,8-tetrachlorodibenzo-p-dioxin on I-compounds in hepatic DNA of Sprague-

Dawley rats: sex-specific effects and structure-activity relationships. Toxicol Appl Pharmacol 103:271

Randerath K, Randerath E, Danna TF (1990c) Lack of I-compounds in DNA from a spectrum of Morris hepatomas. Carcinogenesis 11:1041

Randerath K, Reddy MV, Disher RM (1986) Age- and tissue-related DNA modifications in untreated rats: Detection by ^{32}P-postlabeling assay and possible significance for spontaneous tumor induction and aging. Carcinogenesis 7:1615

Randerath K, Yang P-F, Danna TF, et al. (1991) Bulky adducts detected by ^{32}P-postlabeling in DNA modified by oxidative damage in vitro. Comparison with rat lung I-compounds. Mutat Res 250:135

Randerath K, Zhou G-D, Hart RW, et al. (1993d) Biomarkers of aging: correlation of DNA I-compound levels with median lifespan of calorically restricted and ad libitum fed rats and mice. Mutat Res 295:247

Reddy MV, Randerath K (1986) Nuclease P1-mediated enhancement of sensitivity of ^{32}P-postlabeling test for structurally diverse DNA adducts. Carcinogenesis 7:1543

Sohal RS, Orr WC (1992) Relationship between antioxidants, prooxidants, and the aging process. Ann N Y Acad Sci 663:74

Zar JH (1984) Biostatistical Analysis, 2nd ed. Prentice Hall, Englewood Cliffs, NJ

DNA Polymerase α Function and Fidelity: Dietary Restriction As It Affects Age-Related Enzyme Changes

D. Busbee

College of Veterinary Medicine, Texas A&M University
Texas A&M University Medical Center

S. Miller, M. Schroeder, and V. Srivastava

College of Veterinary Medicine, Texas A&M University

B. Guntupalli

Texas A&M University Medical Center

E. Merriam

College of Veterinary Medicine, Texas A&M University

S. Holt and V. Wilson

Texas A&M University Medical Center

R. Hart

National Center for Toxicological Research

Introduction

DNA polymerase α (pol α) was originally discovered almost 30 years ago and, until the past few years, was thought to be the principal enzyme of both replicative and repair DNA synthesis in eukaryotic organisms. A number of nucleotidyltransferases, polymerases α, β, δ, and ε, are now recognized as different eukaryotic enzymes with interactive functions in nuclear DNA synthesis. DNA pol α and its essential primase subunit have primary functions that include binding to single-stranded DNA (ssDNA) at DNA synthesis origination sites, synthesizing RNA primers necessary for initiating both leading side and Okazaki fragment DNA synthesis, and extending a short DNA segment from the RNA primer to provide an open 3'-OH end for subsequent DNA synthesis (Downey et al. 1988, Weiser et al. 1991). DNA synthesis is continued from the primer with pol δ or ε polymerizing the processive addition of nucleotides on the leading side and a combination of polymerases—β and δ or ε—synthesizing DNA on the lagging side of the eukaryotic replicating fork (Burgers et al. 1990, Linn 1991, Johnson 1992). Since pol α is essential for DNA synthesis initiation occurring at the transition of cells across the G_1/S phase replication boundary in eukaryotic organisms, it is considered to be one of the critical eukaryotic S-phase enzymes (Johnson 1992). Although pol α has been relatively well-characterized in studies of fetal- and newborn-derived cells and transformed cells (Fry and Loeb 1986, Syvaoja et al. 1990), few studies have attempted to characterize the polymerases from essentially amitotic cells of adult animals or from senescing cells in vitro (Linn et al. 1976, Murray 1981, Krauss and Linn 1986, Angello et al. 1987, Collins and Chu 1990, Pendergrass et al. 1991a,b, Srivastava et al. 1991b).

Krauss and Linn (1986) reported that pol α from cells with decreased replicative activity differs chromatographically, based on differences in charge on the molecule, from pol α isolated from actively growing cells. Linn et al. (1976) showed that pol α from senescing cells exhibits decreased activity and fidelity of synthesis, while Murray (1981) demonstrated the existence of active and inactive forms of pol α, showed that enzyme activity in senescing fibroblasts was reduced by two- to fourfold below that of young cells, and proposed that low levels of polymerase activity were correlated with the declining growth rate of senescing cells. Pendergrass et al. (1991a,b) further showed that recoverable pol α activity per unit of total protein declines significantly in senescent cells and proposed a first-order relationship between pol α activity and a cell's capacity to transit the G_1/S replication boundary. These investigations showed that pol α from old cells could be isolated in a variety of forms that were not proteolytic degradation products of the original enzyme and that enzyme(s) isolated from fetal or transformed cells differed from pol α isolated from old cells in their chromatographic characteristics, specific activity, affinity of binding to DNA template, use of synthetic templates, fidelity of synthesis, and capacity to be activated following mitogenic stimulation of the cell.

Human diploid fibroblasts (HDFs) are capable of surviving for a limited number of cell divisions in vitro before entering senescence (Hayflick and Moorhead 1961). Fibroblasts entering senescence enlarge, become less motile, show diminished saturation density (Cristofalo and Sharf 1973, Collins and Chu 1990), and demonstrate decreased DNA synthesis apparently associated with diminished cellular capacity to initiate DNA synthesis (Pendergrass et al. 1991a). The diminished capacity to initiate DNA synthesis correlates with maintenance of a G_1/S phase replication block more typically associated with mitosis-regulating proteins such as Rb or p53 (Collins and Chu 1990, Sager 1992). In spite of their inability to initiate DNA synthesis and transit the G_1/S phase boundary, senescing cells continue to carry out many of the pre-DNA synthesis responses to mitogenic signals (Olashaw et al. 1983, Rittling et al. 1986) and express p53 at levels only slightly higher than younger cells in active mitosis (Rittling et al. 1986).

An effective method of overcoming the senescence-associated inhibition of DNA synthesis and cellular replication is to infect quiescent fibroblasts with SV40 virus (Gorman and Cristofalo 1986). SV40-encoded large tumor antigen (TAg) and two cellular proteins, pol α and replication protein A (RP-A), are necessary and sufficient for initiating DNA synthesis in reconstituted cell-free systems using SV40 DNA as template (Collins and Kelly 1991). In this system TAg binds to the SV40 origin as a double hexamer and, as an ATP-dependent helicase, facilitates "melting" of the double-stranded DNA (dsDNA) sequence. RP-A binds to and stabilizes the resulting single-stranded region of origin DNA. DNA pol α then binds to the stabilized ssDNA, RNA primer synthesis is initiated, and DNA synthesis using the coordinated efforts of polymerases α, β, δ, and ε continues from the initial priming event (Figure 1). TAg also binds to the pol α catalytic subunit, facilitating binding of the pol α-replication complex to single-stranded initiation sites on SV40 DNA (Gannon and Lane 1990, Collins and Kelly 1991). Collins and Chu (1990) found that while senescing cells did not exhibit a decreased pol α content, there was a 100-fold decrease in pol α binding to DNA initiation sites. They proposed that decreased DNA synthesis and cellular proliferation associated with an increased HDF passage number and the onset of senescence may reflect decreased binding affinity of pol α for DNA initiation sites. Srivastava et al. (1993a,c) evaluated pol α from normal and SV40-transformed fetal fibroblasts and from normal and pSV3.neo-transformed adult fibroblasts (pSV3.neo is an SV40-derived plasmid that expresses TAg). They showed that pol α from transformed adult fibroblasts in which the age-related replication block is clearly absent is TAg-positive and exhibits increased levels of both DNA binding and specific activity, while pol α from the corresponding untransformed adult-derived cells exhibits low activity and a somewhat decreased level of binding to DNA. The logical mechanism of TAg action to initiate DNA synthesis suggests the possibility that untransformed cells may have a TAg-analogous protein whose function is correlated with the onset of DNA synthesis. This proposal correlates well with the findings of Dutta et al. (1991), who suggested that

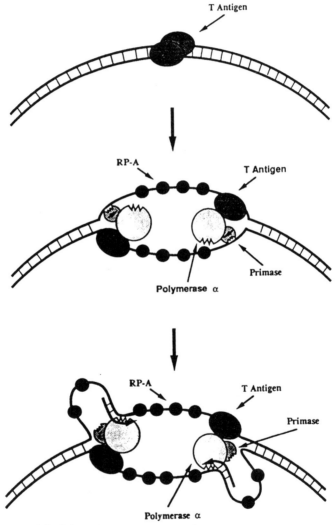

Figure 1. A model of the eukaryotic DNA synthesis initiation reaction using SV40 large tumor antigen (TAg, an ATP-dependent helicase), replication protein A (RP-A, a single-strand DNA stabilizing protein), and DNA polymerase α with its associated primase. Adapted from Collins and Kelly (1991).

normal S-phase cells have a "positive activator of DNA replication" that is absent in G_1 cells and proposed that these proteins in normal cells should perform functions similar to TAg at cellular origins of DNA replication. This suggests that the expression or function of proteins regulating the initiation of DNA replication may be altered in cells from aged animals that show a decline in the initiation of DNA synthesis, and that the increased capacity to initiate DNA synthesis associated with dietary restriction (DR) may also be due to changes in function of cellular proteins required for DNA synthesis.

Goulian and Heard (1990) and Goulian et al. (1990) isolated a pol α accessory protein (aAP) and showed that it binds DNA in the absence of pol α and increases pol α-binding to DNA. Further, they demonstrated that pol α treated with the accessory protein exhibits about a fivefold increase in enzyme activity and a seven- to ninefold increase in processivity (Goulian and Heard 1990). It is not known whether αAP expression is responsive to receptor-mediated extracellular signals or to signal transduction-associated phenomena, how αAP expression and activity are regulated in normal young and normal senescing cells, or whether the expression of αAP, which functions as an activator of DNA initiation, changes with the age or the dietary state of a given organism. In this paper we present data showing that pol a from old cells, but not from fetal or tranformed cells that already have high activity, exhibits a significant increase in specific activity after treatment with exogenous αAP. We further show that αAP is an ATP-dependent helicase and suggest that pol α from fetal and old cells may be differentially associated with endogenous αAP. We propose that intracellular levels of αAP may differ significantly between fetal-derived cells and aged animal-derived cells and that dietary states might influence the expression or function of αAP. We raise the question of whether αAP may have functions in normal cells analogous to those of TAg in SV40-transformed cells.

Materials and Methods

Animals

C57BL/6-NIH male mice (3 and 13 months of age) used in this study were obtained from the specific pathogen-free colony at the National Center For Toxicological Research (NCTR, Jefferson, AR), under a rearing and housing contract from the Biomarkers of Aging project of the National Institute on Aging (NIA). Animals were fed ad libitum (AL) or were subjected to a dietary restriction (DR) regimen beginning at 3 months of

age in which the mice were provided with an amount of food corresponding to 60% of that eaten by the AL animals, as previously described
(Srivastava et al. 1991a).

Purification of DNA Polymerase α

Pol α was purified from HDF, both normal and transformed, using the
procedures of Chang et al. (1984) and Sylvia et al. (1988), as modified by
Srivastava and Busbee (1992). The polymerases were distinguished by
their specific inhibition characteristics and by dot blot analysis using anti-
pol α IgG and anti-TAg (Srivastava et al. 1993a,c).

DNA Polymerase α Assay

The standard pol α assay was carried out as previously reported using a
variety of natural and synthetic substrates (Srivastava and Busbee 1992,
Srivastava et al. 1993a,c).

DNA Polymerase α Fidelity Assays

The reaction mixtures contained 50 mM Tris-HCl, pH 7.8, 1 mM
dithiothreitol, 1 mg/mL bovine serum albumin, 2 μg synthetic template-
primer, 1.5 mM $MgCl_2$, 50 μM complementary deoxynucleotides, 6 μM
noncomplementary deoxynucleotides and enzyme in a total volume of 0.1
mL. Assays were completed in the presence of 0.5 mM $MnCl_2$ for α-poly-
merases and 1 mM $MnCl_2$ with 50 mM KCl for β-polymerases. Each
misincorporation ratio was obtained from a parallel pair of reactions, one
with complementary (^3H) deoxynucleotide (80–100 dpm/pmol) to measure total polymer synthesis, and the other with noncomplementary (^3H)
deoxynucleotide (8–10 dpm/fmol) to measure nonfaithful synthesis. Levels of misincorporation were also measured after treatment with inositol-
1, 4-bisphosphate (1 μg/mL). The complementary and noncomplementary
values are after subtraction of background incorporation of zero-time assays with enzyme but no template or template but no enzyme.

Isolation of DNA Polymerase α Accessory Factor

αAP was isolated from L1210 cells using procedures identical to those
used for pol α (Srivastava and Busbee 1992) followed by additional purification steps as reported by Goulian et al. (1990). Assay of αAP was
completed using the methods of Goulian and Heard (1990).

Cell Culture and Growth Rate of Fibroblasts

An SV40-transformed WI38 cell line (2RA) and a pSV3.neo-trans-
formed GM3529 cell line (3529T) were grown in Dulbecco's modified

Eagle's minimal essential medium (DMEM, Gibco BRL, Gaithersburg, MD), and growth curves were established as previously reported (Srivastava et al. 1993a,c).

Mobility-Shift DNA Binding Assay

Purification of SV40 Origin of Replication. SV40 viral DNA was excised with HindIII and PvuII to produce a 345 bp SV40 origin of replication fragment. The 345 bp fragment was electrophoretically purified using a BioRad Laboratories (Richmond, CA) Prep-A-Gene kit and recovered using the methods of Ansubel et al. (1992).

[32]P-Labeling of SV40 Origin DNA. The 345 bp SV40 DNA fragment (8 µL containing 12 ng of DNA), 2µL [α-^{32}P]-thymidine-5'-triphosphate (20 µCi), 2 µL reaction buffer (40 mM Tris-HCl, pH 7.5, 20 mM $MgCl_2$, 25 mM NaCl), 2 µL dNTP's (7.5 mM GTP, CTP, and ATP), 2 µL of diluted sequenase[R] [United States Biochemical (Cleveland, OH), 1 µL sequenase and 6 µL dilution buffer (10 mM Tris-HCl, pH 7.5, 5 mM dithiothreitol DTT, and 500 µg/mL bovine serum albumen)], and 4 µL of H_2O were incubated at room temperature for 30 minutes. The labeled probe was frozen at −20°C and used within 1 week (Ansubel et al. 1992).

Sample Preparation for Mobility Shift Assay. Each pol α or αAP sample contained 5 µg purified protein, 2.5 µL 0.5 M TNE.1 buffer (10 mM Tris-HCl, pH 7.0, 0.5 mM NaCl, 0.1 mM EDTA), 5 µL 0.01 M TNE.1 + 50% glycerol, 1 mL 10 mM ATP, and 1 µL probe. Samples were incubated for 30 minutes at 25°C and were immediately loaded onto gels that had been prerun for 90 minutes. The control lane contained 5 µL 0.01 M TNE.1 and 5 µL protein buffer (50 mM Tris-HCl, pH 7.5; 2 mM $MgCl_2$, 1 mM DTT, 1 mM EDTA, 0.04 M KCl, and 15% glycerol). Test lanes contained 5 µL pol α peak sample or αAP (Ansubel et al. 1992).

Mobility shift assay. Protein+labeled marker oligonucleotide sequences were loaded onto a low-ionic-strength polyacrylamide slab gel and electrophoresed using the mobility shift methods of Ansubel et al. (1992).

Helicase Assay

Protein Preparation. αAP purified as given above did not exhibit unequivocal helicase activity under the gel shift analysis conditions. Buffer exchange of the αAP preparation was completed by microconcentration using the Amicon Centricon-30 (Amicon, Inc., Beverly, MA) system and methods. αAP was washed three times with helicase buffer without ATP, followed by two washes with 20 mM Tris-HCl, pH 7.6, with 7.5 mM $MgCl_2$. The retentate was dissolved in 20 mM Tris-HCl, pH 7.6, 7.5 mM $MgCl_2$, and concentrated to 20 µL.

Helicase substrate preparation. ^{32}P-labeled 23-base universal primer was obtained using 2 ng M13/pUC forward 23-base sequencing primer, 5X kinase buffer, 0.25 units T4 polynucleotide kinase [dsDNA Cycle Sequencing System (BRL, Gaithersburg, MD)], and 10 μCi of gamma[^{32}P]-ATP (3000 mCi/mmol) at 37°C for 30 minutes. The reaction was stopped by heating to 65°C for 5 minutes, 10 μL M13mp9 ssDNA were added, and the mixture was slow-cooled overnight to room temperature. Excess label and ^{32}P-labeled primer were removed by Centricon-30 spin dialysis. Purity of the substrate was electrophoretically analyzed on an 8% polyacrylamide gel, and, where necessary, the product was further purified by gel filtration on BioGel A-150M agarose beads (BioRad Laboratories).

Helicase assay. The reaction mixture contained 1 μL primed M13mp9 DNA substrate, 20 μL protein or protein buffer, 30 μL helicase buffer containing 20 mM Tris-HCl, pH 7.6, 7.5 mM $MgCl_2$, 1 mM DTT, 1 mg/mL phosphocreatine, 0.1 mg/mL creatinine kinase, and 2 mM ATP (Weikowski et al. 1988). A control containing helicase buffer without ATP was also included. For additional controls, the DNA substrate was heated in boiling water for 15 minutes and cooled on ice for 15 minutes. The reaction mixture without protein and either with or without the heat-denatured DNA served as positive and negative controls. The reaction mixture was incubated at 37°C for 30 minutes, stopped by addition of 0.2 volumes of 1.5% SDS, 0.2 mM EDTA, and 50% glycerol, and cooled on ice. Helicase substrates and products were electrophoretically analyzed and evaluated for ^{32}P-labeled substrate migration by rapid Betascope (Beta-Gen, Waltham, MA) analysis and by exposure to x-ray film (Weikowski et al. 1988).

Results

The contrast between total pol α isolated from approximately 7 grams each of L1210 cells or hepatocytes from either 3-month- or 13-month-old C57BL/6 male mice is given in Figure 2. L1210 cells yielded in excess of 10 times total enzyme activity more than cells from the 3-month-old mice, and in excess of 25 times total activity more than cells from 13-month-old mice. In a separate set of experiments, the specific activity of two hepatic pol α isoforms was found to decline with increased age between 6-, 16-, and 26-month-old animals, with DR animals showing higher activity of both pol α isoforms (Figure 3). The exception to this decline was seen in a single pol α isoform, a_2, from DR animals. A majority of pol α studies have examined mitotically active fetal or transformed cells. The decline in total and specific activity of pol α isolated from less mitotically active cells has been previously reported from this laboratory (Srivastava et al.

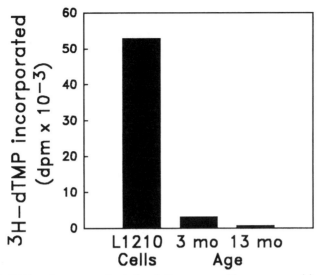

Figure 2. DNA polymerase α (pol α) activity recovered from comparable preparations of L1210 cells or from hepatocytes from 3- or 13-month-old male C57BL/6 mice. L1210 cells or mouse hepatocytes were prepared as given in the Materials and Methods section of the text. Polymerase α was isolated chromatographically from the cells, and the total recoverable pol α activity was determined.

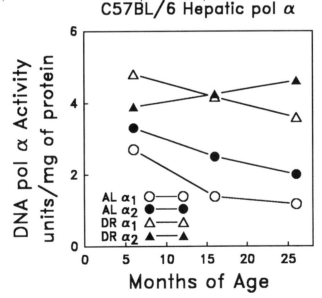

Figure 3. Specific activities of two DNA polymerase α (pol α) isoforms (α_1, α_2,) purified from hepatocytes from 6- 16-, or 26-month-old C57BL/6 mice. Pol α was chromatographically isolated and evaluated as given in the Materials and Methods section of the text. Mice were fed ad libitum (AL) or were maintained on a regimen of dietary restriction (DR).

Table 1. Nucleotide misincorporation frequencies of DNA polymerase fractions from the livers of 6-, 16-, and 26-month-old ad libitum (AL) and diet-restricted (DR) mice using synthetic template primer

Age and dietary regimen[*]	Polymerase fraction	Incorporation dTMP pmol	Incorporation dGMP fmol	Error rate
6-month AL:	α1 Mg	39	8.5	1/4588
	α2 Mg	24.5	9.5	1/2578
6-month DR:	α1 Mg	41	10	1/4000
	α2 Mg	32	12.5	1/2560
16-month AL:	α1 Mg	18.5	14	1/1300
	α2 Mg	26.8	15	1/1800
16-month DR:	α1 Mg	26	14	1/1900
	α2 Mg	22.5	10	1/2300
26-month AL:	α1 Mg	4.5	10	1/450
	α2 Mg	3.5	6	1/583
26-month DR:	α1 Mg	11.5	12	1/958
	α2 Mg	13.5	14	1/964

[*]DNA polymerase α isoforms were purified through the DNA-cellulose affinity column step and were assayed using poly (dA)•poly (dT) as the substrate.

1991a, 1993a,c). These data suggest that the fidelity of pol α isolated from cells of older animals declines (Table 1), with a greater fidelity decline associated with enzyme isolated from cells of AL animals (Srivastava et al. 1991b, Srivastava and Busbee 1992).

When pol α isolated from a fetal-derived HDF (WI38) and an SV40-transformed WI38 corollary HDF cell line (2RA) was contrasted with pol α isolated from HDF established from a 66-year-old donor (GM3529) and a TAg-expressing corollary transformed HDF cell line (2-1) (Figure 4), the enzyme from fetal cells showed approximately five times more total activity with about 2.5 times higher specific activity than pol α from old HDF. The pol α isolated from transformed WI38 was about three times more active than pol α from WI38, with about six times more total enzyme activity. The activity difference between enzyme from GM3529 and 2-1 cells was even greater, with approximately five times more total enzyme activity from 2-1 cells than from GM3529 cells, and with pol α from 2-1, transformed old cells, being about ten times more active than pol α from normal old cells (Table 2). Figure 4 indicates that about 50 times more total pol α activity was isolated from 2-1 cells than from the same wet weight of GM3529 cells.

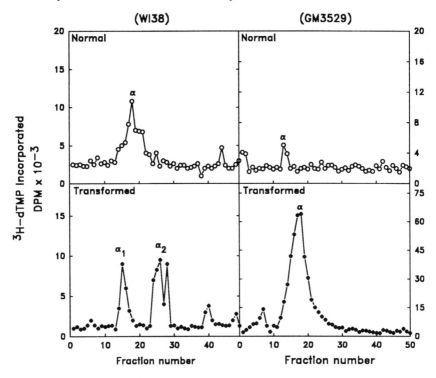

Figure 4. DEAE-52 chromatographic elution profiles of DNA polymerase α (pol α) from WI38 and GM3529 normal human diploid fibroblasts and from SV40-transformed WI38 and pSV3.neo-transformed GM3529 corollary cell lines. The transformed cell lines express SV40 large tumor antigen, and the TAg gene product copurifies with cellular pol α. Adapted from Srivastava et al. (1993a,c).

αAP, which interacts with pol α and increases both its binding to DNA and its specific activity, was isolated from a variety of cells and tissues, including L1210 cells and mouse livers. A purification summary (Table 3) for L1210 αAP shows that the increase in αAP activity is correlated with a decline in total protein (Table 3). Initially, the only assay for αAP was to add the putative αAP to pol α and determine whether it increased pol α activity. Elution fractions from the last αAP purification step, a DNA-cellulose column procedure, were assayed for protein, and the activity of pol α was determined in the presence and absence of aliquots of each αAP fraction. The difference in pol α activity with and without the added elution fractions was plotted to give a profile of αAP activity eluting from the column (Figure 5). In a second experiment, αAP from L1210 cells purified through the DNA affinity chromatography step (Figure 6) gave a

Table 2. Activity of DNA polymerase α (pol α) at different stages during purification. Pol α from WI38 and 2RA fibroblasts and GM3529 and 2-1 fibroblasts were assessed for total protein and enzyme activity at the different stages during purification.

Fraction	Protein (mg)		Total activity[a] (units)		Specific activity[a] (units/mg)	
WI38 and 2RA fibroblasts	WI38	2RA	WI38	2RA	WI38	2RA
Crude extr	168	120	20	37.58	0.12	0.314
DEAE-52						
α1	34.56	16.32	110.76	79.04	3.2	4.84
α2	—[b]	10	—	163.8	—	16.38
DNA cellulose						
α1	0.375	0.225	64.23	72.83	171.28	323.7
α2a	—	0.112	—	68.73	—	613.72
α2b	—	0.15	—	59.67	—	397.8
Glycerol gradient						
α1	0.017	0.01	11	19.45	647.3	1945
α2a	—	0.007	—	18.32	—	2618
α2b	—	0.012	—	24.48	—	2067
GM3529 and 2-1 fibroblasts	GM3529	2-1	GM3529	2-1	GM3529	2-1
Crude extr	29	50	39	125	1.34	2.5
DEAE-52	5.6	9	39	749	7	83
IgG-Sepharose	0.56	0.3	49	283	88	943
DNA cellulose	0.067	0.036	17	86	254	2388

[a]Units are defined as the amount of enzyme activity necessary to catalyze 1 nmol dNMP into TCA precipitable DNA template primer/hour.
[b]Not detected.

total recoverable αAP level stimulating about 1.65×10^6 dpm of ^{3}H-dAMP incorporation into a dsDNA template by pol α. The same pol a preparation was used to assay hepatocyte αAP from 3-month and 13-month C57BL/6 mice (Figures 7 and 8) purified through the hydroxylapatite step. Hepatocyte αAP capable of stimulating pol α to incorporate a total of 6.8×10^3 dpm of ^{3}H-dAMP was recovered from hepatocytes from 3-month-old mice, while αAP stimulating pol α to incorporate 1.35×10^3 dpm of ^{3}H-dAMP was recovered from hepatocytes from 13-month-old mice. These values indicate that more than 240 times as much αAP activity/gram of cells was recovered from L1210 cells than from hepatocytes from 3-month-old mice, and slightly more than 1,200 times as much αAP activity/gram of cells was recovered from L1210 cells than from liver cells from 13-month-old mice.

When αAP was added to a ^{32}P-labeled 345 bp double-stranded sequence of SV40 DNA with multiple TAg binding sites, the electrophoretic mobility of the DNA was shifted significantly, indicating binding of αAP to

Table 3. Summary of the purification of DNA polymerase α (pol α) accessory protein (αAP)

Fraction	Protein	αAP activity	Pol α activity[a]
	mg	units[b]	units/mg
I. S-100 (crude extract)	1827	—[c]	—[c]
II. Phosphocellulose	125.1	—[c]	—[c]
III. Hydroxylapatite	8.4	8000	952.4
IV. DNA-cellulose	1.1	30,000	27,272.8

[a]I.e., the increase in activity of L1210 DNA pol α samples added to the αAP elution fractions.
[b]One unit = amount of αAP sufficient to cause a fivefold increase in the synthesis of poly(dA) by 0.1 unit pol α/primase on a template of unprimed pol (dT) under standard conditions as described in the Materials and Methods section of the text.
[c]Increase in exogenously added pol α activity could not be reliably determined because there was endogenous poly α/primase in Fractions I and II.

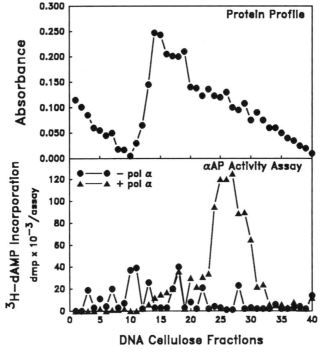

Figure 5. Elution profiles of DNA polymerase α (pol α) accessory protein (αAP) from a DNA-cellulose affinity column. Homogenates of L1210 cells taken through three purification steps were eluted from a DNA affinity column as given in the Materials and Methods section of the text. The total protein in elution fractions was evaluated (upper panel). Aliquots of pol α purified from L1210 cells were added to an aliquot from each αAP elution fraction and assayed for total pol α activity. The presence of αAP in elution fractions was determined by the difference in activity of the pol α fractions before (–pol α) and after (+pol α) addition to the αAP fractions (lower panel).

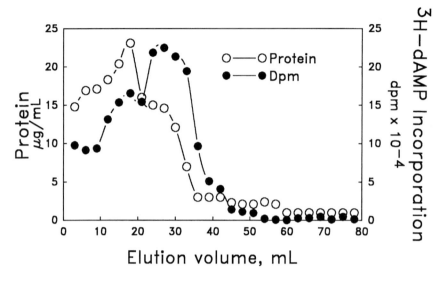

Figure 6. Elution profile of L1210 cell DNA polymerase α (pol α) accessory protein (αAP)from a DNA-cellulose affinity column. L1210 cells homogenized and taken through three purification steps were eluted from a DNA affinity column as given in the Materials and Methods section of the text. Elution fractions were evaluated for total protein and for capacity to increase pol α activity.

TAg binding sites on the DNA (Figure 9). WI38 pol α also exhibited binding to SV40 origin DNA, with an increased mobility shift of the DNA when WI38 pol α and αAP were both added to the DNA mixture. The SV40-transformed WI38 HDF line (2RA) and the pSV3.neo-transformed GM3529 HDF line (2-1) each yielded a very active pol α that was not chromatographically or electrophoretically separable from TAg (Srivastava et al. 1993a,c). In each instance, 2RA and 2-1 pol α initiated a significant shift in mobility of the probe DNA, and in each instance the shift was slightly more pronounced in the presence of L1210 αAP. The HDF line established from a 66-year-old donor (GM3529) yielded pol α that bound the 345 bp SV40 origin dsDNA, giving a definitive mobility shift with discrete embedded DNA bands. In the presence of αAP, GM3529 pol α significantly shifted the electrophoretic mobility of SV40 DNA, indicating a greater degree of pol α binding to the origin DNA. Since pol α was calculated to be in excess in the mobility shift assays, the 2RA and 2-1 gel shifts should indicate the mobility of dsDNA saturated at nearly every binding site. We further showed that pol α in the presence of αAP binds to bovine papilloma virus DNA origin sites and to pUC18 nonspecific dsDNA, while SV40 TAg did not (data not shown).

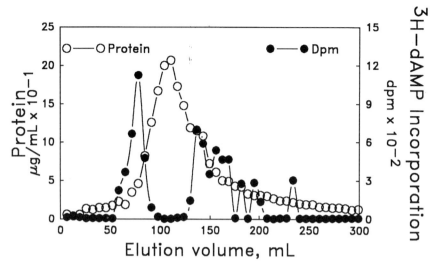

Figure 7. Elution profile of liver cell DNA polymerase α (pol α) accessory protein (αAP) from a hydroxylapatite column. Homogenates of hepatocytes from 3-month-old male C57BL/6 mice taken through two purification steps were evaluated from a DNA affinity column as given in the Materials and Methods section of the text. The affinity elution fractions were evaluated for total protein and for capacity to increase pol α activity.

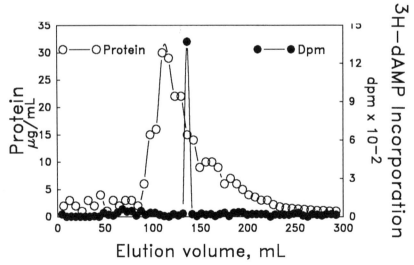

Figure 8. Elution profile of liver cell DNA polymerase α (pol α) accessory protein (αAP) from an hydroxylapatite column. Homogenates of hepatocytes from 13-month-old male C57BL/6 mice taken through two purification steps were eluted from a DNA affinity column as given in the Materials and Methods section of the text. The elution fractions were evaluated for total protein, and aliquots of pol α purified from L1210 cells were added to an aliquot from each elution fraction and assayed for total pol α activity.

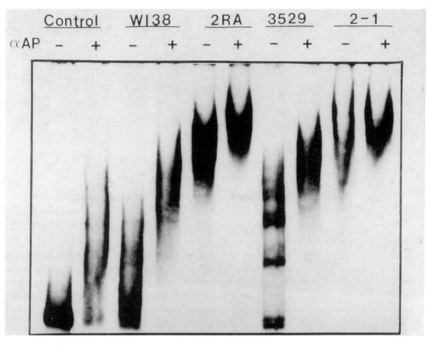

Figure 9. A mobility shift analysis of DNA polymerase α (pol α) binding to SV40 DNA in the presence (+) or absence (–) of a pol α accessory protein (αAP). αAP was added to a ^{32}P-labeled 345 bp DNA sequence containing the SV40 origin of synthesis and electrophoresed in low-ionic-strength polyacrylamide. Migration of the DNA sequence was evaluated by exposure of the gel to x-ray film. Adapted from Srivastava et al. (1993b).

In an attempt to determine the interaction of αAP with dsDNA, M13mp9 ssDNA was ^{32}P-labeled along double-stranded segments, and the partially double-stranded DNA was treated with highly purified αAP dialyzed to ensure the absence of ATP in the preparations. This assay is the classical method used to determine helicase activity in purified protein preparations. The ^{32}P-labeled primer sequences on partially double-stranded M13mp9 DNA did not migrate electrophoretically (Figure 10, lane 1) but did migrate after heat denaturation (lane 2). When αAP was added to the DNA preparation in the absence of ATP, the DNA remained double-stranded, showing no migration of the label (lane 3); however, when αAP and 3 mM ATP were added to the DNA, the labeled primer sequences migrated in the field, indicating that they were single-stranded (lane 4). These data indicate that aAP has helicase activity and that the activity requires ATP.

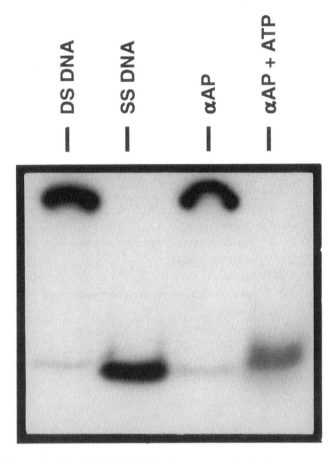

Figure 10. An evaluation of L1210 DNA polymerase α (pol α) accessory protein (αAP) for helicase activity. M13mp9 ssDNA was annealed with ^{32}P-labeled 23-base universal primer obtained as given in the Materials and Methods section of the text using M13/pUC forward 23-base sequencing primer. The M13mp9 DNA was electrophoresed partially double-stranded (lane 1), after heat denaturation (lane 2), after treatment with αAP with no added ATP (lane 3), or after treatment with αAP in the presence of ATP (lane 4).

The interaction of αAP with a variety of pol α from both normal and transformed HDF indicates that only two pol α isoforms showed significantly increased activity in the presence of αAP. GM3529 pol α exhibited very low activity in the absence of αAP and about a sixfold activity increase in the presence of αAP (Figure 11). A second αAP preparation gave only a 2.5-fold increase in GM3529 pol α activity (Figure 12); however, the GM3529 pol α preparation had at that time been frozen for 13 months, and pol α enzymes are labile even when stored at –80°C. The second pol α

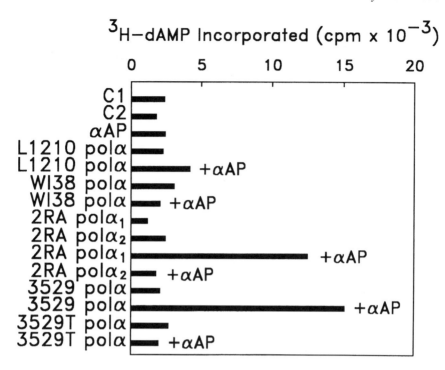

Figure 11. Enhancement of DNA polymerase α (pol α) activity by pol α accessory protein (αAP). Pol α activity isolated from L1210, WI38, 2RA, GM3529, or pSV3.neo-transformed GM3529 cells was determined with or without pretreatment of the enzyme with L1210 αAP. Poly (dT) was used as the template.

isoform activated by αAP (Figure 11) was the α_1 fraction from 2RA that is shown to have a low specific activity (Table 2). The low activity of pol α freshly isolated from hepatocytes from 3-month-old and 13-month-old C57BL/6 mice increased in the presence of αAP (Figure 12).

Discussion

Multiple interacting biochemical and environmental factors, including changed genetic expression, decreased efficiency of protein synthesis, decreased function of essential cellular proteins, and increased levels of somatic mutations appear to interact with caloric intake, available dietary nutrients, dietary antioxidants, and the individual exercise regimen of an animal in the age-dependent decline of physiological processes. The decreased in vitro rates of DNA replication and mitosis in cells derived from young as opposed to old human donors, the significantly longer life spans of DR compared with AL animals, and the tendency of cells from DR old

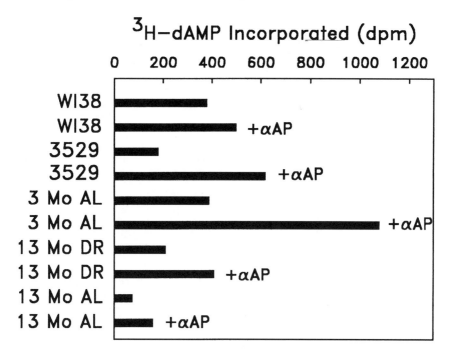

Figure 12. DNA polymerase (pol α) accessory protein (αAP) enhancement of pol α activity isolated from WI38 and GM3529 cells and from hepatocytes from 3-month-old (3 Mo) and 13-month-old (13 Mo) male C57BL/6 3 mice. Mice were fed ad libitum (AL) or were maintained on a regimen of dietary restriction (DR). Poly (dT) was used as the template.

rodents to exhibit increased DNA synthesis initiation typical of cells from much younger animals led us to study the changes in expression and function of DNA pol α in senescing human cells in culture and in AL and DR mice across an age continuum. We investigated the activity, fidelity, and expression of pol α in an attempt to understand, at least in part, the effects of aging on the initiation of DNA synthesis as a critical cellular process affecting life span and how calorie restriction might alter age-related changes in pol α expression and function. We also examined the interaction of pol α with an accessory protein, αAP, as it affects pol α binding to DNA and the onset of DNA synthesis preceding mitosis.

The mechanisms by which increased age of the cell donor or increased passage number of cells in vitro are associated with decreased initiation of DNA synthesis remain unclear. Srivastava et al. (1993b) observed that pol α expression is higher in rapidly dividing transformed cells than in untransformed cells from either fetal- or aged donor-derived HDF but that transcription of pol α is only slightly lower in aged donor- than in fetal donor-derived HDF. The age-related differences in expression of pol α do not appear significant enough to explain the very large differences in total pol α activity isolated from young as opposed to old mice or HDF established from donors of different ages. Our study of pol α isolated from hepatic tissues of C57BL/6 mice indicated, however, that a great deal more total pol α activity/cell can be recovered from tissues of young adult animals than from older adult animals and that enzyme from young adult animals is more active than enzyme from older adult or aged animals.

A number of recent publications have shown that pol α activity is interdependent on a variety of cofactors and auxiliary proteins with which the enzyme may or may not be physically associated, and suggest that interactions between proteins in the eukaryotic replisome, which in turn appear to be dependent on the phosphorylation state of the proteins, are acutely important to determination of pol α activity. Goulian et al. (1990) and Goulian and Heard (1990) reported that a novel pol α accessory protein, which we have called αAP, increased pol α binding to M13mp9 ssDNA and increased both polymerase and primase activities of the enzyme. We show that αAP enhanced pol a binding to SV40 dsDNA, significantly increased the activity of pol α isolated from old donor-derived fibroblasts and aged mice, and that it is an ATP-dependent helicase.

The difference in recoverable αAP between L1210 cells and cells from 3- and 13-month-old mice suggests that there may be a direct relationship between the intracellular content of αAP and the capacity of a cell to initiate DNA synthesis. This relationship is indicated by data showing that approximately four times more αAP can be isolated from livers of 3-month-old than from 13-month-old mice and that as much as 100 times more αAP/cell can be isolated from transformed L1210 cells than from liver cells of 3-month-old mice. Further, when pol α from aging mice or from HDF established from a 66-year-old human donor was treated with L1210 αAP the enzymes responded by showing five- to 10-fold increases in specific activity, while αAP did not appear to greatly enhance the activity of pol α isolated from fetal or transformed cells. This indicates that the activity of relatively inactive pol α isolated from essentially quiescent cells can be increased without the direct stimulatory effects of growth factor-mediated signal transduction. Rather, the increased enzyme activity was apparently a function of the availability of αAP or the capacity of available αAP to bind and destabilize dsDNA, forming ssDNA segments to which pol α may bind to initiate DNA synthesis. The characterization of αAP as an ATP-dependent helicase lends further credibility to the proposal that αAP may be essential for initiating DNA synthesis, leading us to suggest that αAP may interact with pol α in a manner similar to that proposed for

SV40 TAg. If this is the case, the decrease in recoverable intracellular αAP in the tissues of old mice or HDF established from an aged donor could present one possible explanation for the decline in DNA synthesis with increased age.

A decrease in the fidelity of DNA polymerases during either replication or DNA excision repair could present one possible mechanism for the age-associated increased introduction of somatic mutations into eukaryotic cells (Linn et al. 1976, Krauss and Linn 1986, Roberts et al. 1990, Thomas et al. 1991, Srivastava and Busbee 1992). Decreased fidelity of DNA polymerases has been reported for a variety of cell types (Fry and Loeb 1986, Loeb et al. 1986), including enzymes isolated from livers of old rats (Chan and Becker 1979) and HDF established from aged donors (Linn et al. 1976, Thomas et al. 1991). In contrast, no decrease in fidelity of DNA synthesis was reported for pol α from aged avian fibroblasts (Fry and Weisman-Shomer 1976), human lymphocytes from aged subjects (Agarwal et al. 1978), isolated chromatin from livers of aged rats (Fry et al. 1981), or polymerases from livers of partially hepatectomized mice (Silber et al. 1985). The relevance of the fidelity assays on pol α from livers of partially hepatectomized mice is unclear, since liver cells enter an artificially stimulated mitotic state following hepatectomy, which precludes their being compared with quiescent cells. In studies presented here, hepatic pol α from old mice exhibited levels of nucleotide misincorporation into synthetic templates that were significantly higher than those observed for enzymes from young mice. These results are consistent with the concept that alterations in the fidelity of DNA polymerases could be a factor in aging or in the onset of age-related diseases (Linn et al. 1976). Isozymes of hepatic pol α obtained from all age groups of animals maintained under DR conditions copied the synthetic templates with higher fidelity than did enzymes from AL animals. Except for pol α from 26-month-old animals, the poly (dA).poly (dT) homopolymer template was copied with greater accuracy by pol α from old mice, an observation also reported by Krauss and Linn (1980) and Loeb and Kunkel (1982). Although it is not immediately obvious why pol α would show increased fidelity of transcription on defined single-stranded templates in the presence of an ATP-dependent accessory protein expected to bind and destabilize dsDNA, preliminary data indicate that L1210 pol α showed increased fidelity after treatment of the enzymes with αAP. We further show that slowly dividing GM3529 HDF and essentially quiescent hepatic cells from 3- and 13-month-old mice apparently lack significant levels of αAP and that addition of exogenous αAP to pol α from these cells increases the activity of pol α. Whether it also increases the fidelity remains to be seen; however, an increase in fidelity could be consistent with the findings of Goulian and Heard (1990), who reported that αAP accessory protein profoundly affects both the efficiency and processivity of pol α. At this time no data exist to demonstrate a possible correlation between αAP levels and intracellular fidelity of pol α in rapidly dividing as opposed to more quiescent cells. Pol α isolated from AL mice exhibited a greater loss of fidelity with increased age than did pol α from animals maintained under DR conditions. These data are consistent with information suggesting that transient loss of polymerase functional capacities or degradation of polymerases in the cellular pool may contribute to impaired base selection and decreased accuracy of DNA synthesis and that decreased protein synthesis during aging could impede replacement of age-degraded proteins (Richardson 1982, Roberts and Kunkel 1988). If this occurs as a normal process of aging, DR may influence that process, in part, by altering the rate and fidelity of DNA replication.

Data presented here show that DNA pol α expression and activity decline with the age of the HDF donor and that hepatic pol α activity and fidelity of synthesis decrease as a function of aging in mice but do not decrease as significantly in aged DR mice as in old AL mice. A pol α accessory protein (αAP) with ATP-dependent helicase activity was demonstrated to increase not only pol α binding to DNA but both activity and fidelity of the enzyme. The age-related decrease in αAP recovered from cells and the apparent requirement of αAP for initiation of DNA synthesis leads one to speculate that αAP, in addition to pol α, is essential for the G_1/ S phase transition and that regulation of aAP expression and function may be critical to the aging process. Activity and fidelity of pol α are apparently directly correlated with the mitotic state of the cells from which the enzyme was isolated, and it is possible that DR may maintain a level of mitotic competence not typical for aging AL animals. Although the mechanisms of action remain unclear at this time, DR appear to correlate with a decrease in the age-related decline in activity and fidelity of pol α in an experimental animal system. It remains to be seen whether these effects can be demonstrated in human cells.

Acknowledgments

Supported in part by DHHS awards AG06347 and AG07739, by a Research Enhancement Award from the Texas A&M University College of Veterinary Medicine, and by the National Center for Toxicological Research. Parts of this work were completed in collaboration with the Biomarkers of Aging Project.

References

Agarwal SS, Tuffner M, Loeb LA (1978) DNA replication in human lympho-cytes during aging. J Cell Physiol 106:235

Angello JC, Pendergrass WR, Norwood TH, Prothero J (1987) Proliferative potential of human fibroblasts: an inverse dependence on cell size. J Cell Physiol 132:125

Ansubel FM, Brent R, Kingston R, et al. (1992) Short protocols In molecular biology, 2nd ed. John Wiley & Sons, New York, pp 12.5–12.7

Burgers PM, Bambara RA, Campbell JL, et al. (1990) Revised nomenclatures for eukaryotic DNA polymerases. Eur J Biochem 191:617

Chan JY, Becker FF (1979) Decreased fidelity of DNA polymerase activity duringing N-2-fluorenyl acetamide hepatocarcinogenesis. Proc Natl Acad Sci U S A 76:814

Chang LMS, Rafter E, Augl C, Bollum FJ (1984) Purification of a DNA poly-merase-DNA primase complex from calf thymus glands. J Biol Chem 259:14679

Collins JM, Chu AK (1990) Reduction of DNA primase activity in aging but still proliferating cells. J Cell Physiol 143:52

Collins K, Kelly T (1991) Effects of T-antigen and replication protein A on the initiation of DNA synthesis by DNA polymerase alpha-primase. Mol Cell Biol 11(4):2108

Cristofalo VJ, Sharf BB (1973) Cellular senescence and DNA synthesis. Exp Cell Res 76:419

Downey KM, Tan C-K, Andrews DM, et al. (1988) Proposed roles for DNA polymerases a and d at the replication fork. Cancer Cells 6:403

Dutta A, Din S, Brill SJ, Stillman B (1991) Phosphorylation of replication protein A: a role for cdc2 kinase in G_1/S regulation. Cold Spring Harbor symposia on quantitative biology, vol. LVI. Cold Spring Harbor Press, Cold Spring Harbor, NY

Fry M, Loeb LA (1986) Animal cell DNA polymerases. CRC Press, Boca Raton, FL

Fry M, Loeb LA, Martin GM (1981) On the activity and fidelity of chroma-tin-associated hepatic DNA polymerase-b in aging murine species of dif-ferent life spans. J Cell Physiol 106:435

Fry M, Weisman-Shomer P (1976) Altered nuclear deoxyribonucleic acid a-polymerases in senescent cultured chick embryo fibroblasts. Biochemis-try 15:4319

Gannon JV, Lane DP (1990) Interaction between SV40 T antigen and DNA polymerase alpha. New Biol 2:84

Gorman SD, Cristofalo VJ (1985) Reinitiation of cellular DNA synthesis in BrdU-selected nondividing senescent WI-38 cells by simian virus 40 in-fection. J Cell Physiol 125:122

Goulian M, Heard CJ (1990) The mechanism of action of an accessory pro-tein for DNA polymerase a/primase. J Biol Chem 265(22):13,231

Goulian M, Heard CJ, Grimm SL (1990) Purification and properties of an accessory protein for DNA polymerase a/primase. J Biol Chem 265(22):13,221

Hayflick L, Moorhead PS (1961) The serial cultivation of human diploid cell strains. Exp Cell Res 25:585

Johnson LF (1992) G1 events and the regulation of genes for S-phase enzymes. Curr Opin Cell Biol 4:149

Krauss SW, Linn S (1980) Fidelity of fractionated deoxyribonucleic acid polymerases from human placenta. Biochemistry 19:220

Krauss SW, Linn S (1986) Studies of DNA polymerase a and b from cultured human cells in various replicative states. J Cell Physiol 126:99

Linn S (1991) How many pols does it take to replicate nuclear DNA? Cell 66:185

Linn S, Kairis M, Holliday R (1976) Decreased fidelity of DNA polymerase activity isolated from ageing human fibroblasts. Proc Natl Acad Sci U S A 73:2818

Loeb LA, Kunkel TA (1982) Fidelity of DNA synthesis. Ann Rev Biochem 52:29

Loeb L, Liu P, Fry M (1986) DNA polymerase a: enzymology, function, fidelity, and mutagenesis. Prog Nucleic Acid Res Mol Biol 33:57

Murray V (1981) Properties of DNA polymerases from young and ageing human fibroblasts. Mech Ageing Dev 16:327

Olashaw NE, Kress ED, Cristofalo VJ (1983) Thymidine triphosphate synthesis in senescent WI38 cells. Exp Cell Res 149:47

Pendergrass WR, Angello JC, Kirschner MD, Norwood TH (1991a) The relationship between the rate of entry into S phase, concentration of DNA polymerase a, and cell volume in human diploid fibroblast-like monokaryon cells. Exp Cell Res 192:418

Pendergrass WR, Angello JC, Saulewicz AC, Norwood TH (1991b) DNA polymerase a and the regulation of entry into S phase in heterokaryons. Exp Cell Res 192:426

Richardson A (1982) The relationship between age-related changes in gene expression, protein turnover and the responsiveness of an organism to stimuli. Life Sci 31:605

Rittling S, Brooks KM, Cristofalo VJ, Baserga R (1986) Expression of cell-cycle dependent genes in young and senescent WI-38 fibroblasts. Proc Natl Acad Sci U S A 83:3316

Roberts JD, Hamatake RK, Fitzgerald MS, et al. (1990) Effect of accessory proteins on the fidelity of DNA synthesis by eukaryotic replicative polymerases. In Mendelsohn ML (ed), Mutation and the environment: Proceedings of the 5th International Conference on Environmental Mutagens. Wiley-Liss, Inc., part A, pp 91–100

Roberts JD, Kunkel TA (1988) Fidelity of a human cell DNA replication complex. Proc Natl Acad Sci U S A 85:7064

Sager R (1992) Tumor suppressor genes in the cell cycle. Curr Opin Cell Biol 4:155

Silber JR, Fry M, Martin GM, Loeb LA (1985) Fidelity of DNA polymerases isolated from regenerating liver chromatin of aging *Mus Musculus* J Biol Chem 260(2):1304

Srivastava VK, Busbee DL (1992) Decreased fidelity of DNA polymerases and decreased DNA excision repair in aging mice: Effects of caloric restriction. Biochem Biophys Res Commun 182(2):712

Srivastava V, Schroeder M, Busbee D (1993a) Activity of DNA polymerase a in pSV3.neo plasmid transformed human fibroblasts. Int J Biochem 25:385

Srivastava VK, Schroeder MD, Miller SD, et al. (1993b) Age-related changes in cellular levels and functions of DNA polymerase a: effects of calorie restriction. Mutat Res 295:265

Srivastava V, Schroeder M, Miller S, Busbee D (1993c) A comparison of DNA polymerase a from untransformed and SV40-transformed human fibroblasts. Int J Biochem 25:1053

Srivastava V, Tilley R, Hart R, Busbee D (1991a) Effects of aging and dietary restriction on DNA polymerase expression in mice. Exp Gerontol 26:453

Srivastava V, Tilley R, Hart R, et al. (1991b) Effects of aging and dietary restriction on the fidelity of DNA polymerases in aging mice. Exp Gerontol 26:97

Sylvia V, Curtin G, Norman J, et al. (1988) Activation of a low specific activity form of DNA polymerase α by inositol-1,4-biphosphate. Cell 54:651–658

Syvaoja J, Suomensaari S, Nishida C, et al. (1990) DNA polymerases a, d, and e: three distinct enzymes from HeLa cells. Proc Natl Acad Sci U S A 87:6664

Thomas DC, Roberts JD, Sabatino RD, et al. (1991) Fidelity of mammalian DNA replication and replicative DNA polymerases. Biochemistry 30:11,751

Weiser T, Gassmann M, Thömmes P, et al. (1991) Biochemical and functional comparison of DNA polymerases a, d and e from calf thymus. J Biol Chem 266(16):10,420

Wiekowski M, Schwartz MW, Stahl H (1988) Simian virus 40 large T antigen DNA helicase: characterization of the ATPase-dependent DNA unwinding activity and its substrate requirements. J Biol Chem 263(1):436

Oncogene Expression and Cellular Transformation: The Effects of Dietary Restriction

B.D. Lyn-Cook, E.B. Blann, B.S. Hass, and R.W. Hart
National Center for Toxicological Research

Introduction

The beneficial effects of dietary restriction (DR) are well documented. Dietary restriction modulates or delays a number of pathologies and diseases associated with cellular proliferation. In general, DR has been shown to affect processes such as DNA damage, mutation, and synthesis by decreasing the rate of increase in DNA-related genotoxic effects that may lead to various age-associated diseases. The diseased state may be expressed through spontaneous and externally induced mutagenesis, transformation, carcinogenesis, and senescence. Specifically, DR enhances the fidelity of DNA replication (Srivastava et al. 1993), increases the repair of various forms of DNA damage (Lipman et al. 1989), decreases the formation of selected DNA adducts (Chou et al. 1993), increases DNA methylation (Hass et al. 1991, Miyamura et al. 1993, Lyn-Cook et al. 1994), and decreases oncogene expression (Nakamura et al. 1989, Fan et al. 1994). Thus, overall, DR enhances genetic stability.

Of all disease states associated with genetic instability, cancer is of most concern to the general population. Although cancer is considered a chronic disease, it may be induced after only relatively short exposures to certain chemical agents. Since the ad libitum (AL) feeding conditions under which most animal tissue donors are kept results in a tremendous variation in food intake, understanding how differences in food intake or body weight may modulate the incidence and number of tumors that develop within animals is an important component in evaluating the toxic or mutagenetic potential of drugs. There is considerable evidence that aberrant oncogene expression may play a role in the genesis of some neoplasia (Bishop 1985). The initial link between DR and the expression of oncogenes was examined by Nakamura et al. (1989). They observed the suppression of c-*myc* in cells from B6C3F1 mouse liver and further noted that c-*myc* suppression followed a circadian pattern. Very few studies exist using cells in culture after their removal from animals whose dietary histories have been well maintained and documented. Our studies used cell cultures derived from such animals to examine molecular differences in DNA and RNA between such donors. We were interested in addressing three major questions: 1) whether the molecular changes induced by DR in vivo transferred to the in vitro conditions, 2) what the cellular target of programming is, and 3) how these changes are perpetuated long after treatment has been removed and cells are placed in culture.

In order to address these questions, we used pancreatic cells since they, unlike liver cells, replicate well in culture. The pancreas further offered a specific cell type, the acinar cell, that is known to participate in the development of pancreatic cancer (Longnecker et al. 1991). This study was carried out on aged AL and DR animals because pancreatic cancer is primarily a disease of old age and has been shown to be related to diet in humans (Silverberg et al. 1990) and rodents (Longnecker et al. 1981, Roebuck at al. 1993). We have thus examined the impact that DR in vivo has on the results of in vitro studies using in vitro conditions under which there is no restriction of the energy-providing nutrients. In these studies we have analyzed the cellular functions of proliferation, transformation, oncogene expression, and methylation.

Methods

Animals and Diet Regimen

Male Brown-Norway rats were bred and raised at the National Center for Toxicological Research in a specified pathogen-free environment and housed singly in plastic cages. All rats were fed a standard NIH-31 diet and were divided into a control group that was fed AL and a DR group that received 60% of the AL diet using a vitamin-supplemented NIH-31 ration, starting at 14 weeks of age and continuing for the duration of the experiment. Aged (29 months old) AL and DR rats were used in these studies. Rats were sacrificed and pancreatic acinar cells were isolated and placed in tissue culture for transformation and molecular analysis (Hass et. al. 1992).

Transformation Studies

Soft-agar assays were used to evaluate transformation frequency of cells from aged AL animals and those maintained under conditions of DR (Hass et. al. 1992, Hass et. al. 1993).

Methylation Profile and Oncogene Expression

The methylation status of CCGG sequences in DNA from rat pancreatic acinar cells was assessed by cleavage of genomic DNA with the MspI/ Hpa II isoschizomer pair of restriction enzymes coupled with probing of the separated fragments with an H-ras cDNA probe. DNA cleavage by Hpa II is prevented by the presence of a 5-methyl group at the internal C residue on the CCGG recognition sequence, while MspI is able to cleave regardless of the methylation status.

Poly A+ mRNA was obtained using an Invitrogen (San Diego, CA) Fast Track Isolation Kit. Slot-blots of the mRNA were analyzed on a Betagen (Waltham, MA) Betascope after labeling with a ^{32}P H-ras cDNA probe.

Results

Transformation

Transformation can be viewed functionally as the in vitro analog of in vivo carcinogenesis. Using a direct-acting carcinogen, N-methyl-N'-nitro-N-nitrosoguanidine (MNNG), we examined the effects of DR on chemically induced as well as spontaneous transformation. These studies revealed a decrease in chemically induced and spontaneous transformation frequency using soft agar assays (Table 1 and 2). These studies also showed decreased morphological changes induced in vitro.

Methylation and Oncogene Expression

Methyl groups alter the interaction of protein with cellular DNA in two basic ways: 1) by modifying the overall structure of the gene, and 2) by interfering with the binding of protein factors at discrete points on the DNA. Patterns of methylation are inherited by cells and are stable in the genome; however, the loss of methyl groups (hypomethylation) can result from DNA damage. One consequence of hypomethylation of DNA is activation of cellular protooncogenes (Hergersberg 1991, Zapisek et al. 1992). Our studies (Hass et al. 1991), have shown that DNA from pancreatic acinar cells from DR animals is hypermethylated relative to DNA from AL animals. Hypomethylation of the c-H-*ras* gene was noted in aged

Table 1. Soft agar colony frequency of MNNG-induced pancreatic acinar cells from ad libitum (AL) and diet-restricted (DR) Brown-Norway rats[a]

Passage	Group	No. of soft agar colonies/10^5	Transformation[*]
9	AL/Y	8	+
	AL/O	162	+
	DR/Y	0	−
	DR/O	0	−

Y (young) = animals restricted for 13 months; O (old) = animals restricted for 29 months.
[a]Data of Hass et al. (1992).
[*]A corrected chi-square test, conditional on the number of colonies observed in the treated and control groups, was used to test for statistical significance of transformation ($p < 0.05$).

Table 2. Spontaneously induced pancreatic acinar cells from ad libitum (AL) and diet-restricted (DR) Brown Norway rats[a]

Passage	Group	No. of soft agar colonies/10^5	Transformation[*]
9	AL/Y	0	−
	AL/O	0	−
	DR/Y	0	−
	DR/O	0	−
14	AL/Y	13	+
	AL/O	38	+
	DR/Y	3	−
	DR/O	31	+

Y (young) = animals restricted for 13 months, O (Old) = refers to animals restricted for 29 months.
[a]Data of Hass et al. (1992).
[*]A corrected chi-square test, conditional on the number of colonies observed in the treated and control groups, was used to test for the statistical significance of induced transformation ($p < 0.05$).

AL animals (Figure 1). Studies were conducted to determine if hypo-methylation of the c-H-*ras* gene correlated with increased protooncogene expression.

The expression of the protooncogenes c-*myc*, c-H-*ras,* and c-K-*ras* is modulated by DR. DNA from pancreatic acinar cells exhibited increased expression of the H- and K-*ras* oncogenes from aged AL rats when compared to the DNA from aged DR animals (Figure 2). Nakamura et al. (1989) found suppression of c-*myc* gene expression in cells from DR B6C3F1 mice. They further demonstrated that the expression followed a circadian pattern. Although some degree of activation of oncogenes can

Methylation Profile of c-H-ras
in Brown Norway Rats

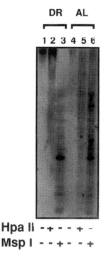

Hpa II - + - - + -
Msp I - - + - - +

Figure 1. Methylation profile of the c-Ha-*ras* gene in pancreatic acinar cells from Brown-Norway rats. Lanes 2 and 5 represent Hpa II-digested DNA from diet-restricted (DR) and ad libitum (AL) cells. Lanes 3 and 6 represent Msp I-digested DNA from DR and AL cells. Lanes 1 and 4 represent uncut DNA from DR and AL animals.

c-H-ras Expression

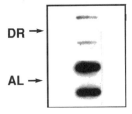

Figure 2. Slot-blot of c-Ha-*ras* mRNA from pancreatic acinar cells from diet-restricted and ad libitum (AL) rats. Note the increased expression of the *ras* gene in pancreatic acinar cells from old AL rats.

be attributed to epigenetic factors, most activation occurs through mutation of normal cellular genes.

Using monoclonal antibodies to the wild-type and mutated c-Ha-*ras* gene product (p21), we observed decreased expression of the mutated phenotype in cells from DR animals (Fan 1994). We have further examined the DNA from these cells using polymerase chain reaction amplification of codons 12 and 13 and oligonucleotide hybridization and found our data to correlate to previous immunological analyses (data not shown).

Conclusion

Our data have clearly shown that the in vivo dietary history of animals from which tissues are derived may be an important factor in interpreting the results of in vitro tests assaying for toxic and genotoxic endpoints. Tests such as growth, mutation, transformation, DNA repair, or altered oncogene expression are all important indicators of the toxicity of a chemical compound, and the results of each of these tests in vitro is significantly altered by the dietary history of the animal from which the tissue was collected.

We further conclude that the effects of DR are transferred from the whole animal to at least some of the individual cells and that the responses observed (decreased proliferation and transformation, depressed oncogene expression and mutation, and increased methylation) constitute cellular and molecular analogs of the in vivo condition. The poor survival rate of rats used in chronic animal bioassays has resulted in some studies being invalidated. Dietary restriction has been suggested as one means by which to avoid this difficulty. Control data obtained from DR studies can enhance current models by separating chemically caused effects from background. Most chemicals under consideration by the government undergo safety- and health-related testing using laboratory rats. Because the rodent's diet (specifically its caloric content) plays a substantial role in the health of the animal and its response to a test chemical, our studies suggest that the animal's diet should be considered a critical variable in evaluating data related to the in vitro transformation, oncogene activation, and DNA methylation of the chemical being tested.

Our studies have used one level of dietary restriction (40%). Future studies should be conducted to determine which level of restriction is optimal for improved health of test animals. If a lower percentage of restriction is found to be adequate, it will be useful in establishing a model diet for testing drugs or food substitutes that will have a direct effect on the animal's weight. Because DR is becoming accepted for use in industrial testing, the current information obtained from DR studies is critical in establishing a model diet for drug or food testing. This diet would optimize the animal's health, provide a standard weight range that would be expected to provide more consistent test responses with implications for the design and interpretation of toxicity and carcinogenicity studies, and decrease the initial number of animals needed in a 2-year study because of increased longevity. Regardless of what level of restriction is ultimately selected for in vivo studies in the future, or even if AL feeding is continued, it will be of utmost importance to address the dietary history of the donor animal in order to achieve reproducibility and to avoid the variability observed in the cellular and molecular responses that can be obtained quickly and efficaciously in in vitro assays.

A growing number of individuals using prescription and over-the-counter drugs are found in the elderly and obese subpopulations. Very little data are available on the possible differential effects of drugs or foods on these people. One can envision using several levels of DR (including AL feeding as a model for obesity) as well as rats of different ages as predictive models for risk assessment in these human subpopulations.

References

Bishop JM (1985) Viral oncogenes. Cell 42:23

Chou MW, Kong J, Chung K-T, Hart RW (1993) Effects of caloric restriction on the metabolic activation of xenobiotics. Mutat Res 295:223

Fan K, Hart RW, Blann E, et al. (1994) Expression of p53 and p21 proteins in dietary restricted rats: biomarkers for susceptibility to carcinogens. Mech Ageing Dev, submitted

Hass BS, Hart RW, Gaylor DW, Lyn-Cook BD (1992) An in vitro pancreas acinar cell model for testing the modulating effects of caloric restriction and ageing on cellular proliferation and transformation. Carcinogenesis 13:2419

Hass BS, Hart RW, Lu MH, Lyn-Cook BD (1993) Effects of caloric restriction in animals on cellular function, oncogene expression, and DNA methylation in vitro. Mutat Res 295:281

Hass BS, Lyn-Cook BD, Poirier L, Hart RW (1991) Differences in growth rate and DNA methylation in c-Ha-ras in pancreatic acinar cells of Brown-Norway rats fed ad libitum (AL) or calorically-restricted (CR) diets. Proc Am Assoc Cancer Res 29:862

Hergersberg M (1991) Biological aspects of cytosine methylation in eukaryotic cells, Experientia 47:1171

Lipman JM, Turturro A, Hart RW (1989) The influence of dietary restriction on DNA repair in rodents: a preliminary study. Mech Ageing Dev 48:135

Longnecker DS, Pettengill OS, Davis BH, et al. (1991) Characterization of preneoplastic and neoplastic lesions in the rat pancreas. Am J Pathol 138:333

Longnecker DS, Roebuck BD, Yager JD, et al. (1981) Pancreatic carcinoma in azaserine-treated rats: induction, classification and dietary modification of incidence. Cancer 47:1562

Lyn-Cook BD, Hass BS, Fan K, Hart RW (1994) Altered methylation profiles and activation of the c-Ha-ras oncogene in pancreatic acinar cells from aged ad libitum fed Brown Norway rats: the effects of caloric restriction. Mech Ageing Dev, Submitted

Miyamura Y, Tawa R, Koizumi A, et al. (1993) Effects of energy restriction on age-associated changes of DNA methylation in mouse liver. Mutat Res 295:63

Nakamura KD, Duffy PH, Lu M-H, Hart RW (1989) The effect of dietary restriction on myc protooncogene expression in mice: a preliminary study. Mech Ageing Dev 48:199

Roebuck BD, Baumgartner KJ, MacMillan DL (1993) Caloric restriction and intervention in pancreatic carcinogenesis in the rat. Cancer Res 53:46

Silverberg E, Boring CC, Squires TS (1990) Cancer statistics. CA Cancer J Clin. 40:9

Srivastava VK, Miller S, Schroeder MD, et al. (1993) Age-related changes in expression and activity of DNA polymerase alpha: Some effects of dietary restriction. Mutat Res, in press

Zapisek WF, Cronin GM, Lyn-Cook BD, Poirier LA (1992) The onset of oncogene hypomethylation in the livers of rats fed methyl-deficient amino-acid defined diets. Carcinogenesis 13:1869

Interactions Between Aging and Moderate Dietary Restriction on the Regulation of Growth Hormone, Insulin-Like Growth Factor-1, and Protein Synthesis

William E. Sonntag and Xiaowei Xu
Wake Forest University

Introduction

During the aging process, a number of cellular and biochemical alterations result in a reduction in metabolic processes and structural and morphological changes in tissues. Abundant empirical and scientific evidence supports the hypothesis that many of these aging processes are closely related to a decline in the capacity of tissues to initiate protein synthesis (Richardson 1981), but the mechanisms responsible for the latter changes remain elusive. One of the earliest investigations into the mechanisms of aging (Brown-Séquard 1889) suggested that alterations in humoral secretions contribute to the decline in tissue function with age, but these effects were quickly disputed. Only recently have studies established that a decline in anabolic hormones occurs with age and that hormone replacement therapy can counteract or delay the physiological effects of aging. One example of the influence of hormones on the aging process is the decrease in estrogen in females, which is known to contribute to the age-related decline in bone mass and the increase in cardiovascular disease. Extensive studies have established that estrogen replacement therapy delays or prevents age-associated diseases (Harman and Talbert 1985). Other hormones, such as growth hormone and insulin-like growth factor (IGF)-1, have recently been proposed to have important actions on the decline in tissue function with age, but their specific actions have not been as extensively documented. Nevertheless, these studies indicate that investigations into the mechanisms of key anabolic hormones will likely yield important information on the etiology of cellular and biochemical alterations with age and possibly the decreased tissue function that accompanies normal aging.

One of the few manipulations that has been reported to consistently influence the rate of biological aging is dietary restriction (DR). This regimen increases both mean and maximal life span (McCay et al. 1943, Yu et al. 1982, Weindruch and Walford 1988, Iwasaki et al. 1988), reduces the appearance of pathological lesions (Weindruch and Walford 1982, Weindruch et al. 1986, Bronson and Lipman 1991), decreases susceptibility to chemical toxicity and increases DNA repair (Turturro and Hart 1984, Lipman et al. 1989, Hart et al. 1990, Djuric et al. 1992, Hart et al. 1992), and increases tissue protein synthetic activity (Birchenall-Sparks et al. 1985, Ricketts et al. 1985, Sonntag et al. 1992). By comparing physiological adaptations in the DR animal to the ad libitum (AL) animal over the life span, this paradigm has been used to identify potential biological mechanisms of aging that could then be pursued in more rigorously controlled studies. We have previously proposed that decreases in growth hormone secretion contribute to biological aging (Sonntag 1987, Sonntag et al. 1992) and noted that many DR actions appear to be similar to the effects achieved by administering growth hormone to aging animals and humans. This report reviews recent studies on the regulation and mechanisms of growth hormone and IGF-1 in aging animals and the compensatory changes resulting from DR that contribute to a delay in physiological processes associated with aging.

In all studies conducted in our laboratory, animals were individually housed and fed an NIH-31 diet. At 14 weeks of age, diet was reduced to 60% compared to AL animals of the same age over a 2-week period. Animals maintained under DR conditions were fed a vitamin- and mineral-supplemented diet once daily, and this regimen was continued throughout the life span.

Protein Synthesis, Dietary Restriction, and Aging

Effects of Age

The age-related decline in protein synthesis is widely recognized as a biochemical correlate of the aging process. In most tissues, including kidney, liver, brain, and cardiac and skeletal muscle (Richardson 1981, Burini et al. 1984, Pluskal et al. 1984), there appears to be a progressive reduction in total protein synthetic activity with age. Several investigators have reported that this decline is related to a reduction in poly A^+RNA synthesis, indicating that deficiencies in transcription of DNA are at least partly responsible for the age-related deficits. Pluskal et al. (1984) reported a 40% decline in the activity of polyribosomes prepared from old as compared to young animals using a cell-free system. There also appears to be a decline in the activity of soluble factors in the pH 5 enzyme fraction prepared from 12- and 22–24-month-old rats. These results suggest not only that there is a progressive, age-related decline in transcription rate, but that the efficiency of the ribosome protein factor complex necessary to initiate and maintain protein synthesis is reduced with age. Although the in vitro studies provide important insight into potential mechanisms of aging, in vivo protein synthesis studies have not revealed entirely consistent decreases in protein synthetic capacity with age. For example, protein synthesis in liver and heart decreases with age (Sonntag et al. 1992) whereas diaphragm exhibits no change (depending on the species) and only the cortex of brain shows age-related decrements (D'Costa et al. 1993). These results suggest that in the basal state some tissues have the ability to maintain protein synthesis (possibly through the induction of compensatory mechanisms). Interestingly, when presented with a stimulus requiring the initiation of protein synthesis, most aging tissues exhibit clear deficiencies. Thus, in vivo studies of protein synthesis are more consistent with in vitro studies when tissues are studied in the "challenged" state. Since the capacity to maintain adequate levels of protein synthesis is intimately related to tissue function, we and others have proposed that decreases in protein synthetic processes are part of the mechanisms responsible for normal aging.

Effects of Moderate Dietary Restriction

The relationship between aging and the decline in protein synthesis has gained additional support in recent years in studies of DR animals. Animals maintained at a caloric level 60% of AL conditions demonstrate a marked decrease in the appearance of age-related pathologies (McCay et al. 1943, Nakagawa et al. 1974, Masoro et al. 1982, Yu et al. 1982, Weindruck and Walford 1988), and tissues from DR animals have increased rates of protein synthesis compared to tissues from AL animals (Birchenall-Sparks et al. 1985, Ricketts et al. 1985, Ward 1988). Although many of these studies have been done in vitro, recent studies from our laboratory (Sonntag et al. 1992) have confirmed these results in vivo. Analysis of relative rates of protein synthesis in liver from AL animals revealed the characteristic decrease in protein synthesis between 5 and 25 months of age after correcting for the specific activity of the amino acid precursor. Protein synthesis in heart and diaphragm increased between 5 and 15 months but decreased thereafter. In each case, moderate DR increased protein synthesis and/or prevented the decline in protein synthesis with age. Although these studies do not address specific functional measures of tissue activity, the decrease in protein synthesis is presumed to have an important role in the decline in tissue function with age.

Growth Hormone, IGF-1 and the Endocrine Regulation of Protein Synthesis in Aging Animals

History

Since the early 1900s, it has been known that a substance present in blood promotes growth. It was only after Li et al. (1945) isolated pure bovine growth hormone from the pituitary gland that the biological effects of growth hormone became evident. Growth hormone was subsequently shown to stimulate amino acid uptake into tissues; DNA, RNA, and protein synthesis; and to have a role in cell division and hypertrophy. Soon after the discovery of growth hormone, highly purified pituitary extracts were used to stimulate growth in growth-hormone-deficient children. An assay for determining circulating concentrations of growth hormone was developed by Utiger et al. (1962), and the amino acid sequence was determined by Li et al. (1969) and revised by Niall (1971). Analysis of growth hormone plasma concentrations demonstrated marked variability in hormone concentrations resulting from discrete growth hormone pulses that increase after the onset of sleep (Finkelstein et al. 1972). Although the precise function of this ultradian pattern remains unknown, the pulsatile release of growth hormone has been confirmed in every species examined to date and is closely related to the biological actions of the hormone.

Other investigations concentrated on specific factors that appeared to mediate the actions of growth hormone. Salmon and Daughaday (1957) discovered a factor that was regulated by growth hormone and promoted incorporation of sulfate into cartilaginous tissue. This led to the purification of the somatomedin family of hormones (Van Wyk et al. 1974), which are peptides of small molecular weight (about 7.5 kDa) that circulate in the blood in high concentrations. These peptide hormones induce mitogenic activity in cultured fibroblasts and fetal cell lines, stimulate anabolic activity in many cell and tissue types, and induce DNA and protein synthesis (Shermer et al. 1987). Somatomedin C, also termed insulin-like growth factor 1 (IGF-1), is structurally related to insulin (Rinderknecht and Humbel 1978) and exhibits similar though less potent effects on glucose regulation. IGF-1 binds with high affinity to the type 1 IGF receptor through which it exerts its actions (Rechler and Nissley 1985); these receptors are found in tissues throughout the body. The type 1 IGF receptor shares 50% amino acid sequence similarity with insulin, and competitive binding and affinity cross-linking studies have demonstrated that IGF-1 binds to the insulin receptor with 1000 times lower affinity than to the IGF-1 receptor (Rechler and Nissley 1985). IGF-1 is synthesized mainly in liver under regulation of growth hormone but is also synthesized in smaller quantities in almost all tissues (Daughaday and Rotwein 1989). Although regulation of the paracrine activity of IGF-1 is poorly understood, alterations in the activity of this hormone or its receptors at the tissue level have a significant effect on many intracellular processes.

IGF-1 circulates in the blood either free (with a half-life of 15–20 minutes) or bound to specific binding proteins that prolong the half-life of the peptide. Six binding proteins have currently been identified (Binoux et al. 1986) and constitute an intricate transport system for the IGFs that regulates their availability to specific tissues. It is now clear that the binding proteins are important regulators of IGF-1 activity and may also prevent hypoglycemia, which can be induced by IGF-1.

Over the past decade there has been increasing interest in determining whether growth hormone acts directly on tissues or whether the effects are mediated through IGF-1. Although investigators initially proposed that all actions of growth hormone were mediated through secretion of IGF-1, other studies provided relatively convincing data that growth hormone could have direct anabolic effects on specific tissues (Beach and Kostyo 1968, Goldspink and Goldberg 1975). Subsequent studies demonstrated, however, that IGF-1 is present in most tissues and that the "local" actions of growth hormone were actually mediated via paracrine secretion of IGF-1. These studies led to the concept that growth hormone stimulates IGF-1 secretion from hepatic tissue, thereby increasing the concentration of IGF-1 in plasma; and that growth hormone may also regulate secretion of IGF-1 from local tissues, thereby influencing the paracrine activity of IGF-1.

Aging

Research over the past decade has clearly established that alterations in the neuroendocrine axis and specifically in growth hormone regulation have an important role in the physiological and biochemical changes normally associated with aging. Early studies in humans indicated that the amplitude of growth hormone pulses and the rise in growth hormone concentration were blunted after insulin-induced hypoglycemia (Laron et al. 1970). Subsequent studies by Sonntag et al. (1980) demonstrated a prominent decrease in the amplitude of growth hormone pulses in old rats and a reduction in growth hormone content in the pituitary gland. The amplitude in the old animals averaged about half of that observed in young animals; there appeared to be no changes in basal concentrations of growth hormone or in the ultradian rhythm. Other investigators reported a progressive reduction in the plasma concentration of IGF-1 in rats (Florini and Roberts 1979, Florini et al. 1981); these results were subsequently confirmed in humans (Johanson and Blizzard 1981, Rudman et al. 1981). Later studies in various strains of rats and mice (Sonntag et al. 1980, Florini et al. 1981, Breese et al. 1991), non-human primates (Kahler et al. 1986), and humans (Carlson et al. 1972, Prinz et al. 1983) consistently confirmed the decline in growth hormone and IGF-1 concentrations with age and suggested that these declines are a robust marker of biological aging in mammalian species.

After a decline in growth hormone pulse amplitude was identified was identified, a number of investigators recognized the potential clinical significance of decreases in growth hormone and initiated replacement regimens. Earlier investigations had shown that purified growth hormone preparations had beneficial effects in old animals and humans, but the measures were unrelated to specific deficiencies that normally appear with age (Asling et al. 1952, Emerson 1955, Everitt 1959, Beck et al. 1960, Root and Oski 1969). Studies by Sonntag et al. (1984) related a specific deficiency, the decline in protein synthetic capacity, to the age-related reduction in the plasma concentrations of growth hormone. Protein synthetic capacity decreased by approximately 40% in diaphragm muscle of Sprague-Dawley rats, and administration of growth hormone over 8 days significantly increased protein synthesis in old animals to values greater than those in young animals. Although this study did not address the question of whether response to growth hormone diminished with age, it clearly indicated that 1) tissue from aged animals could exhibit an increase in protein synthesis, and 2) the decrease in plasma growth hormone concentrations might be a causative factor in the decline with age of protein synthesis.

Growth hormone or IGF-1 administration to old animals partially reverses the decline in immune function (Kelley et al. 1986) and increases the expression of aortic elastin in rats (Foster et al. 1990), while administration of growth hormone alone increases lean body mass, skin thickness and vertebral bone density in elderly men (Rudman et al. 1990). These studies support the concept that the decrease in growth hormone concentration has clinical significance and may be responsible for some of the tissue changes that accompany normal aging.

Effects of Moderate Dietary Restriction on Growth Hormone and IGF-1

Plasma IGF-1

Since protein synthesis is regulated, at least in part, by IGF-1, the hypothesis that protein synthesis increases observed DR in animals may be mediated by altering IGF-1 concentrations in plasma was tested. In all subsequent studies, however, it was found that the IGF-1 concentration was diminished in aged animals in response to DR, and the differences in IGF-1 plasma levels between the DR and AL groups either were not evident or levels were decreased in the DR animals (Breese et al. 1991). These results clearly indicated that the protein synthesis increases in DR animals were not dependent on the concentration of plasma IGF-1 alone.

IGF Binding Proteins

An alternative hypothesis tested the effects of aging and DR on plasma IGF binding proteins that regulate IGF-1 activity. Although there appeared to be an age-related decline in the amount of several binding proteins in AL animals, these changes were generally associated with the concentrations of plasma IGF-1. Dietary regulation did not produce any marked changes in IGF binding proteins that could account for the increase in protein synthesis (Breese et al. 1991).

Type 1 IGF Receptors

Another hypothesis examined type 1 IGF receptors. No consistent changes in the density of these receptors in AL animals were observed with age, but there was a 1.5- to 2.5-fold increase in receptors in the liver, heart, and skeletal muscle from DR animals when compared to the same tissues from AL rats (Figure 1). Since increases in receptor expression are generally associated with increases in tissue response, it was suggested that part of the mechanism for increased protein synthesis in response to moderate DR was an enhanced tissue response to IGF-1 (D'Costa et al. 1993). Although further studies are necessary, this appears to be the first evidence linking DR to an enhanced response to hormones known to regulate protein synthesis. Thus, the lower value for protein synthesis in AL animals is, at least in part, the result of a decline in plasma IGF-1 concentrations; DR may increase protein synthesis by increasing the density of type 1 IGF receptors).

Dietary Restriction

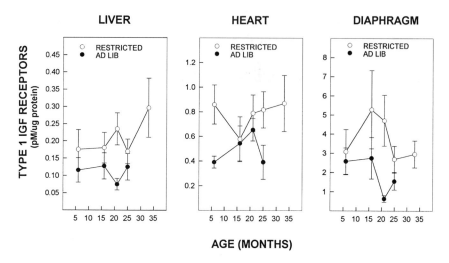

Figure 1. Type 1 IGF receptor densities in liver, heart, and diaphragm of ad libitum (AL, closed circles) or diet-restricted (DR, open circles) male Brown-Norway rats at 5, 16, 21, 26, and 33 months of age (AL animals did not survive to 33 months). IGF receptors were analyzed on crude membrane preparations from each tissue and are corrected to µg protein. No differences in K_a were noted between the age groups. DR was initiated at 14 weeks of age, and animals were maintained on a 60% AL diet. Results indicate that DR increases type 1 IGF receptor density in the tissues analyzed. Data from D'Costa, et al. (1993).

IGF Paracrine Activity

As noted earlier, many tissues express both IGF-1 mRNA and peptide (Murphy et al. 1978, D'Ercole et al. 1984, Noguchi et al. 1987, Wether et al. 1990, Delafontaine et al. 1991). Although regulation of IGF-1 paracrine activity is not fully understood, alterations in the activity of this hormone at the tissue level could have a significant impact on protein synthesis. The results of studies to date suggest hierarchical control of tissue protein synthesis. For example, intrinsic factors within the cell exhibit the initial level of control followed by the action of local (paracrine) factors such as tissue-derived IGF-1, which binds to type 1 IGF receptors. Finally, endocrine control of tissue protein synthesis is manifest by the action of circulating plasma IGF-1 or growth hormone. A complete understanding of the mechanisms responsible for the decline in protein synthesis with age and its reversal by DR requires assessment of the cellular, paracrine, and endocrine factors that regulate protein synthesis.

Growth Hormone

At least part of the reduction in plasma IGF-1 and protein synthesis with age results from a reduction in growth hormone secreted from the

anterior pituitary gland (Sonntag et al. 1980, Sonntag et al. 1984). Because of the decreases in IGF-1 in response to DR and the close association of IGF-1 with growth hormone concentrations (Florini et al. 1985),
the secretory dynamics of growth hormone in DR animals were not initially assessed. However, our results demonstrating an increase in protein
synthesis in DR animals, coupled with increases in the number of type 1
IGF receptors, suggests that alterations in the hormonal milieu could be a
mediating factor. It was initially proposed that alterations in growth hormone secretion might be responsible for the rise in the density of type 1
IGF receptors. Confirming previous studies, it was found that the amplitude of growth hormone secretory pulses decreased with age and, as expected, DR diminished the amplitude of growth hormone pulses in young
animals. However, in older DR animals, growth hormone pulses were similar to those in young AL animals (Figure 2). The increase in growth hormone plasma levels in old DR animals was unexpected, but both the number of detected growth hormone pulses and the concentrations of mean
growth hormone increased. Thus, after an unknown period of adaptation
to DR, these animals exhibited increases in high-amplitude growth hormone secretion that were maintained into old age. Since the plasma concentration of IGF-1 was diminished in the presence of increased tissue
protein synthesis, the studies also suggested that increased amounts of circulating growth hormone may act directly at the tissue level to drive either
paracrine IGF-1 activity or type 1 IGF receptor activity. In either case, the
studies demonstrate that growth hormone secretory dynamics are associated with and may be part of the mechanism for increases in protein synthesis in response to DR.

Neuroendocrine Regulation of Growth Hormone Release

History

Although release of growth hormone in humans is characterized by relatively low-amplitude pulses throughout the day and a large pulse after the
onset of sleep, release in rats is characterized by an ultradian rhythm with
high-amplitude secretory pulses every 3–5 hours; between pulses, growth
hormone concentrations decrease to almost undetectable values
(Tannenbaum and Martin 1976). The regulation of this pattern involves
two different hormones released by the hypothalamus: growth hormone-
releasing hormone (GHRH) (Rivier et al. 1982, Ling et al. 1984), which
increases growth hormone release; and somatostatin (Brazeau et al. 1973),
which inhibits its release. Several studies suggested that both hormones

Dietary Restriction

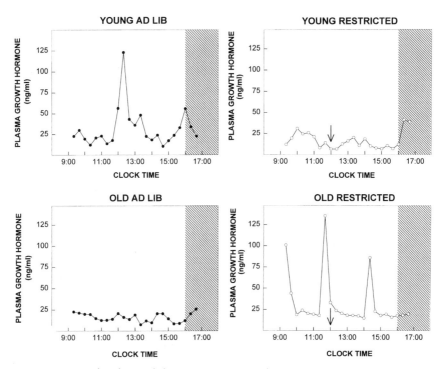

Figure 2. Example of growth hormone secretory dynamics in young (6-month-old) ad libitum (AL) and diet-restricted (DR) animals (top) and old (25-month-old) AL and DR male Brown-Norway rats (bottom). Blood samples from freely moving, cannulated animals were taken at 20-minute intervals from 0920–1640 hours and plasma analyzed for growth hormone by RIA. DR animals were fed at 1200 hours (denoted by arrow). Shaded areas represent the dark phase of the light/dark cycle. Results indicate that the number of pulses, pulse height, and mean growth hormone concentration decrease with age in AL-fed animals. After an initial period of adaptation to the feeding regimen, however, the number of growth hormone pulses, pulse height, and mean growth hormone levels increase in the DR animals.

are secreted in a phasic manner, with GHRH contributing to high-amplitude growth hormone pulses and somatostatin being secreted during trough periods (Tannenbaum and Ling 1984). The dynamic interrelationship between these hypothalamic hormones is responsible for pulsatile growth hormone secretion. A number of other factors contribute to the regulation of growth hormone release by acting directly on the pituitary gland or by regulating hypothalamic somatostatin or GHRH secretion. Both growth hormone and IGF-1 inhibit further growth hormone release (Berelowitz et al. 1981) in a typical feedback relationship at the levels of the hypothalamus and pituitary.

Various neurotransmitters as well as opioid and other neuroactive peptides also influence somatostatin and GHRH release and subsequent growth hormone release from the pituitary. The specific actions of these compounds have been previously reviewed (Sonntag and Meites 1988). More recently there has been increased interest in small peptides, or growth hormone-releasing peptides (GHRPs), that stimulate growth hormone release. One of the peptides, His-D-Trp-Ala-Trp-DPhe-Lys-NH (GHRP-6), stimulates growth hormone release, and although it does not appear to interact with the GHRH receptor, it has been shown to act synergistically with GHRH (Bowers et al. 1990). The exact mechanism of this peptide's action is still unclear, but since it appears to remain active after oral administration, its importance cannot be underestimated.

Somatostatin

To date there have been many studies supporting the conclusion that somatostatin secretion increases with age. Studies of pituitary response of GHRH in vivo with or without passive immunization against somatostatin provide some of the best indirect evidence suggesting increased somatostatinergic tone with aging. In addition, a reported decline in somatostatin receptors in pituitaries from older animals (Spik and Sonntag 1989) suggests a down-regulation in response to increased somatostatin secretion. Analysis of somatostatin concentrations in the pituitary gland has revealed a 40% increase in 22-month-old compared with 6–8-month-old Fischer 344 (F344) rats. In vitro studies also indicate that either K^+ depolarization (Sonntag et al. 1986) or glucopenia increase somatostatin release to a greater extent in old compared with young animals. Analysis of the superfusate revealed greater concentrations of somatostatin-28 than somatostatin-14 in old animals. These results suggest that somatostatin secretion increases with age and that there may be alterations in post-translational processing of the molecule leading to high concentrations of somatostatin-28, which has previously been shown to be more potent than somatostatin-14 in blocking release of growth hormone from pituitary cells (Tannenbaum et al. 1982). Although increased somatostatin secretion plays an important role in the decline of high-amplitude growth hormone pulses in aging animals, it is only one factor involved in this decline.

Because of the varied response to anesthetics with age and the impact of these compounds on neuropeptide secretion, comparing secretion of these hormones into hypophyseal portal blood in young versus old animals has not been attempted. Since there was abundant indirect evidence that somatostatin concentration in plasma increases with age, studies were designed to assess the molecular mechanisms responsible for the increase, assuming a direct relationship between the amount of somatostatin mRNA and expressed peptide. In the initial study, changes in total somatostatin mRNA in hypothalamic neurons were measured, and an unexpected decrease was found in F344 rats with age (Sonntag et al. 1990). Because this finding appeared to conflict with the increase in somatostatin peptide previously reported with age, the amount of somatostatin mRNA in hypothalamic neurons of young and old animals was compared using in situ hybridization. The results demonstrated abundant hybridization of the somatostatin antisense probe to periventricular neuronal cell bodies in young (3–6-month-old) animals and, confirming the previous results, a 40% decrease in hybridization by 24 months of age in male F344 rats.

This experiment was repeated using the Brown-Norway rat. Both total somatostatin mRNA and somatostatin mRNA associated with polysomes were compared in 6-, 17-, and 21–26-month-old rats. The hypothesis was that possible alterations in degradation rates of somatostatin mRNA or recruitment onto polysomes would be observed in aged animals. Total somatostatin mRNA in these animals again revealed a decrease with age, but when somatostatin mRNA associated with polysomes was expressed as a ratio to total somatostatin mRNA, 26-month-old rats exhibited a twofold increase compared with 6-month-old animals.

Although these measurements do not assess the translational efficiency of somatostatin mRNA on the polysome, they provide insight into possible differences in somatostatin mRNA distribution within the cell that could lead to increased production of the somatostatin peptide. Recent evidence suggests that cells contain translational regulatory proteins that bind to mRNA, preventing translation into protein; therefore, it is possible that this type of regulatory control is compromised in aging animals. Since plasma growth hormone and IGF-1 concentrations decrease with age (which should suppress somatostatin mRNA synthesis), the observed decline in somatostatin mRNA in aging animals suggests that cellular regulation of mRNA levels remains intact. By contrast, a higher proportion of somatostatin mRNA bound onto polysomes in the older animal indicates a specific deficiency in translational regulation. We are currently investigating the regulation of somatostatin translational control by neuropeptides and neurotransmitters and its modification in aging animals.

Growth Hormone–Releasing Hormone (GHRH)

Although the studies to date suggest that somatostatin plays an important role in the decline in growth hormone secretion, equally compelling

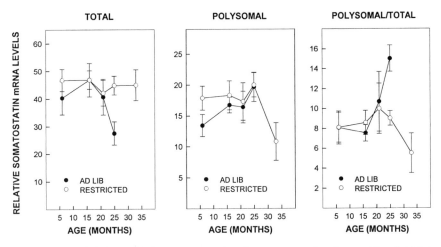

Figure 3. Total (left), polyribosome-associated (center), and polysomal/total (right) somatostatin mRNA in ad libitum (AL) and diet-restricted (DR) male Brown-Norway rats at 6, 16, 21, 26, and 33 months of age (AL animals did not survive to 33 months of age). The hypothalamus was dissected from each animal, homogenized in buffer, and aliquots separated for analysis of total and polysome-associated somatostatin mRNA by dot-blot hybridization to an antisense somatostatin probe. Total/polysomal values were derived after correction to polyA⁺RNA levels. Data represent mean ± SEM for 10 animals/group.

data suggest that GHRH secretion decreases with age. In a series of important experiments, it was reported that GHRH mRNA decreased with age (DeGennaro Colonna et al. 1989), and more importantly, that the feedback relationship between growth hormone and GHRH mRNA was deficient (DeGennaro Colonna et al. 1993). Thus, the emerging data suggest that decreases in growth hormone secretion with age result from deficiencies within the hypothalamus involving both transcriptional regulation of GHRH and translational regulation of somatostatin.

Moderate Dietary Restriction

The previous results demonstrating that moderate DR increases growth hormone pulse amplitude led us to compare amounts of total and polysomal-bound somatostatin mRNA in AL and DR animals. The strategy was to determine whether DR could modify the recruitment of somatostatin mRNA onto polysomes and whether the modification was associated with the growth hormone pulse increases observed in these animals. It was found that DR prevented both the age-related decline in total somatostatin mRNA and the recruitment of mRNA onto the polysome (Figure 3). These data suggest that DR decreases the synthesis and release of somatostatin-14 and somatostatin-28, which then leads to an increase in the amplitude of growth hormone pulses. Unfortunately, no data are currently available to assess the effects of moderate DR on GHRH mRNA levels.

The DR model may provide further insight into both the normal mechanisms for regulating growth hormone pulses and the age-related perturbations in this system. Additional studies to assess the importance of translational regulation of somatostatin mRNA are currently in progress. It is clear, however, that the decrease in the total amount of somatostatin mRNA with age is most likely a physiological response to the decrease in growth hormone concentration in plasma. With increasing age, the feedback signal from circulating growth hormone remains intact, and hypothalamic neurons respond appropriately by decreasing transcription of somatostatin mRNA. The increase in somatostatin secretion likely results from an inability of cellular regulatory factors to suppress the active translation of somatostatin mRNA into protein, resulting in increased somatostatin synthesis and reduced amplitude of growth hormone pulses.

Conclusions

The studies mentioned here have given new insights into the biological mechanisms of aging. First, it appears that alterations within the endocrine axis result in a cascade of events that culminate in decreased amplitude of growth hormone pulses and the amount of IGF-1 secreted (and a subsequent decline in the capacity of tissues to synthesize protein). The broad range of growth hormone actions and ubiquitous distribution of IGF-1 suggest that we are only beginning to decipher the physiological events of aging that result from decreases in growth hormone concentration. Second, DR increases protein synthesis, and this effect may be mediated by an increase in type 1 IGF receptor density in tissues and possibly by an increase in tissue response to IGF-1. The increase in growth hormone secretion in response to DR and its contribution to IGF-1 paracrine activity (and subsequent protein synthesis) remain to be determined. Finally, the influence of translation regulation and its contribution to increased somatostatin release may provide additional insights into biological mechanisms of aging and the regulation of hormone secretion in general.

Acknowledgments

This work was supported by NIH grant AGO7752.

References

Asling CW, Moon HD, Bennet LL, Evans HM (1952) Relation of the anterior hypophysis to problems of aging. J Gerontol 9:292

Beach RK, Kostyo JL (1968) Effect of growth hormone on DNA content of muscles of young hypophysectomized rats. Endocrinology 82:882

Beck JC, McGarry EE, Dyrenfurth I, et al. (1960) Primate growth hormone studies in man. Metabolism 9:699

Berelowitz M, Szabo M, Frohman LA, et al. (1981) Somatomedin-C mediates growth hormone negative feedback by effects on both the hypothalamus and the pituitary. Science 212:1279

Binoux M, Hossenloop L, Hardouin S, et al. (1986) Somatomedin (insulin-like growth factors)-binding proteins: molecular forms and regulation. Hormone Res 24:141

Birchenall-Sparks MC, Roberts MS, Staecker J, et al. (1985) Effect of dietary restriction on liver protein synthesis in rats. J Nutri 115:944

Bowers CY, Reynolds GA, Durham D, et al. (1990) Growth hormone releasing peptide stimulates growth hormone release in normal men and acts synergistically with growth hormone-releasing hormone. J Clin Endocrinol Metab 70:975

Brazeau P, Vale W, Burgus R, et al. (1973) Hypothalamic polypeptide that inhibits the secretion of immunoreactive pituitary growth hormone. Science 179:77

Breese CR, Ingram RL, Sonntag WE (1991) Influence of age and long-term dietary restriction on plasma insulin-like growth factor-1 (IGF-1), IGF-1 gene expression, and IGF-1 binding proteins. J Gerontol 46:B180

Bronson RT, Lipman RD (1991) Reduction in rate of occurrence of age-related lesions in dietary restricted laboratory mice. Growth Devel Aging 55:169

Brown-Séquard CE (1889) Expérience démontrante la puissance dynamogénique chez l'homme d'un liquide extrait de testicules d'animaux. Arch Physiol Norm Path 5:651

Burini RC, Pluskal MG, Wei IW, Young VR (1984) Protein synthesis studies in skeletal muscle of aging rats II. In vitro studies with 0.5M potassium chloride washed polyribosomes. J Gerontol 39:392

Carlson H, Gillin J, Gorden P, Snyder F (1972) Absence of sleep-related growth hormone peaks in aged normal subjects and in acromegaly. J Clin Endocrinol Metab 34:1102

Daughaday WH, Rotwein P (1989) Insulin-like growth factors I and 11. Peptide, messenger ribonucleic acid and gene structures, serum and tissue concentrations. Endocr Rev 10:68

D'Costa AP, Lenham JE, Ingram RL, Sonntag WE (1993) Moderate caloric restriction increases type 1 IGF receptors and protein synthesis in aged rats. Mech Aging Devel 71:59

De Gennaro Colonna V, Fidone F, Cocchi D, Muller EE (1993) Feedback effects of growth hormone on growth hormone releasing hormone and somatostatin are not evident in aged rats. Neurobiol Aging 14:503

De Gennaro Colonna V, Zoli M, Cocchi D, et al. (1989) Reduced growth hormone releasing factor (GHRF) immunoreactivity and GHRF gene expression in the hypothalamus of aged rats. Peptides 10:705

Delafontaine P, Bernstein KE, Alexander RW (1991) Insulin-like growth factor-1 gene expression in vascular cells. Hypertension 17:689

D'Ercole AJ, Stiles AD, Underwood LE (1984) Tissue concentrations of somatomedin-C: further evidence for multiple sites of synthesis and paracrine or autocrine mechanisms of action. Proc Natl Acad Sci U S A 81:935

Djuric Z, Lu MH, Lewis SM, et al. (1992) Oxidative DNA damage levels in rats fed low-fat, high fat, or calorie restricted diets. Toxicol Appl Pharmacol 115:156

Emerson JD (1955) Development of resistance to growth hormone action of anterior pituitary hormone. Am J Physiol 181:390

Everitt AV (1959) The effect of pituitary growth hormone on aging mate rats. J Gerontol 14:415

Finkelstein J, Roffwarg H, Boyar R, et al. (1972) Age-related changes in the twenty-four hour spontaneous secretion of growth hormone. J Clin Endocrinol Metab 35:665

Florini J, Harned J, Richman R, Weiss J (1981) Effect of rat age on serum levels of growth hormone and somatomedins. Mech Ageing Dev 15:165

Florini JR, Prinz PN, Vitiello MV, Hintz RL (1985) Somatomedin-C levels in healthy young and old men: relationship to peak and 24-hour integrated levels of growth hormone. J Gerontol 40:2

Florini JR, Roberts SB (1979) Effect of age on blood levels of somatomedin-like growth factor. J Gerontol 35:23

Foster JA, Rich CB, Miller M, et al. (1990) Effect of age and IGF-1 administration on elastin gene expression in rat aorta. J Gerontol 45:B113

Goldspink DF, Goldberg AL (1975) Influence of pituitary growth hormone on DNA synthesis in rat tissues. Am J Physiol 228:302

Harman SM, Talbert GB (1985) Reproductive Aging. In Finch CE, Schneider EL (eds), Handbook of the biology of aging, 2nd ed. Van Nostrand Reinhold, New York, pp 457–510

Hart RW, Leakey JE, Chou M, et al. (1992) Modulation of chemical toxicity by modification of caloric intake. In Jacobs MM (ed), Exercise, fat and cancer. Plenum Press, New York, pp 73–81

Hart RW, Turturro A, Pegram RA, Chou MW (1990) Effects of caloric restriction on the maintenance of genetic fidelity. In Sutherland BM, Woodhead AD (eds), DNA damage and repair in human tissues. Plenum Press, New York, pp 351–361

Iwasaki K, Gielser CA, Masoro EJ, et al. (1988) Influence of the restriction of individual dietary components on longevity and age-related disease of

Fisher rats: the fat component and the mineral component. J Gerontol 43:B13

Johanson A, Blizzard R (1981) Low somatomedin-C levels in older men rise in response to growth hormone administration. Johns Hopkins Med J 149:115

Kahler LW, Gleissman P, Craven J, et al. (1986) Loss of enhanced nocturnal growth hormone secretion in aging rhesus monkeys. Endocrinology 119:1281.

Kelley K, Brief S, Westly H, et al. (1986) Growth hormone pituitary adenoma cells can reverse thymic aging in rats. Proc Natl Acad Sci U S A 83:5663

Laron A, Doron M, Arnikan B (1970) Plasma growth hormone in men and women over 70 years of age. Med Sports Phys Act Aging 4:126

Li CH, Dixon JS, Liu WK (1969) Human pituitary growth hormone, XIX. The primary structure of the hormone. Arch Biochem Biophys 133:70

Li CH, Evans HM, Simpson ME (1945) Isolation and properties of the anterior hypophysial growth hormone. J Biol Chem 159:353

Ling N, Esch F, Bohlen P, et al. (1984) Isolation, primary structure, and synthesis of human hypothalamic somatocrinin: growth hormone-releasing factor. Proc Natl Acad Sci U S A 81:4302

Lipman JM, Turturro A, Hart RW (1989) The influence of dietary restriction on DNA repair in rodents: A preliminary study. Mech Ageing Dev 48:135

Masoro EJ, Yu BP, Bertrand HA (1982) Action of food restriction in delaying the aging process. Proc Natl Acad Sci U S A 79:4239

McCay CM, Sparling G, Barnes LL (1943) Growth, aging, chronic diseases and life-span in rats. Arch Biochem Biophys 2:469

Murphy LJ, Bell GI, Freisen HG (1978) Tissue distribution of insulin-like growth factor I and II messenger ribonucleic acid in the adult rat. Endocrinology 120:1279

Nakagawa I, Sasaki A, Kajimoto T, et al. (1974) Effect of protein nutrition on growth, longevity, and incidence of lesions in the rat. J Nutr 104:1576.

Niall HD (1971) A revised primary structure for human growth hormone. Nature 230:90

Noguchi T, Kurata LM, Sugisaki T (1987) Presence of a somatomedin-C immunoreactive substance in the central nervous system: immunohistochemical mapping studies. Neuroendocrinology 46:277

Pluskal MG, Moreyra M, Burini RC, Young VR (1984) Protein synthesis studies in skeletal muscle of aging rats I. Alterations in nitrogen composition and protein synthesis using a crude polyribosome and pH 5 enzyme system. J Gerontol 39:385

Prinz PN, Weitzman ED, Cunningham GR, Karacan I (1983) Plasma growth hormone during sleep in young and aged men. J Gerontol 38:519

Rechler MM, Nissley PS (1985) The nature and regulation of the receptors for insulin-like growth factors. Ann Rev Physiol 47:425

Richardson A (1981) The relationship between aging and protein synthesis. In Florini JR (ed), Handbook of biochemistry in aging. CRC Press, Boca Raton, FL pp 70–105

Ricketts WG, Birchenall-Sparks MC, Hardwick JP, Richardson A (1985) Effect of age and dietary restriction of protein synthesis by isolated kidney cells. J Cell Physiol 125:492

Rinderknecht E, Humbel RE (1978) The amino acid sequence of human insulin like growth factor-1 and its structure homology with proinsulin. J Biol Chem 253:2769

Rivier J, Spiess J, Thorner M, Vale W (1982) Characterization of a growth hormone-releasing factor from a human pancreatic islet tumor. Nature 300:276

Root AW, Oski FA (1969) Effects of human growth hormone in elderly males. J Gerontol 24:97

Rudman D, Axel GF, Nagraj HS, et al. (1990) Effects of growth hormone in men over 60 years old. N Engl J Med 323:1

Rudman D, Kutner MH, Rogers CM, et al. (1981) Impaired growth hormone secretion in the adult population. J Clin Invest 67:1361

Salmon WD, Daughaday WH (1957) A hormonally controlled serum factor which stimulates sulfate incorporation by cartilage in vitro. J Lab Clin Med 49:825

Shermer J, Raizada MK, Masters BA, et al. (1987) Insulin-like growth factor I receptors in neural and glial cells: characterization and biological effects in primary culture. J Biol Chem 262:7693

Sonntag WE (1987) Hormone secretion and action in aging animals and man. Rev Biol Res Aging 3:279

Sonntag WE, Boyd RL, Booze RM (1990) Somatostatin gene expression in hypothalamus and cortex of aging male rats. Neurobiol Aging 11:409

Sonntag WE, Gotschall PE, Meites J (1986) Increased secretion of somatostatin-28 from hypothalamic neurons of aged rats in vitro. Brain Res 380:229

Sonntag WE, Hylka VW, Meites J (1985) Growth hormone restores protein synthesis in skeletal muscle of old male rats. J Gerontol 40:689

Sonntag WE, Lenham JE, Ingram RL (1992) Effects of aging and dietary restriction on tissue protein synthesis: relationship to plasma insulin-like growth factor 1. J Gerontol 47:B159

Sonntag WE, Meites J (1988) Decline in growth hormone secretion in aging animals and man. Interdis Topics Gerontol 24:111

Sonntag WE, Steger RW, Forman LJ, Meites J (1980) Decreased pulsatile release of growth hormone in old male rats. Endocrinology 107:1875

Spik KW, Sonntag WE (1989) Increased pituitary response to somatostatin in aging male rats: relationship to somatostatin receptor number and affinity. Neuroendocrinology 50:489

Tannenbaum GS, Ling N (1984) The interrelationships of growth hormone releasing factor and somatostatin in generation of the ultradian rhythm of growth hormone secretion. Endocrinology 115:1952

Tannenbaum GS, Ling N, Brazeau P (1982) Somatostatin-28 is longer acting and more selective than somatostatin-14 on pituitary and pancreatic hormone release. Endocrinology 111:101

Tannenbaum GS, Martin J (1976) Evidence for an endogenous ultradian rhythm governing growth hormone secretion in the rat. Endocrinology 98:562

Turturro A, Hart RW (1984) DNA repair mechanisms in aging. In Schiapelli D, Migaki G (eds), Comparative biology of major age-related diseases. AR Liss, New York, pp 351–361

Utiger RD, Parker ML, Daughaday W (1962) Studies on human growth hormone, I. A radioimmunoassay for human growth hormone. J Clin Invest 41:254

Van Wyk JJ, Underwood LE, Hintz RL, et al. (1974) The somatomedins: a family of insulin-like hormones under growth hormone control. Recent Prog Hormone Res 22:259

Ward WF (1988) Enhancement by food restriction of liver protein synthesis in the aging Fisher 344 rat. J Gerontol 43:B50

Weindruch R, Walford RL (1982) Dietary restriction in mice beginning at 1 year of age: effects on lifespan and spontaneous cancer incidence. Science 215:1415

Weindruch R, Walford RL (1988) The retardation of aging and disease by dietary restriction. Charles C. Thomas, Springfield, IL

Weindruch R, Walford RL, Fligel S, Guthrie D (1986) The retardation of aging in mice by dietary restriction: longevity, cancer, immunity and lifetime energy intake. J Nutr 116:641

Wether GA, Abate M, Hogg A, et al. (1990) Localization of insulin like growth factor I mRNA in rat brain by in situ hybridization—relationship to IGF-1 receptors. Mol Endocrinol 4:773

Yu BP, Masoro EJ, Murata I, et al. (1982) Life-span study of SPF Fischer 344 males fed ad libitum or restricted diets: longevity, growth, lean body mass and disease. J Gerontol 37:130

In Vivo Cell Replication Rates and In Vitro Replication Capacity Are Modulated by Caloric Intake

Norman S. Wolf and William R. Pendergrass
University of Washington

Introduction

A number of studies have shown that reduced cell replication rates in vivo, reduced cell proliferative potential in vitro, and accumulation of cell-based pathologies are accompaniments of advancing age in several animal species (Schneider et al. 1981, Martin 1977, Cameron and Thrasher 1976, Lu et al. 1991, Peacocke and Campisi 1991, Martin 1979, Pendergrass et al. 1993). Thus, it might be said that aging in the whole animal is consonant with, and at least partly a result of, the decline of multiple cellular functions in the body's several organ systems. If one assumes that a decline in function and in self-replacement activities by the several cell systems is thus causative, then any means that preserves normal cellular replicative capacity and function should extend life span. The one method that has been repeatedly shown to accomplish life-span extension is restriction of total caloric intake (as opposed to restriction of a single component of the diet) over a significant period of the animal's life span (McCay et al. 1939, Maeda et al. 1985, Weindruch et al. 1986, Weindruch and Walford 1988). We refer to this form of dietary restriction, i.e., the proportional reduction of all dietary components other than vitamins, as caloric restriction (CR). CR in our studies was 60% of same-age ad libitum (AL) intake. We show in the present studies that this level of restriction in mice also slows cellular replication in a number of organs early in life while preserving the replication capacity of these cell systems over time. The latter is demonstrable late in life by moving the mice onto AL intake. By doing so, adequate energy intake is provided to fuel a relative increase in replication rate, which is superior to that of either long-term CR or long-term AL animals. CR also delays the loss of in vitro replicative capacity that accompanies aging in these tissues.

In the studies described here, we used male B6D2F1 and female B6C3F1 mice. As furnished to us by the National Center for Toxicological Research (NCTR, Jefferson, AR), the mean and maximum life spans (last 10%) of the AL B6D2F1 males were 138 weeks and 171 weeks, respectively. These were extended by 36% and 20%, respectively, by CR. The mean and maximum life spans for AL B6C3F1 females were 132 and 158 weeks and were extended by 36% and 30%, respectively.

We chose the thymidine analogue 5-bromo-2-deoxyuridine (BrdU) to measure the cellular replication rates in vivo because it is not reused in the body (Pera and Mattias 1976, King et al. 1982) and the required period of infusion was quite long (2 weeks). We were also equipped to use a dye-coupled antibody technique for image analysis, which was convenient and accurate. As an alternative measurement of cellular capacity, we used a well-established clone size distribution method as a sensitive measure of the in vitro replicative capacity of several cell types from non–BrdU-infused mice (Smith et al. 1978, Schneider et al. 1981, Pendergrass et al. 1993). This provided two quite different approaches to measuring the cellular replicative capacity of mice on AL and CR diets at several ages.

Materials and Methods

Animals and Conditions

(C57BL/6 × DBA/2)F1 (B6D2F1) male mice or (C57BL/6 × C3H)F1 (B6C3F1) female mice, ages 6 to 28 months, were obtained from the National Institute on Aging (NIA) rodent colony maintained by the NCTR. These are the two hybrid long-lived mouse strains maintained for aging studies by the NIA. These barrier-maintained animals are routinely tested and found to be free of all testable mouse pathogens. Calorie-restricted mice were gradually moved to 60% of the AL feed intake between 14 and 16 weeks of age and were so maintained thereafter on the basis of 60% relationship to the mean weighed intake of pelleted food by AL mice of the same age. The AL diet consisted of full access to NIH-31 pellets at all times. The CR diet consisted of NIH-31 diet with sufficient vitamin supplementation to equal the vitamin intake of AL animals of the same age. Similar conditions were maintained after their arrival at the University of Washington; a 2-week acclimation period was standard upon their arrival. The 4.35 kcal/g diet fed was NIH-31 in pellet form with CR mice receiving a preweighed pellet on the cage bottom according to their weight, thus allowing complete access to the pellet content. Refeeding (RF) diet, which consisted of abruptly moving CR mice onto the AL diet, was commenced at the ages indicated (see Refeeding Schedules, below) and maintained thereafter. Animals were on a 12-hours-on/12-hours-off light cycle, with darkness commencing at 3:30 p.m. and with feed for the CR mice given as a single pellet at 4:00 p.m. The CR mice received a pellet weighing either 3.0 or 3.5 g daily, on the basis of their weight, which was within ± 6% of the 60% target, since adult CR animal weights fell within a narrow range. All mice were individually housed in standard plastic mouse box caging with filter bonnets under standard SPF conditions.

In Vivo BrdU Delivery

The 5-bromo-2-deoxyuridine (BrdU) was dissolved in a 1% ammonia solution in distilled water and placed in 2-week continuous delivery osmotic minipumps (Alzet Corp., Palo Alto, CA) at a concentration calculated to deliver 2 µg/g body weight/hour. Adequate delivery of BrdU was assured by preliminary measurements of serum and blood cell accumulations. The pumps were placed subcutaneously at the dorsal midline, just posterior to the scapulae, under metophane anesthesia and with surgical clip wound closure.

Refeeding Schedules

Some groups of CR mice were placed on the AL diet either 2 or 6 weeks before implantation of the BrdU pump; this diet switch is referred to as refeeding (RF). The RF diet was then continued for the additional 2 weeks of pump time until sacrifice (total of either 4 or 8 weeks on RF).

Tissue Preparation

At time of sacrifice, the right kidney and a piece of midbelly skin were fixed in 10% buffered neutral formalin. Following standard dehydration and paraffin embedding the tissues were sectioned at 4 µm, placed on slides, and hydrolyzed in 1 N HCl for 30 minutes at room temperature to denature DNA, and then were neutralized with 0.1 M sodium borate for 15 minutes. The sections were blocked with 10% normal goat serum (Vector Labs, Inc., Burlingame, CA) for 1 hour, then incubated in mouse monoclonal anti-BrdU (Becton-Dickinson, Inc., San Jose, CA) diluted 1:100 in 3% bovine serum albumin (BSA) for 4 hours at room temperature. The slides were washed in phosphate-buffered saline (PBS) and incubated in biotin-conjugated goat antimouse IgG1 (Southern Biotechnology Associates, Inc., Birmingham, AL) diluted 1:300 in 3% BSA for 2 hours at room temperature. Following a PBS rinse, labeled cells in the section were detected by the presence of the enzyme marker alkaline phosphatase and associated dye substrate (Vector Labs) that combines with the biotinylated secondary antibody (techniques as described by supplier). Unlabeled cell nuclei were also counted in the tissues, which were counterstained with hematoxylin.

The bone marrow endothelial-like cells were treated differently. These cells were obtained by extruding the contents of both femurs with PBS containing 2% fetal calf serum (FCS). Following dispersion, cells were allowed 24 hours to adhere to tissue culture plastic. Following removal of nonadherent cells by washing, the adherent cells were freed with 0.5% trypsin in PBS. The newly suspended cells were exposed to magnetic beads 4.5 μm in diameter (M450, Dynal, Inc., Great Neck, NY) for 2 hours, followed by magnetic selection of cells that had ingested beads. These cells were grown in minimal essential medium (MEM)-alpha containing 15% FCS, 10^{-5} M 2-mercaptoethanol (ME), and 200 units of murine GM-CSF (gift from Amgen, Inc., Thousand Oaks, CA). The colony cell type was identified as endothelial-like by vWF polyvalent rabbit antibody (DAKO, Inc., Carpinteria, CA), with secondary goat antirabbit antibody and immunogold label (Amersham, Inc., Arlington Heights, IL) using the secondary antibody alone for controls (according to supplier's instructions). Before final plating, the cells from the donors were divided into two aliquots, one of which was exposed for 15 seconds to 150 ergs of near-ultraviolet light, wavelength 254 nm at a distance of 88 cm, with fluid depth of 2 mm (Pietryzk et al. 1985). The second aliquot was sham exposed. Both aliquots were then light-protected and plated at sufficient dilution to produce clonal growth. Control studies showed that this level of near-UV exposure to non–BrdU-treated cells of this type did not reduce clonal growth when compared to cells not so exposed (Pietrzyk et al. 1985). When clone numbers (regardless of size) were counted at the end of 10 days of incubation (7% O_2, 7% CO_2, 86% N_2), the relationship between the UV-exposed and nonexposed clone numbers represented the percentage of cells that had entered at least one S-phase while the BrdU pumps were in place in the donor animals. Cells exposed to UV are unable to replicate if they have taken BrdU into their DNA during a previous S-phase (Hagan and MacVittie 1981, Pietrzyk et al. 1985).

Cell Counts in Tissue Sections

BrdU-labeled and unlabeled cells from kidney tubules or dermal fibroblasts of intact belly skin were counted in the fixed sections (see above) using an Olympus CUE-2 Image Analysis system (Olympus Corp., Lake

Success, NY) for monitor screen projection and grid sector onlay but were enumerated by eye (found to be most accurate). The labeling index was determined by counting a minimum of 2000 cells per tissue from each mouse at 400× magnification. Ten to 15 randomized fields from each of four consecutive sections from each tissue were counted. The percentage of labeled cells/total cells counted were reported as mean rate of cell replication, with the caveat that a proportion of labeled cells might have proceeded through S-phase more than once during the 2 weeks of continuous BrdU delivery.

In Vitro Clone Size Distribution Measurements

The percentage of large clones (more than four cell doublings) among the total number of clones grown was determined using a modification of the clone size distribution method described by Smith et al. (1978). We and others have found this method to be more sensitive for cell growth differences than mass culture (Pendergrass et al. 1993). Skin from the ear pinna was used for cell culture studies because it was free of hair. Briefly, the skin from which dermal fibroblasts were to be grown was cut into 1–2 mm pieces, washed in PBS, and digested in Dulbecco's minimal essential medium (DMEM) containing 0.1% collagenase/dispase supplemented with 10% fetal bovine serum (FBS). In the case of the kidney pieces, the enzymes were omitted. In either case, the mixture was placed in a sterile bag and gently macerated by rolling on the outside of a roller bottle at 40 revolutions per hour for 45 minutes. The cell preparations were then washed with versene buffer to inactivate the proteases, and the freed cells were filtered through 20 μm nylon screen mesh to remove clumps. The cell preparations were immediately plated into 25-mL flasks containing 7 mL of cloning medium (F12 medium supplemented with 10% FBS, 50 units/mL penicillin, 50 μg/mL streptomycin, and 12.5 μg/mL amphociterin B and 10^{-4} M ME) for two days. After two days, the dermal cells were trypsinized (0.05%), transferred into 25-mL flasks (300–400 cells/flask) containing the above cloning medium, and incubated in a 7% CO_2, 7% O_2, 86% N_2 incubator at 37°C. The clones forming under these conditions were found to be fibroblasts by morphology. Kidney cells again had special handling, being cloned immediately following filtration of the primary preparation at 500 cells/25 mL flask in the above medium supplemented with 20 ng/ mL of epidermal growth factor. After 7 or 14 days the clones were fixed and stained with crystal violet, and the number of cells per clone was enumerated by eye using a dissecting microscope. The plating efficiency of the cells from all tissues was 20–40%. Clones of inappropriate morphology were not counted (<10% for either tissue). The endothelial-like

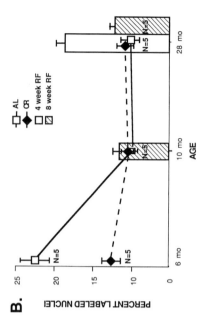

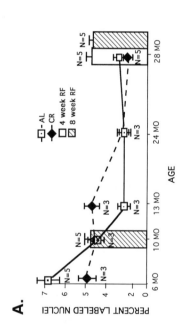

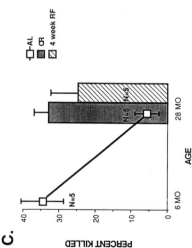

Figure 1. Comparison of the effects of diets on percentage of nuclei BrdU-labeling in B6D2F1 male mice and reported as mean rate of cell replication (MRCR) in the text. N = number of animals. Mean and SE are shown. A. Kidney tubular cells labeled in tissue sections. Significant and near-significant values are 4-week RF: 28-month RF versus 28-month AL, $p < .05$; 28-month RF versus 28-month CR, $p < .05$. 8-week RF: 28-month RF versus 28-month AL, $p < .01$; 28-month RF versus 28-month CR, $p < .05$. 6-month AL versus 6-month CR = .14; 13-month RF versus 13-month AL, $p = .06$. B. Effect of RF on dermal fibroblasts labeled in tissue sections. Twenty-eight-month-old CR mice were placed on an AL diet for either 4 or 8 weeks. 4-week RF: 28-month RF versus 28-month AL, $p < .05$; 28-month RF versus 28-month CR, $p < .05$. 8-week RF: 28-month RF versus 28-month AL or CR, N.S. 6-month AL versus 6-month CR, $p < .05$. C. Cultured bone marrow stromal endothelial-like cells from animals treated previously in vivo with BrdU. The proportion of cells that have replicated are shown by the percentage killed by near-ultraviolet irradiation (see Materials and Methods section). The RF was for a 4-week period. 6- versus 28-month AL, $p < .001$; 28-month RF versus 28-month AL, $p < .001$; 28-month RF versus 28-month CR, $p < .001$.

cells from the marrow stroma were obtained and cultured as described above after being dispersed in 35-mm flasks at 100–200 cells per flask. In this instance we were able to repeat the first experiment using B6C3F1 females with a second experiment using male B6D2F1 mice.

Statistical Analysis

Student's two-tailed t-test was used for all analyses. For in vivo studies, the percent of BrdU-positive cells (evidence of entry into at least one S-phase) was determined and reported as the mean rate of cell replication. For the clone size analysis of in vitro maximal proliferative potential, the percentage of large clones present in triplicate flasks was determined.

Results

In Vivo Measurements of Cell Replication Rates

B6D2F1 male mice were used in these studies. Shown in Figure 1A is the mean rate of cell replication (MRCR) among kidney tubular cells, as extrapolated from the BrdU-labeling percentage. The percent of labeled cells was higher for the AL than for the CR mice at 6 months of age but fell to a level even lower than that of the CR mice by 13 months. The values for both diet groups remained low through 28 months, except in the group in which the 28-month-old CR mice were placed on the RF diet for 4 weeks, where a significantly higher MRCR occurred during the last 2 weeks of the RF period. This difference remained but was diminished when measured during the last 2 weeks of the 8-week RF protocol. Essentially the same pattern is shown for dermal fibroblasts (Figure 1B) and for endothelial-like cells obtained from femoral marrow stroma (Figure 1C), although in this last study the MRCR of the CR mice was also greater than that of the AL mice at 28 months. It is notable that when CR was carried out to early midlife (at 10 months, see Figures 1A and 1B), the MRCR differences between AL and RF mice were not present.

In Vitro Measurements of Cell Growth Potential

B6C3F1 female mice were used in all but one of these studies. As seen in Figure 2A, the proliferative potential of kidney tubular epithelial cells determined by the fraction of large clones produced in vitro under optimal growth conditions was similar in 6- or 13-month-old mice on either AL or CR diets. When measured at 22 months, however, cells from the CR mice performed significantly better. This advantage was apparently lost by 30 months for this cell type. A similar pattern was present for ear skin fibroblast, as seen in Figure 2B. As seen in Figures 2C and 2D, however, bone marrow endothelial-like cells from both B6C3F1 female and B6D2F1 male CR mice displayed a superior clone size distribution as early as 13 months, which remained even at 30 months.

Discussion

When measured in vivo by the proportion of cells that have entered S-phase in several different tissue types and locations, we found that young (6-month-old) AL animals have a more rapid MRCR than those on CR and thus have accrued a greater number of cell divisions within the given cell population at this time. However, this differential disappeared at 10 months of age. If at 28 months of age the CR animals were switched to an AL diet (RF), the MRCR of the RF animal then increased to one that was more rapid than that of either AL or CR mice of the same age. This advantage was most remarkable after 4 weeks of RF status and was often still present, but reduced, after 8 weeks on RF. The response was present in tissues that do not participate in digestion or excretion that may be affected by increased food intake, i.e., dermal fibroblasts and marrow stromal endothelial-like cells. Therefore, the results suggest that the nearly lifelong condition of CR had in some manner preserved the proliferative potential in these cell populations, either through an early reduction of the number of cell doublings in populations with a limited doubling potential (Hayflick 1965) or by a limitation of gradually accruing damage, perhaps by glycation or oxidative radical reactions (Adelman et al. 1988, Krystal and Yu 1992). Supporting these concepts was the finding that the CR-to-AL diet switch, which increased MRCR when instituted late in life, usually did not alter the rate of cell replication if performed early in life, i.e., at 10 months, at which time CR had been in effect for only 6 months and the AL mice had not yet completed their decline in MRCR. In addition to the kidney studies, this finding was present for hepatocytes (data not shown).

A similar husbanding of the capacity for maximal cellular replication with CR was found in "early" old age (22 months) in our in vitro clone size distribution studies using these same tissue types. In this instance RF was unnecessary, since cells from both AL and CR animals were immediately provided with a maximal energy source when plated in the high serum-containing growth medium. The decrease at 30 months of age in clonal proliferation by the nonhematopoietic tissues (kidney epithelial cells and skin fibroblasts) from CR animals to the level of those same tissues taken from AL animals suggests that very late in life the cells protected by the

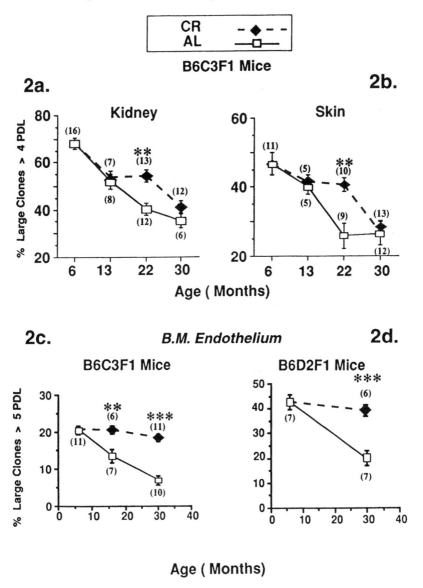

Figure 2. In panels 2a and 2b, the percentage of large clones (over 4 population doublings, abbreviated PDL) produced at 1 week by cells from kidney and skin of mice on CR and AL diets is compared as a function of age in (a) fibroblast-like cells from ear skin and (b) epithelial cells from kidney from female B6C3F1 mice (see Materials and Methods section in text). In panels 2c and 2d, the fraction of large clones (attaining over 5 PDL in 9 days) formed by bone marrow stroma-derived endothelial-like cells is shown for cells from B6C3F1 mice (2c) and B6D2F1 mice (2d). $^{**}p < .01$ and $^{***}p < .001$ for comparisons at ages shown. Number of mice is shown in parentheses. Mice were individually studied.

Dietary Restriction

CR diet have used up their reserve of cell doublings or have finally accumulated enough damage to fall to a lower level of growth capacity. An interesting exception to this later loss of potential was the bone marrow endothelial-like cells, which had a significantly higher percentage of large clones, even in CR animals at 30 months of age, as compared to AL controls. This cell type also demonstrated a beneficial effect of CR in the in vivo MRCR studies (Figure 1C). In those studies, cells from CR mice maintained a superior MRCR over those from AL mice, i.e., RF was not required. This suggests that this cell type, which undergoes replacement at a surprisingly rapid rate, is highly responsive to the conditions provided by CR. One may conclude that CR would affect studies of marrow depressive or stimulatory agents, especially those that act on the marrow stroma. In relation to our findings, it is noteworthy that Duffy et al. (this volume) have found a much higher resistance to the physiopathological effects on the cardiovascular system of certain drugs by animals on long-term CR than by AL animals. Leaky et al. (this volume) and Feuers et al. (this volume) reported that CR markedly altered the metabolism of carcinogens and intermediate metabolic processes, respectively. Our studies provide a cautionary note that agents that affect cells in active growth, or even tissues with a slow rate of cell replication, must initially be evaluated with both AL feeding and CR if CR animals are to be used in physiopathological effect and dose tolerance evaluations.

In summary, B6D2F1 mice placed on CR consisting of 60% of AL intake of the same food source lived significantly longer and displayed reduced MRCR in early life (6 months of age) in the several organs measured. Following this, the early higher MRCR in the AL animals fell to that of mice on CR by 10 months of age and was indistinguishable from it for the remainder of the life span (measured to 28 months). However, if CR mice were switched to an AL diet late in life for 4 or 8 weeks (RF), the MRCR of these animals increased, often significantly, relative to that of either the AL or CR mice of the same age. In some cases the MRCR approached that of 6-month-old AL mice. These in vivo findings were corroborated by in vitro studies in which the capacity of cells from AL or CR mice of several ages was measured for maximum clonal replication under high serum conditions. We suggest that long-term CR slows accruing damage to cells in vivo, possibly by limiting cellular responses to mitogenic stimuli and/or by reducing damage accrued during metabolic activity. These studies indicate that CR produces profound alterations in metabolism, cellular replication rates, and the accrual of cellular damages, thereby affecting life span of the animal and its expected responses to drug stimuli.

Acknowledgments

This study was supported by NIH grant AG07724. The data presented in this report have been submitted in more detail for publication elsewhere.

References

Adelman R, Saul RL, Ames BN (1988) Oxidative damage to DNA: relation to species metabolic rate and life span. Proc Natl Acad Sci U S A 85:2706–2708

Cameron IL, Thrasher JD (1976) Cell renewal and cell loss in the tissues of aging mammals. Interdisciplinary Topics Gerontol 10:108–129

Hagan MP, MacVittie TJ (1981) CFU-S kinetics observed in vivo by bromodeoxyuridine near UV-light treatment. Exp Hematol 9:123–130

Hayflick L (1965) The limited in vitro lifetime of human diploid cell strains. Exp Cell Res 37:614–636

King MT, Wild D, Gocke E, Eckhardt K (1982) 5-bromodeoxyuridine tablets with improved depot effect for analysis in vivo of sister-chromatid exchanges in bone-marrow and spermatogonial cells. Mutat Res 97:117–129

Krystal BS, Yu BP (1992) An emerging hypothesis: synergistic induction of aging by free radicals and Maillard reactions. J Gerontol Biol Sci 47:B107–B114

Lu MH, Henson WG, Turturro A, et al. (1991) Cell cycle analysis in bone marrow and kidney tissues of dietary restricted rats. Mech Ageing Dev 59:111–121

Maeda H, Gleiser CA, Masoro EJ, et al. (1985) Nutritional influences on aging of Fischer 344 rats: II. pathology. J Gerontol 40:671–688

Martin GM (1977) Cellular aging—clonal senescence. A review (part I). Am J Pathol 89:484–511

Martin GM (1979) Genetic and evolutionary aspects of aging. Fed Proc 38:1962–1967

McCay CM, Maynard LA, Sperling G, Barnes LL (1939) Retarded growth, life span, ultimate body size and age changes in the albino rat after feeding diets restricted in calories. J Nutr 18:1–13

Peacocke M, Campisi J (1991) Cellular senescence: a reflection of normal growth control, differentiation or aging? J Cell Biochem 45:147–155

Pendergrass WR, Li Y, Jiang D, Wolf NS (1993) Decrease in cellular replicative potential in "giant" mice transfected with the bovine growth hormone gene correlates to shortened life span. J Cell Physiol 156:96–103

Pera F, Mattias P (1976) Labelling of DNA and differentiated sister chromatid staining after BrdU treatment in vivo. Chromasoma 57:13–18

Pietrysk ME, Priestley GV, Wolf NS (1985) Normal cycling patterns of hematopoietic stem cell subpopulations: an assay using long-term in vivo BrdU infusion. Blood 66:1460–1462

Schneider EL, Monticone R, Smith J, et al. (1981) Skin fibroblast cultures derived from members of the Baltimore Longitudinal Study: A new resource for studies of cellular aging. Cytogenet Cell Genet 31:40–46

Smith JR, Pereira-Smith O, Schneider EL (1978) Colony size distributions as a measure of in vivo and in vitro aging. Proc Natl Acad Sci U S A 75:1353–1356

Weindruch R, Walford RL (1988) The retardation of aging and disease by dietary restriction. Charles C. Thomas, Springfield, IL, pp 231–298

Weindruch R, Walford RL, Fligiel S, Guthrie D (1986) The retardation of aging in mice by dietary restriction: longevity, cancer, immunity and lifetime energy intake. J Nutr 116:641–654

Possible Role of Apoptosis in Immune Enhancement and Disease Retardation with Dietary Restriction and/or Very Low Doses of Ionizing Radiation

Takashi Makinodan

West Los Angeles VA Medical Center
University of California at Los Angeles

S. Jill James

National Center for Toxicological Research

Introduction

Apoptosis, or physiologic cell death, is an endogenous process whereby damaged, senescent, or superfluous cells are selectively deleted within normal tissues. Agents or conditions that increase the incidence of apoptosis will protect the organism from DNA damage-inducing agents and disease by eliminating mutated or diseased cell precursors (Wyllie 1992). In this report, we propose the novel hypothesis that apoptotic cell death is enhanced in tissues of diet-restricted animals and that this mechanism may contribute to increased life span and disease retardation in this model. In our initial studies of immunologic enhancement and disease retardation with very low doses of ionizing radiation (LDR), we discovered a remarkable similarity between the effects of dietary restriction (DR) and LDR. For example, both manipulations have been independently associated with increased longevity (Lorenz 1950, Masaro 1985, Neafsey 1990), immune enhancement (Good et al. 1991, James et al. 1990), increased DNA repair (Turturro and Hart 1991, James et al. 1991), and delay or abrogation of disease progression (Good et al. 1991, Fernandes et al. 1976, James et al.

1990). Based on these considerations, we examined the independent and additive effects of DR and LDR on the progression of spontaneous autoimmune disease in the lpr/lpr mouse and spontaneous mammary tumor development in the C3H/He mouse. These two models are shown to provide indirect support for our hypothesis. In the autoimmune-prone C57BL/ 6 lpr/lpr mice, DR and LDR independently and additively enhanced immune function by promoting selective deletion of the pathologic autoreactive cells (James et al. 1990). In the C3H/He mice, DR alone was effective in retarding tumor growth and incidence; however, the combined regimen of DR and LDR additionally stimulated tumor regression (Kharazi et al. 1994). In the regressing tumors, we observed massive infiltration of $CD8^+$ cytolytic tumor-infiltrating lymphocytes, which kill tumor cells by triggering apoptosis in the tumor cells. In a third model, we provide more direct support for our hypothesis by examining the relative rates of apoptosis and proliferation in livers of B6C3F1 mice, a model prone to develop spontaneous hepatoma (James and Muskhelishvili 1994). A significant increase in the rate of apoptosis was found to be associated with a decrease in the rate of proliferation in hepatocytes from the DR B6C3F1 mice relative to ad libitum (AL) mice. We conclude with a discussion of the potential implications for caloric modulation of apoptosis in terms of the chronic rodent bioassay and in risk assessment analyses of carcinogenic or toxic compounds.

Apoptosis

Apoptosis is an evolutionarily conserved, highly regulated physiologic process whereby single cells are selectively deleted in the midst of healthy tissues. It is the homeostatic complement of mitosis in the normal maintenance of tissue size and shape (Wyllie 1992). Apoptosis occurs normally during embryogenesis, tissue cell turnover, immunologic cytotoxicity, and after withdrawal of trophic hormones or growth factors. The process does not appear to be random but preferentially eliminates senescent, preneoplastic, DNA-damaged, or superfluous cells within normal tissues (Thompson et al. 1992, Schulte-Hermann et al. 1993). The hormonal and metabolic alterations induced by DR are consistent with increased incidence of apoptosis. These alterations include 1) increased circulating levels of glucocorticoids, 2) decreased circulating levels of trophic factors such as insulin and glucose, 3) decreased basal rate of proliferation, and 4) decreased tissue concentrations of the trophic hormones estrogen, prolactin, testosterone, and growth hormone (Ruggeri et al. 1989; Ames and Shigenaga 1992).

Hypothesis

Based on the results obtained from the three experimental studies described below, it is our hypothesis that the rate of apoptotic cell death is increased in tissues of DR animals relative to AL animals. The experimental observations support a protective role for apoptosis by providing a mechanism for the deletion of diseased or DNA-damaged cells. A relative increase in the elimination of cells with potentially deleterious genetic or biochemical lesions with DR may contribute to an increased functional capacity in the remaining cells, a reduced risk of disease development, and an extension of natural life span.

Experimental Mouse Models

Spontaneous Autoimmune Disease

These studies were initiated to explore mechanisms of immune potentiation that occur with chronic in vivo exposure to very low doses of ionizing radiation. This counterintuitive paradigm of a beneficial effect of very low doses of an otherwise toxic agent has many examples in nature and is referred to as "hormesis" (Sagan 1989). Radiation hormesis in the immune system was evaluated in normal C57BL/6 mice and the congenic autoimmune-prone strain C57BL/6 lpr/lpr (James and Makinodan 1988, James et al. 1990). Mice bearing the lpr/lpr gene develop progressive lymphadenopathy and a lupus-type autoimmune syndrome. The genetic defect in the lpr model of autoimmune disease has recently been elucidated and is due to a mutation in the Fas (or APO-1) gene, which encodes a membrane antigen that is essential for T-cell negative selection to occur in the thymus (Itoh et al. 1991). As a result of this defect, immature thymocytes are resistant to Fas-activated apoptosis and exit the thymus inappropriately to accumulate peripherally in the spleen and lymph nodes (Watanabe-Fukunaga et al. 1992). The resulting splenomegaly is due to the accumulation of these cells, which are unique in that their immature phenotype lacks both CD4 and CD8 T-cell surface antigens. These pathologic cells are therefore referred to as "double-negative" (CD4$^-$-CD8$^-$) cells and can be quantified by flow cytometry.

At two months of age, C57BL/6 normal and lpr/lpr mice were randomly assigned to either the AL or DR feeding regimens. Dietary restriction was included in the experimental design as a positive control group because it is a well-defined model of immunologic enhancement (Walford et al. 1974) and of autoimmune disease retardation in autoimmune-prone mice (Good et al. 1991). The semi-purified diets were obtained commercially from

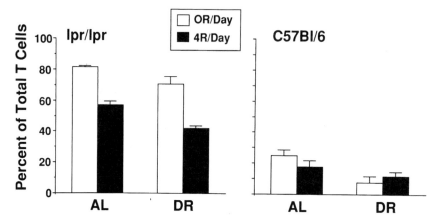

Figure 1. The independent and additive effects of dietary restriction and low dose radiation (4 rads/day) on proportion of pathologic "double-negative" splenic T cells in autoimmune-susceptible lpr/lpr and normal C57BL/6 mice (adapted from James et al. 1988 and James 1990).

Teklad Test Diets (Harlan-Sprague Dawley, Madison, WI). Dietary restriction to 70% of the AL food consumption was induced over a 3-week period in a progressive step-wise manner (10% restriction). The DR diet was supplemented with vitamins and minerals such that micronutrient intake between both groups was identical. One week after DR was established, both the AL and DR groups were subdivided into LDR and sham-LDR subgroups. The LDR groups were exposed to 0.04 Gy/day, 5 days a week, for 4 weeks. Three days after the last radiation exposure, the mice were sacrificed and their spleens were assessed for total spleen cell numbers, T-cell subsets, and mitogenic proliferative capacity as previously described in detail (James and Makinodan 1988, James et al. 1990).

The results indicated that DR alone in the normal C57BL/6 mice significantly reduced the absolute number of spleen cells and total organ weight with further reduction of splenomegaly with the combined regimen of DR and LDR (James et al. 1990). In Figure 1, a reduction in the proportion of the pathologic "double-negative" cells in the lpr/lpr mice is apparent with DR alone and augmented with combined DR/LDR. The LDR effect on the elimination of double-negative cells would appear to indicate that these cells are highly radiosensitive to apoptotic cell death. The effect of DR in decreasing the double-negative cell population and associated splenomegaly is most consistent with a DR-induced elevation of glucocorticoids, which is well known to induce apoptotic cell death in immature T cells (Kerr and Harmon 1991). Thus, DR and LDR independently and additively reduce

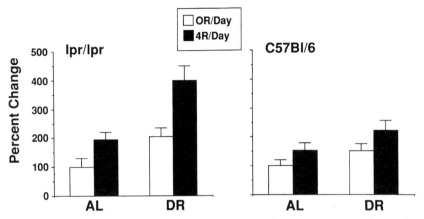

Figure 2. Effect of dietary restriction and low dose radiation on percent change in T cell mitogen-stimulated proliferative response of splenocytes of autoimmune-susceptible lpr/lpr and normal male C57BL/6 mice (adapted from James et al. 1990).

splenomegaly in autoimune-prone mice by eliminating the immunologically abnormal double-negative cells. We conclude that both interventions induce apoptosis, but by independent mechanisms (James et al. 1990). Associated with the elimination of double-negative cells was an increase in the proliferative response of spleen cells to mitogen stimulation (Figure 2) and a relative increase in immunologically normal CD4[+] and CD8[+] T lymphocytes as indicated by flow cytometric analysis (James and Makinodan 1988). Therefore, the physiologic outcome of DR or combined DR/LDR was an increase in immune responsiveness and a decrease in autoimmune manifestations. These results can be explained by the selective elimination of diseased cells by apoptosis resulting in the functional enhancement of the remaining cells as predicted by our hypothesis.

Spontaneous Mammary Carcinogenesis

The C3H/He mouse develops spontaneous mammary tumors by 12 months of age. Previous studies have shown that DR significantly delays tumor latency and tumor incidence (Fernandes et al. 1976). Since both DR and LDR enhance immune responsiveness in normal and autoimmune-prone mice (James and Makinodan 1988), we reasoned that enhanced cytolytic T-cell activity may contribute to tumor retardation in this model. At 7 months of age, before the appearance of spontaneous mammary tumors, half of the C3H/He mice were adapted to DR (70% of AL food consumption) over a 3-week period, and the other half were maintained on AL feeding. Within each dietary group, half the mice were subjected to LDR

Dietary Restriction

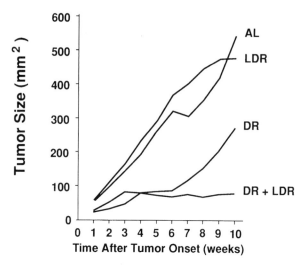

Figure 3. Effect of dietary restriction and low dose radiation on growth patterns of spontaneous mammary tumor in tumor-susceptible female C3H/He mice (adapted from Kharazi et al. 1994).

(0.04 Gy/day for 4 weeks) and the other half to sham-LDR. The mice were palpated for appearance of mammary tumors 5 days a week for the subsequent 12 months. The DR and DR/LDR groups were maintained in the DR protocol throughout the study.

Tumor growth rate was monitored weekly over a 10-week period following initial appearance of each tumor in mice of the AL, AL/LDR, DR, and DR/LDR groups (Figure 3). The results have been previously published in detail (Kharazi et al. 1994) and showed that LDR was ineffective in down-regulating the growth rate of tumors in AL mice. In contrast, DR (initiated at 7 months of age) was effective in delaying or reducing tumor growth rate for approximately 6 weeks. Thereafter, however, tumors of DR mice grew rapidly, at a rate comparable to that observed for AL and AL/LDR mice. These observations are consistent with those of other investigators (Masaro 1985). The most striking effect on tumor growth rate was the combined regimen of DR/LDR where the mean tumor growth rate approached zero. The reduced mean growth rate in this group reflects the fact that a significant percentage of tumors in this group exhibited regression. In Figure 4, the percent of tumors exhibiting growth and regression within each group is presented. Perhaps the most interesting observation from this study was the high percentage of tumors observed to undergo regression with the combined regimen of DR/LDR. Malignant tissue is well known to exhibit a high apoptotic cell

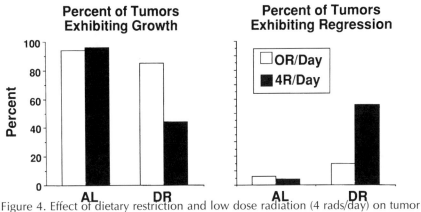

Figure 4. Effect of dietary restriction and low dose radiation (4 rads/day) on tumor growth and regeneration (adapted from Kharazi et al. 1994).

death rate (Schulte-Hermann et al. 1988). Tumor growth rate reflects the balance between tumor cell proliferation and tumor cell death. In rapidly growing tumors, the cell proliferation rate far exceeds the cell death rate. The slow growth of tumors reflects a proliferation rate that only marginally exceeds the apoptosis rate. Tumor regression occurs when the death rate exceeds the proliferation rate. In this present study, immunohistologic examination of the regressing tumors revealed large clusters of CD8+ tumor-infiltrating lymphocytes (Kharazi et al. 1994). Both the slow growth rate and frank tumor regression in the DR and DR/LDR groups are consistent with an enhanced apoptotic rate with DR and provide indirect support for our hypothesis.

Spontaneous Hepatocarcinogenesis

The indirect evidence in our previous studies suggested that DR may retard autoimmune disease and spontaneous mammary tumorigenesis in mice by enhancing selective cell deletion through apoptosis. To more directly pursue this possibility, we quantified the spontaneous apoptotic rate in liver sections from DR and AL 12-month-old male B6C3F1 mice, a murine strain known to develop a high incidence of spontaneous liver tumors by 18 months of age. In this study, half of the mice were randomly allocated to receive 60% of the food consumption of the AL group, which was instituted in a step-wise procedure over a 2-week period. The animals were fed an NIH-31 open-formula diet (Purina Mills, Inc., Richmond, IN) and were maintained as part of the Program on Caloric Restriction at the National Center for Toxicological Research (Jefferson, AR) as previously

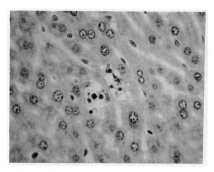 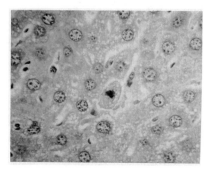

Figure 5. *Left:* Representative example of in situ end labeling (ISEL) immun-ohistochemical staining of an apoptotic body in a liver section from a diet-restricted B6C3F1 mouse counterstained with methyl green (× 345). *Right:* Hepatocytes exhibiting necrotic morphology in liver section from ad libitum B6C3F1 mouse. Note cytoplasmic swelling and vacuoles and nuclear pycnosis (× 345).

described (Witt et al. 1991). The restricted diet was supplemented with vitamins to ensure equal micronutrient consumption between groups. At 12, 18, 24, 30, and 36 months of age, 15 mice from each group were fasted for 24 hours before sacrifice at 10:00 a.m. The fasting procedure and timing of sacrifice was essential to assure that animals from both dietary groups were metabolically and hormonally synchronized in terms of food-induced circadian rhythms. All animals were examined grossly and microscopically for hepatocellular adenomas and carcinomas. The incidence of spontaneous hepatic neoplasms as a function of age and diet in these mice has been previously published (Witt et al. 1991). At 12 months of age, neither group had developed tumors; however, between 18 and 36 months, the AL mice developed consistently more tumors, with a 53% incidence at 36 months compared to no tumors in the DR group. The 12-month-old age group was selected for immunohistochemical analysis because it represents a preneoplastic stage at which evaluation of cell proliferation and cell death rates may provide predictive indices for subsequent tumor development.

Formalin-fixed paraffin-embedded liver sections from 12-month-old AL and DR mice (14/group) were processed for in situ end labeling (ISEL) of 3'OH DNA strand breaks localized in apoptotic bodies (ABs) using the Apoptag detection system (Oncor, Gaithersburg, MD). A representative liver section demonstrating the utility the ISEL technique in identifying ABs is presented in Figure 5A. Approximately 50,000 hepatocytes per animal were counted, and the rate of apoptosis was expressed as the percent incidence per hour (James and Muskelishvili 1994) using the formula proposed by Bursch et al. (1990). Proliferating hepatocytes in S phase were quantified by immunohistochemical analysis of proliferating cell nuclear antigen (PCNA) using a standard biotin-streptavidin peroxidase detection system. Only cells with dark brown stained nuclei and clear cytoplasm were scored as PCNA⁺s cells from approximately 50,000 hepatocytes (200 microscopic fields) per mouse. The proportion of proliferating cells was expressed as the relative incidence of PCNA⁺ cells/100 hepatocytes. The proliferation per hour was obtained by dividing the percent of incidence by the duration of the S phase in the mouse (~8 hours). Analysis for statistically significant differences between means was performed by using the nonparametric Mann-Whitney U-test and Student's t-test.

In Table 1, the percent incidence of ABs and S phase cells and the relative rates per hour in livers of AL and DR mice are presented ($n = 14$/group). In Figure 6, the relationship between the rates of cell proliferation and apoptosis in livers of DR and AL mice is presented and related to balanced tissue homeostasis (when cell death = proliferation). The error bars around each point represent the potential variation ($\pm$ 1 hr) in the

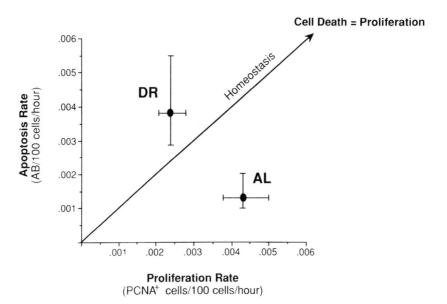

Figure 6. The relationship between rates of cell proliferation and cell death measured in livers of diet-restricted (DR) and ad libitum (AL) B6C3F1 mice as related to balanced tissue homeostasis (when cell death = cell proliferation). Presented in this manner, it is apparent that caloric intake can alter the balance between cell death and proliferation in a direction consistent with cancer promotion in AL mice and cancer protection in the DR mice. Error bars represent uncertainty in the estimate of measurement duration ($\pm$ 1 hour).

Table 1. Rates of apoptosis and proliferation in livers of 12-month-old ad libitum (AL) and diet-restricted (DR) B6C3F1 mice

Diet group	n	Apoptosis[a]		Proliferation[a]	
		Incidence AB/100 nuclei	Rate[b] AB/100 nuclei/hr	Incidence PCNA+ nuclei/100 cells	Rate[b] PCNA+ nuclei/100 cells/hr
AL	14	0.008 ± 0.0006	0.0013 ± 0.0001	0.0300 ± 0.0014	0.0043 ± 0.0002
DR	14	0.023[b] ± 0.0006	0.0038[*] ± 0.0001	0.0170[**] ± 0.0001	0.0024[**] ± 0.0001

[a]Mean values ± SEM.
[b]Based on a mean duration of the histological stage of apoptosis (ABs) of 3 hours (Bursch et al. 1990) and a mean duration of S phase of 7 hours (Rotstein et al. 1988).
[*]$p < 0.01$ relative to AL.
[**]$p < 0.05$ relative to AL.

estimate of the mean S phase duration (Rotstein et al. 1986) and the mean duration of apoptotic bodies (Bursch et al. 1990). The data indicate that a significant imbalance exists in the rates of apoptosis and proliferation in livers of the AL mice, which subsequently developed a high incidence of spontaneous hepatoma as reported by Witt et al. (1991). Within the livers of the DR mice, the rates of apoptosis and proliferation were more closely balanced and associated with low incidence of spontaneous hepatoma over a 36-month period. When the comparison is made between diet groups, the proliferation rate was increased in the AL mice relative to DR mice, and the rate of apoptosis was decreased ($p < 0.05$ and $p < 0.01$, respectively). An increase in cell proliferation associated with a decrease in apoptosis would be consistent with aspects of tumor promotion. Despite differences in proliferation and apoptotic rates, tissue size homeostasis was remarkably constant in both diet groups as indicated by constant liver weights and liver weight body weight with age (Table 2). Although difficult to quantify, single cells exhibiting necrotic morphology, such as swelling, blebbing of cytoplasm, and nuclear pycnosis, were histologically visible in liver sections from the AL mice (Figure 5B) but were nonexistent in livers of DR mice (James and Muskhelishvili 1994). Therefore, we postulate that low-level or single-cell necrotic cell death occurs in livers of AL mice such that total cell death and proliferation are in approximate balance. An increase in necrotic cell death in AL mice would explain the observed higher proliferation rate in terms of a regenerative proliferative stimulus not present or necessary in DR mice.

In summary, at 12 months of age, before the development of spontaneous hepatic neoplasms, an imbalance between rates of proliferation and apoptosis was apparent in livers of the AL mice. These data provide experimental evidence for recent hypotheses that suggest that dysregulation

Table 2. Changes in liver weights and body weights of B6C3F1 mice with age and diet[a]

Age/diet group	Body weight (g)	Liver weight (g)	Liver weight/g body weight
12 months			
AL	38.3 ± 0.29	1.628 ± 0.018	0.042 ± 0.0002
DR	26.5 ± 0.14	1.070 ± 0.064	0.040 ± 0.0001
18 months			
AL	37.9 ± 0.33	1.666 ± 0.013	0.044 ± 0.0002
DR	26.1 ± 0.14	1.142 ± 0.006	0.043 ± 0.0002
24 months			
AL	39.3 ± 0.32	1.593 ± 0.018	0.041 ± 0.0002
DR	27.4 ± 0.17	1.188 ± 0.013	0.043 ± 0.0003
30 months			
AL	32.3 ± 0.39	1.477 ± 0.020	0.045 ± 0.0002
DR	23.5 ± 0.10	1.059 ± 0.007	0.045 ± 0.0003

AL = ad libitum, DR = diet-restricted.
[a]Values are the means ± SEM in non-tumor-bearing mice.

in the homeostatic balance between proliferation and apoptosis contributes to the progression of multistage carcinogenesis (McDonnell 1993). Relative to the AL mice, livers of DR mice exhibited an increased rate of apoptosis, a reduced rate of proliferation, and decreased subsequent hepatoma development. Chronic DR is well known to reduce basal levels of trophic hormones such as insulin, estrogen, testosterone, and growth hormone (Ruggeri et al. 1989). Reduced availability of trophic factors has been shown to increase the incidence of apoptosis (Collins et al. 1994) and may provide a partial explanation for the increased spontaneous apoptosis with DR. Taken together, the results suggest that caloric intake modulates the balance between apoptosis and proliferation in a direction consistent with a cancer-promoting effect in AL mice and a cancer-protective effect in DR mice.

Conclusions

Tissue Size Homeostasis

The homeostatic maintenance of cell populations in normal tissues is the result of a highly regulated balance between cell proliferation and cell death. Although the focus in cancer research has been on understanding the dysregulation of physiologic signals and mechanisms that control cell proliferation, recent evidence suggests that knowledge of apoptosis or physiologic cell death may be of equal importance in elucidating multi-stage carcinogenesis (Moolgavkar and Luebeck 1992). Apoptotic cell death, as opposed to necrotic cell death, provides a protective mechanism by removing senescent, DNA-damaged, or diseased cells that could potentially interfere with normal function or lead to neoplastic transformation (Thompson et al. 1992). The relative efficiency or dysfunction of the cell death program would, therefore, have a direct impact on the risk or susceptibility to both degenerative and neoplastic diseases. Abnormalities in cell death can promote cancer development by allowing the "inappropriate survival" of DNA-damaged or mutated cells or by allowing cell proliferation rates to exceed cell death rates, leading to clonal expansion of preneoplastic cells. Several lines of investigation have recently suggested that initiated cells may be more sensitive to apoptosis than normal cells (Schulte-Hermann et al. 1993). Thus, agents or conditions that promote apoptosis would delay or prevent the development of multistage carcinogenesis. We hypothesized that the hormonal and metabolic alterations that occur with DR may increase the rate of apoptosis and enhance elimination of potentially neoplastic cells. An increase in the spontaneous level of apoptotic cell death may contribute to the mechanisms of disease resistance and prolongation of life span by promoting natural selection at the cellular level.

Relevance to Optimizing the Bioassay: Modulation of Apoptosis and Proliferation by Level of Caloric (Energy) Intake

A major challenge in attempting to optimize the rodent carcinogen-testing chronic bioassay is identifying quantitative determinants to maximize carcinogen sensitivity while minimizing spontaneous tumorigenesis. Parameters that modulate the homeostatic maintenance of cell numbers may offer mechanistic determinants that can be used to optimize the signal-to-noise ratio in the chronic bioassay. During normal cell turnover kinetics, the rate of apoptotic cell death is equal to the rate of cell proliferation. Under pathologic conditions, such as exposure to genotoxic and nongenotoxic carcinogens, an imbalance between these rates may be predictive of subsequent risk of tumorigenesis. Accordingly, it could be predicted that a relatively high level of endogenous proliferation would promote increased sensitivity to the test carcinogen because of the well-known promotional aspects of increased proliferation. However, a higher rate of proliferation would for the same reasons tend to promote a high background of spontaneous tumorigenesis, which would be undesirable in the design of rodent carcinogen testing. Under pathologic conditions, an increased rate of apoptosis can provide a protective mechanism by eliminating potentially neoplastic cells. Conversely, agents or conditions that inhibit apoptosis would be expected to enhance neoplastic transformation. Thus, although a low rate of apoptosis in the rodent bioassay would be predicted to increase sensitivity to carcinogens, it would also increase the background level of spontaneous carcinogenesis.

As presented in Figure 6, our data suggest that rates of apoptosis and proliferation may be modulated by caloric intake. Plotted in this manner, hypoplastic tissues would be expected to fall above the region of homeostasis, and hyperplastic tissues would tend to fall below the region of homeostasis (McDonnell 1993). Using this approach, it should be possible to use measurements of the relative rates of apoptosis and proliferation to determine the appropriate level of caloric intake that would increase sensitivity (increase proliferation) and decrease background neoplasms (increase apoptosis). The observed imbalance between proliferation and apoptosis with AL feeding (enhanced proliferation and reduced apoptosis) would predict a model with increased sensitivity to potential carcinogens accompanied by a high background of spontaneous neoplasms relative to DR animals. This prediction is consistent with experimental observations. In the DR mice, rates of proliferation and apoptosis were in approximate balance, with apoptosis slightly exceeding proliferation rate. Thus, it would be predicted that 40% DR would reduce "noise" or background level of spontaneous neoplasms; however, sensitivity to carcinogen testing would also be reduced. Using this approach with different levels of caloric intake, it should be possible to predict the level of caloric intake that would be "optimal" for the purpose of testing both genotoxic and nongenotoxic carcinogens.

Acknowledgments

The authors gratefully acknowledge the significant contibutions of Dr. Levan Muskhelishvili for immunohistochemical analyses of apoptotic bodies and PCNA$^+$ cells and for data interpretation. This work was supported by VA Medical Research Funds and an American Cancer Society Grant (SJJ, #CN-76B).

References

Ames BN, Shigenada MK (1992) Oxidants are a major contributor to aging. Ann N Y Acad Sci 663:85

Bursch W, Paffe S, Putz B, et al. (1990) Determination of the length of histological steps of apoptosis in normal liver and in altered hepatic foci of rats. Carcinogenesis 11:847–853

Collins MKL, Perkins GR, Rodriguez-Tarduchy G, et al. (1994) Growth factors as survival factors: regulation of apoptosis. Bioessays 16:133–138

Fernandes G, Yunis EJ, Good RA (1976) Suppression of adenocarcinoma by the immunological consequences of caloric restriction. Nature 263:504

Good RA, Lorenz E, Engelman RW, and Day NK (1991) Chronic energy intake restriction: influence on longevity, autoimmunity and cancer in autoimmune-prone mice. In Fishbein L (ed), Biological Effects of Dietary Restriction. Springer-Verlag, New York, pp 147–156

Itoh N, Yonehara S, Ishii A, et al. (1991) The polypeptide encoded by the cDNA of human cell surface antigen *fas* can modulate apoptosis. Cell 66:2332

James SJ, Enger SM, Peterson WJ, et al. (1990) Immune potentiation after fractionated exposure to very low doses of ionizing radiation and/or caloric restriction in autoimmune-prone and normal C57B1/6 mice. Clin Immunolog Immunopathol 55:427

James SJ, Enger SM, Makinodan T (1991) DNA strand breaks and DNA repair response in lymphocytes after chronic in vivo exposure to very low doses of ionizing radiation in mice. Mutat Res 249:255

James SJ, Makinodan T (1988) T cell potentiation in normal and autoimmune-prone mice after extended exposure to low doses of ionizing radiation and/or caloric restriction. Int J Radiat Biol 53:137

James SJ, Muskhelishvili L (1994) Rates of apoptosis and proliferation vary with caloric intake and may influence incidence of spontaneous hepatoma in B6C3F1 mice. Cancer Res 54:5508–5510.

Kerr JFR, Harmon BV (1991) Definition and incidence of apoptosis: an historical perspective. In Tomei DL, Cope FO (eds), Apoptosis: the molecular basis of cell death. Cold Spring Harbor Laboratory Press, Plainview, NY, pp. 5–29

Kharazi AI, James SJ, Taylor JMG, et al. (1994) Combined chronic low dose radiation-caloric restriction: a model for regression of spontaneous mammary tumor. Int J Radia Oncol Biol Phys 28:641

Lorenz E (1950) Some biologic effects of long continued irradiation. Am J Roentgenol Radium Ther Nucl Med 63:176

Masaro EJ (1985) Nutrition and aging—a current assessment. J Nutr 115:842

Moolgavkar SH, Luebeck EG (1992) Risk assessment of non-genotoxic carcinogens. Toxicol Lett 64/65:631

McDonnell TJ (1993) Cell division versus cell death: a functional model of multistep neoplasia. Molecular Carcinog 8:209–213

Neafsey PJ (1990) Longevity hormesis: a review. Mech Ageing Dev 51:1

Rotstein J, Sarma DSR, Farber E (1986) Sequential alterations in growth control and cell dynamics of rat hepatocytes in early precancerous steps in hepatocarcinogenesis. Cancer Res 51:67–73

Ruggeri BA, Klurfield DM, Kritchevsky D, Furlanetto RW (1989) Caloric restriction and 7,12 dimethylbenzanthracene-induced mammary tumor growth in rats: alterations in circulating insulin, insulin-like growth factors I and II, and epidermal growth factor. Cancer Res 49:4130–4134

Sagan LA (1989) On radiation, paradigms, and hormesis. Science 245:574

Schulte-Hermann R, Bursch W, Fesus L, Kraupp B (1988) Cell death by apoptosis in normal, preneoplastic and neoplastic tissue. In Feo F, Pani P, Columbano A, Garcia R (eds), Chemical carcinogenesis: models and mechanisms. Plenum Press, New York, pp 263–274

Schulte-Hermann R, Bursch W, Kraupp-Grasl B, et al. (1993) Cell proliferation and apoptosis in normal liver and preneoplastic foci. Environ Health Perspect 101:87–90

Thompson HJ, Strange R, Schedin PJ (1992) Apoptosis in the genesis and prevention of cancer. Cancer Epidemiol Biomarkers Prev 1:597–602

Turturro A, Hart RW (1991) Caloric restriction and its effects on molecular parameters especially DNA repair. In Fishbein L (ed), Biological effect of dietary restriction. ILSI Monographs. Springer Verlag, New York, pp 185–190

Walford RL, Liu RK, Gerbase-DeLima MN, et al. (1974) Longterm dietary restriction and immune function in mice: response to sheep red blood cells and to mitogenic agents. Mech Ageing Dev 2:447

Watanabe-Fukunaga R, Brannan CI, Copeland NG, et al. (1992) Lympho-proliferation disorder in mice explained by defects in fas antigen that mediates apoptosis. Nature 356:314

Witt WM, Sheldon WG, Thurman JD (1991) Pathological end points in dietary restricted rodents—Fischer 344 rats and B6C3F1 mice. In Fishbein L (ed), Biological effects of dietary restriction. ILSI Monographs. Springer-Verlag, New York, pp 73–86

Wyllie AH (1992) Apoptosis and the regulation of cell numbers in normal and neoplastic tissues: an overview. Cancer Metastasis Rev 11:95.

Effects of Dietary Restriction on Endogenous and Exogenous Retrovirus Expression

Donna M. Murasko and Deneen R. Stewart
Medical College of Pennsylvania

Kenneth J. Blank
Hahnemann University School of Medicine

Background

An increased incidence of lymphomas with advancing age in most rodent species has been recognized (Turturro et al. 1994), although the etiology of the development of these tumors remains unknown. However, development of lymphomas in certain inbred strains of mice early in life has been shown to be dependent on the expression of endogenous retroviruses (Weiss et al. 1982). Most noticeably in the AKR mouse, ecotropic endogenous retrovirus expression begins at 17 days of gestation and continues throughout the mouse's life, with lymphomas developing at about 11 months of age. In strains of mice exhibiting a low incidence of early lymphoma development, however, the expression of infectious, endogenous virus has been reported to be sporadic in young animals (Copeland et al. 1983) and has not been extensively investigated in aged animals. The lack of endogenous virus production in young mice appears to be the result of small defects in the endogenous, ecotropic retrovirus genome (Weiss et al. 1982). In the case of DBA/2 mice, this defect has been localized to a point mutation in the *gag* gene of the virus that encodes the p30 core molecule (Copeland et al. 1984, 1988). Since endogenous retroviruses are known to be the etiologic cause of murine lymphomas, it is possible that similar retroviruses are important in lymphoma development in aged mice. For virus-induced leukemogenesis to occur in aged mice, however, it may first be necessary for endogenous retrovirus production to occur at some point during the life of these mice.

Since dietary restriction (DR) implemented at weaning extends both median and maximum life spans of rodents (Yu et al. 1989) and decreases or at least postpones development of lymphomas (Turturro et al. 1994), the effect of DR on endogenous retrovirus needed to be explored. In addition to endogenous retrovirus expression, there are few reports concerning the ability of aged mice to support exogenous infection with retroviruses. One study suggested that aged mice may allow more extensive proliferation of Friend virus leukemia (Cinader et al. 1987); however, this report has not been extended by other investigators. It has been reported (Fernandez et al. 1992) that DR can limit pathogenesis and disease induced by the murine retrovirus LP-BM5. The effect of DR on the virus itself has not been investigated, however.

The purpose of our studies was to explore the effect of age on the expression of both endogenous and exogenous retroviruses and to determine whether DR postponed any age-associated changes in that expression.

Materials and Methods

Mice

DBA/2 and (C57BL/6 × DBA/2)F1 mice were obtained from the National Center for Toxicological Research (NCTR, Jefferson, AR). BALB/c mice were obtained from the Jackson Laboratory (Bar Harbor, ME). Mice were maintained individually in micro-isolator units in an environmentally controlled animal facility according to AAALAC regulations. Autoclaved NIH-31 food and water were provided according to dietary constraints. All animal handling was performed in laminar flow hoods. Tail snips were taken from the mice 5 days after their arrival at the animal facility. Tail and spleen samples from C57BL/6 mice were obtained from Dr. William Sonntag (Bowman Gray School of Medicine). Dr. Sonntag originally obtained these mice from the NCTR and subsequently provided us with frozen tails and spleens.

DR was initiated at weaning as outlined in Thurman et al. (1994). Basically, through a sequential and increasing reduction in total calories while maintaining all appropriate vitamins and minerals, mice were stabilized on the 60% DR regimen, i.e., 60% of ad libitum (AL) intake, by four months of age. Mice were maintained on the DR regimen by providing weighed amounts of food between 8–9 a.m. each morning.

Cells

SC-1 cells (Rowe et al. 1970) obtained from the American Type Culture Collection (Rockville, MD) were maintained in Dulbecco's Eagle's medium (DMEM), supplemented with 10% fetal calf serum (FCS), 2 mM L-glutamine, and 50 μg/ml Gentamicin.

Virus

NB-tropic Friend murine leukemic virus (FV), originally obtained from Dr. Frank Lilly (Albert Einstein College of Medicine), was obtained from the spleens of 2- to 3-month-old CBA/Ca mice that were infected with undiluted FV. Spleens were collected at 9 days post-infection and a 10% weight/volume (w/v) homogenate was prepared. The titer, as determined by spleen focus forming units (SFFU) assay (Axelrod and Steeves 1964), in 2-month-old DBA/2 mice was 1.2×10^6/mL.

Spleen Weight and Spleen Focus Forming Units (SFFU)

Mice were infected intravenously with various dilutions of FV. Spleens were removed, weighed, and placed in Bouin's solution to visualize virus-induced foci at 9 days post-infection. Virus titers were determined by counting the number of spleen foci with the unaided eye and were expressed as SFFU per spleen. One SFFU is the amount of FV required to induce an average of one focus/spleen. The amount of SFFU reflects not only the amount of virus but the ability of the host cells to proliferate in response to virus.

The SFFU assay also was performed to assess the effect of age and diet on the amount of spleen focus forming virus (SFFV) produced after inoculation of a standard preparation of FV. Spleen homogenates from FV-infected mice are injected into susceptible young mice. Comparing the amount of SFFU in the various preparations, therefore, reflects only the amount of virus inoculated rather than variations in host control. In the experiments described, 8-month and 25-month-old (C57BL/6 × DBA/2)F1 mice were infected with a 10^{-2} dilution of FV. Spleens were removed at 9 days-post infection, and a 10% w/v homogenate was prepared. The homogenate was clarified by centrifugation at 1xg at 4°C, diluted, and injected intravenously into BALB/c mice. Spleens from the BALB/c mice were removed at 9 days post-infection, weighed, and placed in Bouin's solution. SFFU/spleen was determined as described earlier.

Preparation of Tails and Spleens

Spleens and tails were homogenized in 1 mL of cold DMEM media supplemented with 10% FCS, 2 mM L-glutamine and gentamicin. Samples were clarified by centrifugation at 1xg for 10 minutes at 4°C, and 100 μl was placed on subconfluent SC-1 cells (1 × 10³ cell per well) in a 24-well plate. Cells were observed daily, and supernatants were collected from the wells when the cells were 80% confluent.

Reverse Transcriptase Activity

The presence of reverse transcriptase (RT) activity in the supernatants of the infected SC-1 cells was detected by the production of ³²P-labeled double-stranded DNA (Taylor et al. 1976). Each sample was assayed in triplicate. Positive controls (supernatants from FV-infected SC-1 cells) and negative controls (supernatants from uninfected SC-1 cells) were run with each assay. Activity was detected by both beta scintillation counting and autoradiography. Results were expressed in CPM/g of spleen or tail. A sample was considered positive when the counts were 50% above the negative control count after the background was subtracted.

Statistics

Statistically significant p differences (< 0.05) were determined using analysis of variance and the Student t-test, as appropriate.

Results

Effect of Age and Dietary Restriction on Infectious Endogenous Retrovirus Expression

Tail snips and spleen samples were obtained from two different strains of mice, DBA/2 and C57BL/6. To determine whether these samples contained replication-competent endogenous retrovirus, homogenates of the samples were placed on SC-1 cells. After 5 days, the supernatants from the SC-1 cells were assessed for the presence of RT activity. As shown in Figure 1, there was RT activity in supernatants of tail homogenates of DBA/2 mice, demonstrating that progeny infectious, endogenous ecotropic virus was being shed by the infected SC-1 cells. By 12 months of age 65% of tail samples of AL DBA/2 mice demonstrated RT activity. The level of expression in tail samples of AL DBA/2 mice increased to 100% by 24 months of age. This expression was postponed by DR, with only 30% of tail samples demonstrating RT activity at 12 months. In order to demonstrate that this phenomenon was not limited to DBA/2 mice, the presence of infectious, ecotropic virus was assessed in C57BL/6 mice. Similar to the DBA/2 mice, there was an age-associated increase in the expression of endogenous retroviruses in both tail and spleen cell homogenates (Figure

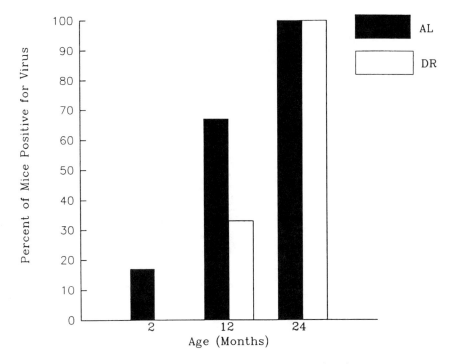

Figure 1. Effect of dietary restriction (DR) on expression of endogenous ecotropic retrovirus in DBA/2 mice. Results are expressed as the percentage of mice positive for virus. Tails from ad libitum (AL) and DR DBA/2 mice at the stated ages were homogenized and assayed for reverse transcriptase activity. Number of mice per group varied from 5 to 9.

2). Even at 28 months of age, however, only 50–60% of the spleens of C57BL/6 mice demonstrated infectious retrovirus activity. DR diminished this increase in virus expression, with only 10–20% of the mice demonstrating RT activity in splenocyte homogenates by 28 months of age. It is interesting to note that spleen (Figure 2, panel B) appears to be a more sensitive indicator than tail (Figure 2, panel A) of expression of infectious endogenous retroviruses.

Association of Endogenous Infectious Retrovirus with Late-Onset Lymphomas

Lymphomas were obtained from two aged AL DBA/2 mice (20 and 24 months of age). Homogenates of the lymphomas were assessed for RT activity. As a control, spleens of age-matched AL animals were harvested for assessment of the retrovirus activity level that would normally be present

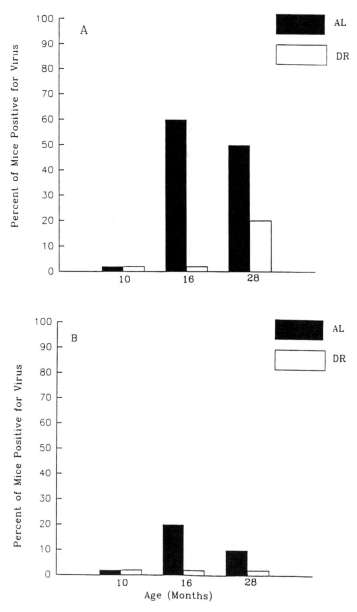

Figure 2. Effect of dietary restriction (DR) on expression of endogenous ecotropic retrovirus in C57BL/6 mice. Results are expressed as the percentage of mice positive for virus. Tails (A) and spleens (B) from ad libitum (AL) and DR C57BL/6 mice at the stated ages were homogenized and assayed for reverse transcriptase activity. Number of mice per group ranged from 5 to 10. Tails and spleens were obtained from the same mice.

Cell Line Virus

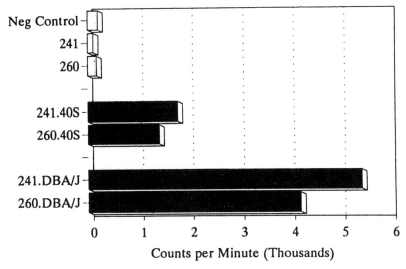

Figure 3. Detection of infectious retrovirus in lymphomas of aged ad libitum DBA/2 mice. Homogenates of spleens and lymphomas from two separate DBA/2 mice were assayed for reverse transcriptase activity. 20-month-old: 260; 24-month-old: 241. S = spleen samples. DBA/J = lymphoma samples.

in animals of that age. As shown in Figure 3, although there was RT activity in the spleens of both aged mice, the level of RT expression in the lymphomas of the aged DBA/2 mice was considerably higher.

The cell-free homogenate was then inoculated into sublethally irradiated BALB/c-H-2^b (BALB.B) mice. Within 3 months of inoculation, all of the mice (3/3) inoculated with the tumor homogenate demonstrated splenomegaly. The cells obtained from these enlarged spleens were assessed for major histocompatibility antigen expression by staining with monoclonal antibody and assessment by flow cytometry to determine whether the tumor was of host or donor origin. Consistent with the hypothesis that the tumor was induced by inoculation of the homogenate, the induced lymphoma was H-2^b, homologous to the BALB.B recipient mice, rather than being H-2^d, like the DBA/2 donors.

Effect of Age and Dietary Restriction on Exogenous Retrovirus Infection

FV was used to assess the effect of age and DR on exogenous retrovirus expression. FV is a complex of two retroviruses. One, a spleen focus forming virus (SFFV), is an acute transforming virus that is replication-defective and induces formation of foci of transformed cells on the surface of

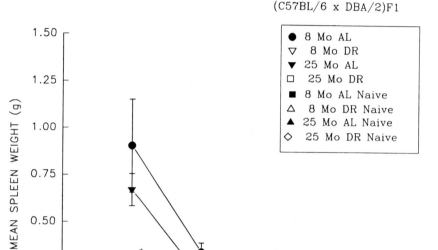

Figure 4. Effect of dietary restriction (DR) on spleen weight in infected (C57BL/6 × DBA/2)F1 mice. Ad libitum (AL) and DR (C57BL/6 × DBA/2)F1 mice at 8 and 25 months of age were injected intravenously with 0.2 mL NB-tropic Friend murine leukemia virus (FV) at varying dilutions. Spleens were removed and weighed at 9 days post-infection. Each group contained three mice. * = statistical significance ($p < 0.05$) between 8-month-old AL FV-infected mice and indicated groups within the dilution as determined by ANOVA followed by a Scheffe post hoc test.

the infected mouse's spleen; the other, F-MuLV, is a replication-competent "helper" virus necessary for replication of SFFV. As shown in Figure 4, the size of the spleen in response to FV inoculation was considerably smaller in aged compared to young AL (C57BL/6 × DBA/2) F1 mice. Similarly, the extent of FV infection, as expressed in SFFU per spleen, was decreased in the aged versus young mice (young => $1.0 × 10^5$; aged = $8.5 × 10^3$) (Figure 5).

It was hypothesized that DR would allow a level of virus replication in 24-month-old DR mice that was comparable to that observed in young AL mice. Contrary to expectations, splenomegaly in aged FV-infected mice maintained under DR was comparable to the splenomegaly in aged AL mice. In fact, mice maintained under DR, regardless of age, demonstrated little increase in spleen weight after inoculation with FV (Figure 4). In

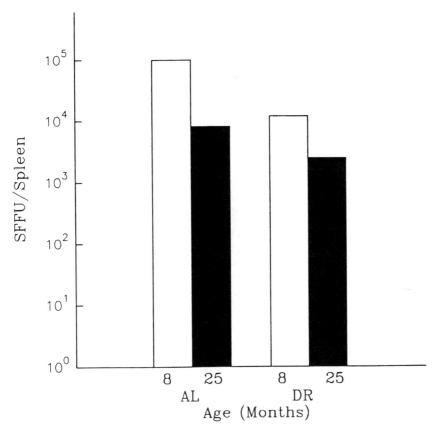

Figure 5. Effect of dietary Restriction (DR) on spleen focus forming units (SFFUs). Ad libitum (AL) and DR (C57BL/6 × DBA/2)F1 mice at 8 and 25 months of age were injected with 0.2 mL of NB-tropic Friend murine leukemia virus at varying dilutions. Spleens were placed in Bouins solution for foci determination at 9 days post-infection. The values represent the average number of foci at the dilution that is countable by the unaided eye. Number of mice per group: four to six.

addition, similar to aged AL mice, young and aged mice maintained under DR exhibited fewer SFFU after FV infection than did young AL mice (Figure 5).

Since both spleen weight and SFFU reflect the extent of FV-induced disease and thus host response to the virus, it was necessary to determine the effect of age and DR on FV replication alone. This was assessed by quantitating RT activity in supernatants of SC-1 cells infected with homogenates of spleens from mice inoculated with FV. All four groups of mice, regardless of age or diet, demonstrated similar levels of RT activity per spleen (Figure 6). This result suggested that while FV disease was

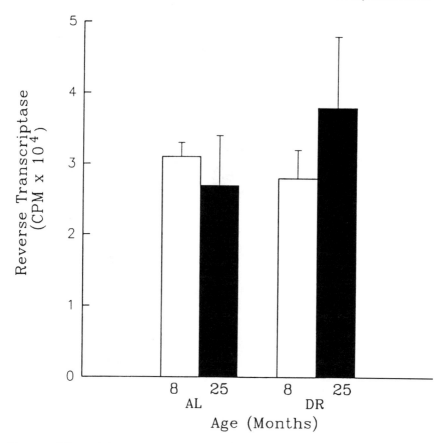

Figure 6. Effect of dietary restriction (DR) on reverse transcriptase (RT) in spleens of (C57BL/6 × DBA/2)F1 mice. Ad libitum (AL) and DR (C57BL/6 × DBA/2)F1 mice at 8 and 25 months of age were injected with 0.2 mL of a 10^{-2} dilution of NB-tropic Friend murine leukemia virus. Spleens were removed, quick frozen, and assayed for RT activity at 9 days post-infection. Each group contained three mice. Differences among groups were not statistically significant ($p > 0.05$) as determined by ANOVA.

limited by age and DR, the level of virus expression was comparable in all four groups. To further address this discrepancy, the amount of infectious SFFV in FV-infected animals was determined in a secondary spleen focus-forming virus assay. Homogenates from spleens of mice of each of the four groups were inoculated into susceptible young BALB/c mice, and the amount of SFFV in each group was determined. While both aged and young AL mice had comparable levels of SFFV, splenic homogenates from both young and aged mice maintained under DR had one to two log fewer SFFV than did homogenates from AL mice (Figure 7).

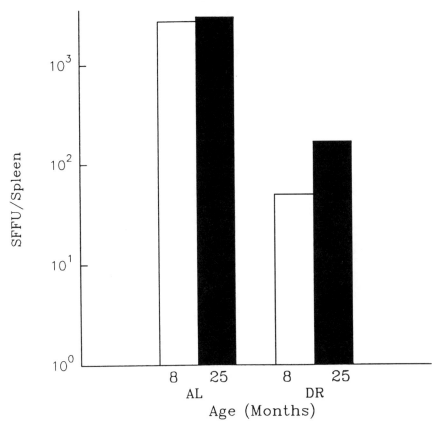

Figure 7. Effect of dietary restriction on spleen focus forming virus. Spleen homogenates (10% w/v) from Friend virus-infected 8- and 25-month-old (C57BL/6 × DBA/2)F1 mice were injected intravenously into BALB/c mice. Spleens were placed in Bouins solution for foci determination at 9 days post-infection. There were three mice in each group.

Discussion

The current studies clearly demonstrate that in at least two strains of mice, DBA/2 and C57BL/6, there is an age-associated increase in infectious, endogenous retrovirus expression. Additional studies in our laboratory have indicated that expression of replication-competent retrovirus in aged DBA/2 mice is the result of recombination between an ecotropic retroviral sequence and another endogenous retrovirus sequence (Bartman et al. 1994). This expression of replication-competent retrovirus was postponed in DBA/2 mice and reduced in C57BL/6 mice by DR. The role of the expression of infectious, endogenous retroviruses in the development of late-onset lymphoma is yet to be established. Our data suggest, however, that at least two lymphomas of aged DBA/2 mice contained infectious, ecotropic retrovirus and that inoculation of homogenates from late-onset lymphomas induces splenomegaly. The relationship of the virus in the lymphoma extract to the endogenous virus expressed at middle and late age can only be accomplished through molecular analysis of the various virus isolates. Such studies are currently under way.

Based on these results, which indicate that there was an increased expression of retrovirus with increasing age and those of Cinader et al. (1987), it was postulated that there would be an increased expression of exogenous retrovirus infection in aged mice. Contrary to the hypothesis, there was a decrease in FV-induced disease in aged mice as demonstrated by reduced splenomegaly and SFFU compared to young mice. There was, however, little or no effect on either RT activity or the amount of SFFV detectable upon inoculation of cell-free homogenates from aged and young AL FV-infected animals into susceptible young mice. These results can be interpreted to mean that while virus expression was comparable in young and old mice, the effect of the virus on cell proliferation and the resulting disease was decreased in aged AL mice. This is consistent with the work of Jiang et al. (1992), who have found a reduced proliferative potential of aged versus young cells (see also Wolf and Pendergrass, this volume). Since both splenomegaly and SFFU are dependent on the ability of host cells to proliferate in response to the FV p55 protein and host growth factors (Bernstein et al. 1990, Aizawa et al. 1990), the decreased splenomegaly and focus formation may be due to this decreased proliferative potential.

It had been originally postulated that DR would return the level of FV disease to the level seen in young mice. Again, contrary to expectation, the level of FV-induced disease in both young and aged DR mice was comparable to that seen in aged AL mice. More importantly, DR not only limited FV-induced disease but also limited expression of FV. It has been reported that constitutive proliferation of cells of aged mice maintained under DR is similar to that of cells from aged AL mice. Upon appropriate stimulation, however, the high level of proliferation that is possible in young AL or DR mice can be obtained in aged DR mice (Grossmann et al. 1990). One possible mechanism, therefore, for the decreased FV disease in DR mice is the lack of the appropriate stimulus to induce proliferation of FV-infected cells. This is a reasonable explanation, because our data indicate that there is less virus produced after infection in DR mice and therefore there is less stimulus for proliferation. The mechanism of this FV regulation in DR mice itself needs to be explored.

These results suggest that both age and diet can significantly affect both exogenous and endogenous retrovirus expression, with DR limiting virus replication to a much more substantial degree than age alone. Therefore, although DR may help maintain a young phenotype in mice, one cannot assume that physiologic parameters are comparable in young AL mice and those maintained under DR.

Acknowledgments

This work was supported by AG07719, AG10905, and a grant from the American Cancer Institute for Cancer Research.

References

Aizawa S, Suda Y, Furuta Y, et al. (1990) Env-derived gp 55 gene of Friend spleen focus forming virus specifically induces neoplastic proliferation of erythroid progenitor cells. EMBO J 9:2107–2116

Axelrod AA, Steeves RA (1964) Assay for Friend leukemia virus: rapid quantitative method based on enumeration of macroscopic spleen foci in mice. Virology 24:513–518

Bartman T, Murasko DM, Sieck TG, et al. (1994) A murine leukemia virus expressed in aged DBA/2 mice is derived by recombination of the Emv-3 locus and another endogenous gag sequence. Virology, in press

Bernstein A, Chabott B, Dubrcuil P, et al. (1990) The mouse W/c-kit locus. Ciba Found Symp 148:158–172

Cinader B, Van der Gaag HC, Koh SW, Axelrod AA (1987) Friend virus replication as a function of age. Mech Ageing Dev 40:181–191

Copeland NG, Bedigian HG, Thomas CY, Jenkins NA (1984) DNAs of two molecularly cloned endogenous ecotropic proviruses are poorly infectious in DNA transfection assays. J Virol 49:437–444

Copeland NG, Jenkins NA, Lee BK (1983) Association of the lethal yellow (Ay) coat color mutation with an ecotropic murine leukemia virus genome. Proc Natl Acad Sci U S A 80:247–249

Copeland NG, Jenkins NA, Nexo B, et al. (1988) Poorly expressed endogenous ecotropic provirus of DBA/2 mice encodes a mutant Pr65gag protein that is not myristylated. J Virol 62:479–487

Fernandez G, Tomar V, Venkatreman MN, Venkatreman JT (1992) Potential of diet therapy on murine AIDS. J Nutr 122:716–722

Grossmann A, Maggio-Price L, Jinneman JC, Wolf NS (1990) The effect of long-term caloric restriction on function of T cell subsets in old mice. Cell Immunol 131:191–204

Jiang D, Fei RG, Pendergrass WR, Wolf NS (1992) An age-related reduction in the replicative capacity of two murine hematopoietic stroma cell types. Exp Hematol 20:1216–1222

Rowe WP, Pugh WE, Hartley JW (1970) Plaque assay techniques for murine leukemia viruses. Virology 42:1136–1139

Taylor JM, Illmensee R, Summers J (1976) Efficient transcription of RNA into DNA by avian sarcoma virus polymerases. Biochem Biophys Acta 442:324–330

Thurman JD, Bucci T, Hart R, Turturro A (1994) Survival body weight and spontaneous neoplasms in ad libitum and diet restricted fed Fisher 344 rats. Toxicol Pathol 22:1–9

Turturro A, Blank KJ, Murasko DM, Hart R (1994) Mechanism of caloric restriction affecting aging and disease. Ann N Y Acad Sci 719:159–170

Weiss R, Teich N, Varmus H, Coffin J (eds) (1982) RNA tumor viruses, 2nd ed. Cold Spring Harbor Laboratory, Cold Spring Harbor, NY

Yu BP (1989) Why dietary restriction may extend life: a hypothesis. Geriatrics 44:87–90

Reproductive Effects Associated with Dietary Restriction

B.A. Schwetz
National Center for Toxicological Research

Introduction

Malnutrition or other variations in the intake or composition of food have been shown to adversely affect all stages of the reproductive process in humans, livestock, and experimental animals where these effects have been examined. Much of the experimental literature dealing with diet optimization has been contributed because of the importance of diet for the livestock industry, particularly because of problems of fertility, growth, and survival in situations where the diet has been somewhat different from optimal composition. Therefore, much of the knowledge gained on optimizing diet constituents for maximum reproductive performance has come from studies related to animal husbandry and the livestock industry. The literature also reflects the fact that sensitivity to variations in the diet and the manifestations that result from dietary alterations are a function of species, strain, gender, the nature of the dietary change, and the duration of the change in diet.

While extensive literature relates changes in diet to many aspects of the reproductive process, most of the literature has presented descriptive rather than mechanistic studies. There has, however, been a recent increase in the amount of mechanistic information, because new research tools have been applied to nutrition-related questions. While the early literature focused primarily on classic endocrine changes as they relate to homeostasis, the more recent literature describes effects on paracrine and autocrine regulation; to some extent, new information regarding the impact of dietary composition changes on gene product regulation is also coming into the literature (Bernstein et al. 1991; Dwyer and Stickland 1992). A thorough, recent review of the literature on dietary restriction and reproduction was not found, but a review of the historic evidence for an effect of nutrition on reproductive ability was published by Frisch (1978).

Effects of Dietary Restriction on Development

There are many papers in the literature that summarize the effect of dietary restriction (DR) and other changes in diet on development of the embryo, fetus, neonate, and young adult. This review summarizes studies that are representative of the literature or that report key effects of DR on development.

Dietary restriction of the pregnant female has significant effects on postnatal growth and function; it therefore is an important toxicological effect as well as a confounder in interpreting developmental toxicity studies as they are conducted today for risk assessment and regulatory purposes. The effect of DR during pregnancy on growth of offspring is expected to be related to the time of pregnancy during which food is restricted. Malnutrition during the early part of pregnancy impedes cell division, and there is little recovery in the decreased growth of offspring (Pond et al. 1990). Malnutrition during late pregnancy has been associated with decreased growth of offspring, but recovery to control weights is more likely to occur compared to the effect of malnutrition during early pregnancy. In a study by Jones and Friedman (1982), pregnant Sprague-Dawley (SD) rats were restricted to 50% of their pre-pregnancy food intake during the first two weeks of pregnancy. The animals were allowed to deliver their offspring, and there was no effect on offspring body weight at birth or at time of weaning. There was also no change in the motor activity of these offspring. However, male offspring became hyperphagic and obese as adults as a result of food deprivation of their mothers. The fat-pad weights of these male offspring were 2–3 times greater than those of control animals, and fat cells were about 25% larger than cells in control animals. A somewhat different effect was observed in female offspring in that they did not overeat and were not obese. The fat-pad weights of female offspring were more variable than controls but were not significantly increased. Fat cells were 10–25% larger in female offspring of DR mothers, but again the difference was not statistically significant.

In contrast to the results reported by Jones and Friedman (1982), Anguita et al. (1993) used a similar experimental design of restricting food of Wistar rats during the first two weeks of pregnancy to 50% of pre-pregnancy intake. They observed a body weight decrease in male offspring up to 53 days of age and reported no change in their food intake level. They did, however, report a marked fat accumulation by 53 days of age among female offspring.

The effect of DR on lung growth and function of offspring has been the subject of considerable study. The mortality rate in human neonates weighing less than 2500 grams is 40 times that of larger infants during the first month after birth. Smaller neonates show a higher incidence of neonatal respiratory distress (summarized by Shapiro et al. 1980, Hack et al. 1980). One of the early experimental approaches to evaluate the effect of maternal DR on lung growth and function of offspring was reported by Faridy (1975), who used 50% or 100% DR of Holtzman rats during gestation days 3–7, 9–13, or 17–21. Faridy concluded that the later deprivation caused more adverse toxic effects: lungs were hypocellular with decreased relative dry and wet weight, and there was a significant decrease in lung lecithin levels expressed relative to lung DNA. The lecithin levels recovered postnatally, but the lungs remained small and hypocellular. Several informative studies have also been conducted in guinea pigs. Lechner (1985) restricted guinea pigs to 50% of their normal food consumption during gestation days 45 through term (gestation days 67–69 in guinea pigs), or from delivery until postnatal day 21 or 42. There was a significant reduction in lung tissue mass as well as a decrease in the alveolar and capillary surface areas and pulmonary diffusing capacity. There was recovery if restriction only occurred postnatally, but values for these parameters did not recover among offspring of those mothers that were restricted prenatally. Using a similar experimental design, Lin and Lechner (1991) restricted guinea pigs to 50% of their food consumption from gestation day 45 through term. Offspring were significantly lighter than control animals, and there was a significant decrease in the amount of lung surfactant. There was a change in the surfactant phospholipid content and retardation of the maturation of secretory type II cells (those cells that produce surfactant). However, the authors concluded that the changes in surfactant phospholipid and maturation of secretory cells may or may not have been related to the poor growth and survival of the offspring. Effects on other cellular elements were also considered to be important for the development of normal lung function.

Another important manifestation of DR during pregnancy is the effect on the development of immune function. Miller (1991) summarized studies in mice ([C57BL/6J × CBA/H-T6J]F$_1$) where restriction of food was initiated around the time of weaning. Food consumption in the restricted mice ranged from 55–63% of normal intake; body weights ranged from 61–69% of normal values at 6, 18, and 30 months of age. Dietary restriction caused significant changes in the composition of the peripheral T-cell pool. Mitogen-sensitive cells were replaced by mitogen-insensitive T cells. There was also a significant decrease in cells that were identified by the surface markers CD4 and CD8 in memory T cells.

The studies summarized above support the conclusion that DR can significantly affect the structural and functional development of the embryo, fetus, and neonate. Because of this effect and the fact that developmental toxicity studies are often designed using dose levels that cause some decrease in food intake at least at the highest dose level, DR can potentially have a significant impact on the outcome and interpretation of developmental toxicity studies. Decreased food intake is often a confounder in these studies because the level of decreased food intake is not uniform throughout the experimental groups and is not observed in the control animals. Thus, in most experiments, there is no corresponding DR control to help interpret the results at those dose levels where food was restricted by virtue of palatability or toxicity.

Dietary restriction can also be a cofactor in causing developmental toxicity. Two approaches have been used to further evaluate DR as a cofactor in demonstrating developmental toxicity. One is the use of appropriately restricted control animals to determine if restriction of controls to the same level observed in experimental groups accounts for the developmental toxicity observed in animals treated with a chemical agent. Another approach involves use of whole-embryo culture techniques where embryos are exposed to the agent directly without the modulating effect of the maternal organism. Some studies have reported a significant potentiation, that is, effects that are in excess of an additive effect of the chemical alone or the same level of DR observed in the chemical-treated animals. Thus, the effect of DR ranges from a confounding effect in many experiments to potentiation of developmental toxicity in others.

Among the studies reporting an effect of DR on the toxicity of drugs or other chemicals, Fujii et al. (1977) reported that teratogenicity of hydroxyurea was significantly enhanced in SD rats by maternal fasting. Beall and Klein (1977) found that DR of CD rats significantly enhanced the teratogenicity of aspirin. The effect of decreased food consumption on the teratogenicity of ethanol was studied by Goad et al. (1984). C3H mice were given ethanol in the diet; the experiment involved several different levels of reduced dietary intake. The combination of DR and ethanol caused a significant inhibition of fetal growth and development. Ethanol effects were not dependent on maternal dietary intake. Decreased food intake, but not ethanol consumption, resulted in decreased fertility and maternal pregnancy weight gain and litter size. Thus, maternal nutrition played an important role, depending on the endpoint measured. In a recent study (Singh et al. 1993), the teratogenicity and developmental toxicity of carbon monoxide was evaluated in protein-deficient CD-1 mice. Control mice were given diets that were 27% protein, while experimental groups received diets with 16%, 8%, or 4% protein. Carbon monoxide levels ranged from 65 to 500 ppm. Exposure was on gestation days 8–18. The authors concluded that carbon monoxide was teratogenic under protein-deficient conditions and that protein deficiency had an additive effect on carbon monoxide teratogenicity and a synergistic effect on fetal mortality.

Among the studies that elucidate how DR might adversely affect development, many have focused on insulin-like growth factors (IGFs). Insulin-sensitive tissues are more sensitive to undernutrition than other organs, as reflected by the significant effect of DR on muscle fibers and liver mass. There is less of an effect of under-nutrition on the development of bone, heart, and brain (Dwyer and Stickland 1992). As a result of this selective effect of malnutrition, the authors hypothesized that insulin-like growth factors may mediate the effects of undernutrition. Using guinea pigs that were 40% DR (i.e., food intake restricted to 60% of that of control animals) from gestation day 25 to birth, Dwyer and Stickland (1992) reported that maternal and fetal IGF-I was reduced, while IGF-II was reduced only in the fetal compartment. They concluded, therefore, that maternal IGF-I may mediate the effects on fetal growth through its impact on the development of the fetal-placental unit. A similar conclusion about the relative importance of IGF-I and IGF-II was reached by Jones et al. (1990), who also used a DR regimen in guinea pigs and observed that decreases in fetal growth correlated with changes in plasma glucose (insulin) and IGF-I, but not with changes in IGF-II. In contrast, Bernstein et al. (1991) restricted the feed of SD rats on gestation days 18–21 and found that IGF-I was a good marker for growth retardation on the basis of changes in IGF-I levels in fetal plasma, liver, and placenta. They also concluded that placental IGF-I was very closely associated with fetal size and was important in the classic endocrine fashion in the sense that it is formed in the placenta and has an effect on fetus development.

Effects of Diet on Reproduction

Dietary restriction also affects other aspects of the overall reproductive process, including age at sexual maturity, mating rate and fertility, ovulation rate and cycle length, implantation, resorptions, litter size, birth weight, and neonatal growth and survival (summarized by Krackow 1989). There are gender-specific DR reproduction effects that go beyond the obvious physiological differences in the sexes. For example, sperm production is not impaired when adult males are exposed to levels of energy intake known to inhibit ovarian function (summarized by Blank and Desjardins 1985). In a series of studies conducted by Hamilton and Bronson (1985, 1986) in mice that were feed-restricted after weaning, reproductive development capabilities proceeded independent of body growth in males but not in females.

As in the case of developmental toxicity studies, DR is a significant confounder in the design and interpretation of reproductive toxicity studies. The most extensive studies in rats and mice designed to evaluate the role of DR on the interpretation of reproductive studies are those by Chapin et al. (1993a,b). In these studies, SD rats and Swiss mice were food-restricted to maintain groups at 90%, 80%, or 70% of control body weight and were evaluated for the effects on reproduction in an experimental design similar to that used for evaluating chemical toxicity. The restriction levels were selected to bracket the level of decreased food intake commonly observed in reproductive toxicity studies. Observations on females included effects on fertility, the number and live/dead status of uterine implants, resorptions, numbers of corpora lutea, and weights of the body and selected organs. In the case of males, mating trials were conducted during weeks 8 and 15 of DR to assess the number of litters per male, live pups per litter, and live pup weight. Necropsy data included weight of the cauda epididymis and testis, measures of sperm motility and density, and percent of abnormal sperm. Effects on rats (Chapin et al. 1993a) were found only in the group maintained at 70% of control body weight and were limited to females. There was a significant increase in estrous cyclicity and a decrease in the number of corpora lutea. There were no adverse effects on male rats at any of the three DR levels. In comparison, there were significant adverse effects on reproductive performance of mice at 70% and 80% of control body weight, and there were marginal effects at 90% of control body weight (Chapin et al. 1993b). These adverse effects are summarized in Table 1, where it is evident that a number of endpoints were adversely affected in both males and females. Females were affected to a greater extent than males in that there were adverse effects at 80% and 90% of control body weight. The studies in mice were conducted in two laboratories to evaluate the repeatability of the observations. The effect observed in the 90% group, a decrease in the number of live pups per litter, was observed in only one of the two laboratories. This probably

Table 1. Comparison of effects of feed restriction in rats and mice[a]

Control body weight (%)	Swiss mice			Sprague-Dawley rats		
	90	80	70	90	80	70
Female						
Fertility	—	—	—	—	—	—
Estrous cycle length	—	—	Increased	—	—	Increased
Copulatory plugs	—	—	Decreased	—	—	—
Corpora lutea/dam		Not measured in mice	—	—	—	Decreased
Implants/dam	—	Decreased	Decreased	—	—	—
Live pups/litter	Decreased*	Decreased	Decreased	—	—	—
Relative ovary weight	—	—	Decreased*	—	—	Decreased
Male						
Fertility	—	—	Decreased*	—	—	—
Live pups/litter	—	—	Decreased*	—	—	—
Epididymal sperm density	—	—	Decreased	—	—	—
Sperm motility	—	—	—	—	—	—
Abnormal sperm %	—	—	—	—	—	—
Testicular testosterone	—	—	Decreased	—	—	—
Relative testis weight	Increased	Increased	Increased	—	—	Increased

[a]Summarized from Chapin et al. 1993a,b.
[b]— = no change from control value, "increased" or "decreased" = statistically significant difference from control values.
*Difference from control was significant only in one of two laboratories where mouse studies were conducted.

reflects the fact that the marginal effect in the 90% group is the "tail end" of the dose-response curve for the effect that was significant in the 80% and 70% groups. In the males, there was also an effect in one of the two laboratories on fertility in the 70% group. This is a significant observation in view of the robust breeding capability of these mice.

Significant adverse effects have also been reported on the lactation process. Two studies were reported from one laboratory (Young and Rasmussen 1985, Fischbeck and Rasmussen 1987) in which SD rats were restricted to 75%, 60%, 50%, or 40% of the ad libitum food consumption level from 28 days before breeding to the time of the second lactation. The body weights of dams and litters were significantly decreased in all groups. Milk yield, milk protein, and lactose concentrations were significantly affected in the 60% restriction groups and greater.

Many studies were designed to identify the site and mechanism through which DR disrupts reproductive function. The most extensively reported mechanism is the effect of DR on neuropeptide Y. Both neuropeptide Y and beta-endorphin are important in the mediation of DR effects as well as normal reproductive function through their secretion from the hypothalamus. The secretion of these factors regulates the secretion of luteinizing-hormone-releasing hormone (LHRH), which in turn controls the formation and release of luteinizing hormone (LH), which, in the plasma, regulates reproductive function in females. Significant changes in plasma LH levels caused by DR have been shown to delay the onset of puberty and to adversely affect follicle maturation in the ovary. The role of neuropeptide Y in this sequence of events has been the subject of considerable research. Prasad et al. (1993) used an animal model in which lambs with intact ovaries were studied. Under DR conditions, lamb weight was maintained at <20 Kg by providing 20% of the proteins and calories required by the National Research Council to achieve 200g growth/day. The authors reported a decreased level of beta-endorphin with no change in neuropeptide Y release. Under these conditions, there was decreased amplitude in the release of LHRH (no change in the frequency of secretory surges), and plasma levels of LH were decreased. In contrast, McShane et al. (1992), using ovariectomized lambs, concluded that the link between undernutrition and reproductive function involved a neuropeptide Y-mediated effect on LHRH. In studies conducted in SD rats, Gruaz et al. (1993) concluded that neuropeptide Y is involved in inhibition of sexual maturation.

Conclusions

Most if not all parts of the reproductive process are adversely affected by certain levels of dietary component manipulation or DR. The mechanisms by which the adverse effects might be caused are complex and not yet well understood. Little commonality has been used in the experimental approaches to evaluate the role of DR or diet changes on reproduction and development. As a result, it is not possible at this time to draw firm conclusions about specific dietary changes or DR in specific periods of the overall reproductive process, because many of the studies have not been replicated. Even when similar approaches were used for one aspect of the study, other aspects varied so widely that it is not possible to draw conclusions based on replicated experimental findings. There is little doubt that dietary component manipulation or DR may enhance or suppress expression of toxicity of substances that are commercially important or found in the environment. Chemical-related changes in food intake in reproductive and developmental toxicity studies may confound the interpretation of these studies. In some cases, the effects attributed to the agent under test may, to some extent, be attributable to the effect of DR.

References

Anguita RM, Sigulem DM, Sawaya AL (1993) Intrauterine food restriction is associated with obesity in young rats. J Nutr 123:1421–1428

Beall JR, Klein MF (1977) Enhancement of aspirin-induced teratogenicity by food restriction in rats. Toxicol Appl Pharmacol 39:489–495

Bernstein IM, DeSouza MM, Copeland KC (1991) Insulin-like growth factor I in substrate-deprived, growth-retarded fetal rats. Pediatr Res 30(2):154–157

Blank JL, Desjardins C (1985) Differential effects of food restriction on pituitary-testicular function in mice. Am J Physiol 248:R181–R189

Chapin RE, Gulati DK, Barnes LH, Teague JL (1993a) The effects of feed restriction on reproductive function in Sprague-Dawley rats. Fund Appl Toxicol 20:23–29

Chapin RE, Gulati DK, Fail PA, et al. (1993b) The effects of feed restriction on reproductive function in Swiss CD-1 mice. Fund and Appl Toxicol 20:15–22

Dwyer CM, Stickland NC (1992) The effects of maternal undernutrition on maternal and fetal serum insulin-like growth factors, thyroid hormones and cortisol in the guinea pig. J Devel Physiol 18:303–313

Faridy EE (1975) Effect of maternal malnutrition on surface activity of fetal lungs in rats. J Appl Physiol 39(4):535–540

Fischbeck KL, Rasmussen KM (1987) Effect of repeated reproductive cycles on maternal nutritional status, lactational performance and litter growth in ad libitum-fed and chronically food-restricted rats. J Nutr 117:1967–1975

Frisch RE (1978) Population, food intake, and fertility. Science 199(6J):22–30

Fujii O, Ikeda Y, Sukegawa J (1977) Effects of maternal fasting on teratogenicity of hydroxyurea in rats. Teratology 16(1):103

Goad PT, Hill DE, Slikker Jr. W, et al. (1984) The role of maternal diet in the developmental toxicology of ethanol. Toxicol Appl Pharmacol 73:256–267

Gruaz NM, Pierroz DD, Rohner-Jeanrenaud F, et al. (1993) Evidence that neuropeptide Y could represent a neuroendocrine inhibitor of sexual maturation in unfavorable metabolic conditions in the rat. Endocrinology 133(4):1891–1894

Hack M, Merkatz IR, Jones PK, Fanaroff AA (1980) Changing trends of neonatal and postneonatal deaths in very-low-birth-weight infants. Am J Obstet Gynecol 137:797–800

Hamilton GD, Bronson FH (1985) Food restriction and reproductive development in wild house mice. Biol Reprod 32:773–778

Hamilton GD, Bronson FH (1986) Food restriction and reproductive development: male and female mice and male rats. Am J Physiol 250:R370–R376

Jones AP, Friedman MI (1982) Obesity and adipocyte abnormalities in offspring of rats undernourished during pregnancy. Science 215:1518–1519

Jones CT, Lafeber HN, Rolph TP, Parer JT (1990) Studies on the growth of the fetal guinea pig. The effects of nutritional manipulation on prenatal growth and plasma somatomedin activity and insulin-like growth factor concentrations. J Devel Physiol 13:189–197

Krackow S (1989) Effect of food restriction on reproduction and lactation in house mice mated post partum. J Reprod Fert 86:341–347

Lechner AJ (1985) Perinatal age determines the severity of retarded lung development induced by starvation. Am Rev Respir Dis 131:638–643

Lin Y, Lechner AJ (1991) Surfactant content and type II cell development in fetal guinea pig lungs during prenatal starvation. Pediatr Res 29(3):288–291

McShane TM, May T, Miner JL, Keisler DH (1992) Central actions of neuropeptide Y may provide a neuromodulatory link between nutrition and reproduction. Biol Reprod 46:1151–1157

Miller RA (1991) Caloric restriction and immune function: developmental mechanisms. Aging 3(4):395–398

Pond WG, Yen J-T, Mersmann HJ, Maurer RR (1990) Reduced mature size in progeny of swine severely restricted in protein intake during pregnancy. Growth Dev Aging 54:77–84

Prasad BM, Conover CD, Sarkar DK, et al. (1993) Feed restriction in prepubertal lambs: Effect on puberty onset and on in vivo release of luteinizing-hormone-releasing hormone, neuropeptide Y and beta-endorphin from the posterior-lateral median eminence. Neuroendocrinology 57:1171–1181

Shapiro S, McCormick MC, Starfield BH, et al. (1980) Relevance of correlates of infant deaths for significant morbidity at 1 year of age. Am J Obstet Gynecol 136:363–373

Singh J, Aggison Jr. L, Moore-Cheatum L (1993) Teratogenicity and developmental toxicity of carbon monoxide in protein-deficient mice. Teratology 48:149–159

Young CM, Rasmussen KM (1985) Effects of varying degrees of chronic dietary restriction in rat dams on reproductive and lactational performance and body composition in dams and their pups. Amer J Clin Nutr 41:979–987

Interspecies Variations in Physiologic and Antipathologic Outcomes of Dietary Restriction

Richard Weindruch

University of Wisconsin—Madison
William S. Middleton VA Hospital

Joseph W. Kemnitz and Hideo Uno

University of Wisconsin—Madison

Introduction

As evidenced by these proceedings, the dietary restriction (DR) paradigm can no longer be viewed as an interesting curiosity being studied, in large part, by a few gerontologists. Instead, the clear and easily obtained anti-disease and anti-aging actions of this simple dietary change have attracted widespread attention and now raise important new questions germane to safety assessment. Because the majority of bioassay studies conducted by the drug industry use rats (most often males from the Sprague-Dawley strain), whereas mice (male B6C3F1) serve as the main model for the National Toxicology Program (NTP), it is important to consider the variations observed among different species in response to DR. Because these animals serve as models for humans, it is also appropriate to examine data on the physiologic effects of DR in nonhuman primates.

The retardation of aging and diseases by long-term DR has been most closely studied in mice and rats, usually by imposing DR at 40–60% of the ad libitum (AL) caloric intake level on young (< 3-month-old) animals (Weindruch and Walford 1988). Moderate DR (~70% of AL intake) gradually imposed on middle-aged (12-month-old) male mice from two strains (B10C3F1 and C57BL/6) is also quite effective at opposing the development of lymphoma and increasing maximum life span by 10–20%

(Weindruch and Walford 1982). This finding and the observation that DR started at 6 months of age is just as effective as that started at 6 weeks of age in attenuating the development of nephropathy and increasing the life span of male Fischer 344 (F344) rats (Yu et al. 1985) show that a significant part of DR's actions is independent of growth inhibition.

The rodent subjected to DR provides an excellent model to study the biology of aging. Because DR attenuates the development of several diseases, the gerontologist can study aging in healthy old animals, which often live to ages exceeding earlier estimates of the species-specific maximum life span. Likewise, the toxicologist could use DR as a way to study toxicity in healthy old animals. DR is the only intervention repeatedly shown to extend maximum life span in mammals and to retard a broad spectrum of age-associated physiologic changes. It is important to realize that, as long as malnutrition is avoided, these outcomes become increasingly more overt as the severity of DR increases (Figure 1). This direct relationship between life span and the severity of energy intake restriction supports the importance of studying mitochondrial energy and free radical metabolism in animals on long-term DR (Weindruch and Walford 1988, Yu 1990, Weindruch et al. 1993, Feuers et al. 1993).

A major gerontologic issue is whether DR will exert actions in primates similar to those in rodents. This topic is clearly germane to safety assessment. One would hope that extensive use of rodents as models both for the study of aging and for safety evaluations is not a situation where scientists are assuming too high a level of biologic similarity between rodents and primates. We are now in the fifth year of a study of adult-onset, moderate DR in male rhesus monkeys, with the long-term goal of assessing the influence of DR on the rate of aging (Kemnitz et al. 1993). The study entails longitudinal evaluation of 30 male rhesus macaques, half of which have been subjected to a 30% reduction in dietary intake. The animals are regularly assessed across several categories of measurements that are purported indicators of functional age (loosely referred to as "biomarkers of aging"). These assessments include measures of body size and composition, energy expenditure, glucose tolerance and insulin sensitivity, immunologic function, and visual accommodation. The battery of assessments is performed on each animal at intervals of 6 months, or annually in the cases of body composition measurement and ocular exams.

In this report, comparisons are made of selected physiologic and antipathologic outcomes observed in mice and rats subjected to long-term DR. The available information on the physiologic effects of DR in rhesus monkeys is also considered. A main conclusion is that, for certain physiologic measures and diseases, the results of DR are repeatedly (but not

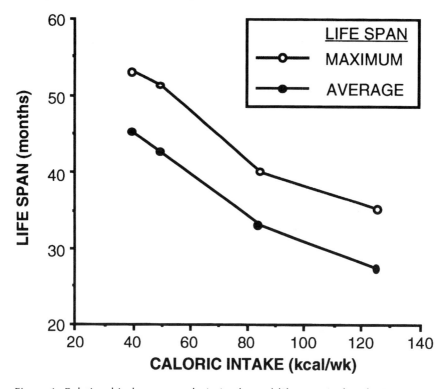

Figure 1. Relationship between caloric intake and life span in female C3B10RF1 mice. Dietary restriction was initiated at 3 weeks of age. Four cohorts of mice (*n* = 49–71) were fed either 125 (the average ad libitum intake), 85, 50, or 40 kcal/week. "Maximum life span" is defined as the average of the cohort's longest-lived decile. Adapted from Feuers et al. (1993).

always) more striking in the mouse rather than the rat studies. The few physiologic responses measured to date in rhesus monkeys subjected to DR also resemble those of mice and rats on restricted energy intakes.

Physiologic Outcomes

Three physiologic outcomes (metabolic rate reduction, body temperature reduction, and improved glucoregulation) are discussed here. These physiologic adaptations to DR, although not universally observed, were selected because they have been hypothesized to be important in explaining the retardation of aging and diseases by DR (discussed by Sacher 1977, Weindruch and Walford 1988, Masoro and McCarter 1991). Accordingly, these adaptations may prove to be important biomarkers of DR and may also be associated with disease prevention in old rodents on restricted diets. Summaries of some of the available data on body fatness (an indicator of DR severity), blood glucose levels, and body temperature in rodents maintained on DR are provided in Tables 1 (rats) and 2 (mice).

Metabolic Rate Reduction

Metabolic rate has long been regarded as a potentially important regulator of the rate of aging. Both Sacher (1977) and Harman (1981) suggested that retardation of aging by DR can be explained as being a result of a reduction in metabolic rate induced by low-calorie intake. Studies by McCarter et al. (1985) and McCarter and McGee (1989) in rats subjected to DR at 6 weeks of age have, however, led to the view that DR only transiently decreases the metabolic rate (oxygen consumption) per lean body mass and that any reductions of metabolic rate by DR vanish after a few weeks. As a result, Masoro (1992) and others have argued strongly that a reduction of metabolic rate by DR is uninvolved in DR's actions on aging processes. Others have questioned this conclusion (Weindruch and Walford 1988, Lynn and Wallwork 1992), and newly published data in rats show that DR reduced the metabolic rate (Gonzales-Pacheco et al. 1993b), which led to a new round of debate on the topic (McCarter 1993, Gonzales-Pacheco et al. 1993a, Lynn 1993).

Metabolic rate has been less often studied in mice on long-term DR. Duffy et al. (1991) continuously monitored the 24-hour oxygen consumption of 28-month-old female B6C3f1 mice and found that DR (61–67% of the AL intake started at 14 weeks of age) reduced rates of whole-body oxygen consumption. Like the findings in F344 rats by McCarter et al. (1985, 1989), however, metabolic rate per lean body mass was not altered by DR.

Our studies in monkeys show that DR reduces oxygen consumption independent of the mode of data expression, i.e., per whole monkey or per lean body mass (Kemnitz et al., in preparation). The difference between groups remains statistically significant even when the contribution of lean body mass is accounted for. We imposed DR on adult animals, and it is quite possible that age of onset may influence the effect of DR on metabolic rate. We are continuing to monitor metabolic rate to determine the persistence and magnitude of any effect of DR on this parameter.

Body Temperature Reduction

Body temperature decreases with DR in rodents (Tables 1 and 2). Figure 2 plots core body temperatures in 18-month-old female F344 rats and in 28-month-old B6C3F1 mice subjected to DR (60% of AL intake) from 14 weeks of age (Duffy et al. 1990, 1991). The results indicate

Table 1. Indicators of the physiologic severity of dietary restriction in rats (all values except ratios are averages)

Measure	DR	AL	DR/AL	Body weight (g) DR/AL[a]	Food intake DR/AL (kcal/day)[a]	Reference[b]
Body fat (%)	20	44	0.46	467/881 (0.53)	54/90 (0.60)	1
	24	32	0.75	330/500 (0.66)	37/59 (0.63)	2
Body glucose (mg/dl)[c,d]	130	150	0.87	284/519 (0.55)	35/59 (0.59)	3[e]
	120	140	0.86	284/519 (0.55)	35/59 (0.59)	3[e,f]
	60	78	0.77	280/450 (0.62)	36/60 (0.60)	4[h]
Body temperature (°C)[c,g]	35	36	0.97	190/300 (0.63)	29/48 (0.60)	5[h]

DR = dietary restriction, AL = ad libitum, DR/AL = ratio of DR value to that of the control.
[a]DR/C ratio is in parentheses; several of these values were not reported but were calculated.
[b]1 = Nolen 1972: male Sprague-Dawley, ~1 month old at diet initiation, 24 months old when studied. 2 = Rikimaru et al. 1988: male Wistar, 4.5 months old at diet initiation, 17.5 months old when studied. 3 = Masoro et al. 1992: male F344, 1.5 months old at diet initiation, 15–19 months old when studied. 4 = Feuers, personal communication: male F344, 3 months old at diet initiation, 9 months old when studied. 5 = Duffy et al. 1990: female F344, 3 months old at diet initiation, 18 months old when studied.
[c]Blood glucose levels and body temperature vary over the day.
[d]Glucose data were obtained ~2 hours after feeding.
[e]Higher glucose values probably reflect the assay (glucose oxidase) versus a glucose analyzer used in the other studies.
[f]Values were the lowest of the day.
[g]Lowest body temperatures observed during the day (see Figure 2 for complete daily profiles).
[h]Food intakes and body weights derived from Witt et al. (1991).

Table 2. Indicators of the physiologic severity of dietary restriction in mice (all values except ratios are averages)

Measure	DR	AL	DR/AL	Body weight (g) DR/AL[a]	Food intake DR/AL (kcal/day)[a]	Reference[b]
Body fat (%)	13	22	0.59	20/30 (0.67)	7/11 (0.67)	1
Blood glucose (mg/dl)[c,d]	74	104	0.72	25/40 (0.63)	12/19 (0.60)	2[g]
	52	96	0.54	25/40 (0.63)	12/19 (0.60)	2[e,g]
Body temperature (°C)[c,f]	25	36	0.69	20/40 (0.50)	7/14 (0.50)	3
	28	35	0.80	24/38 (0.63)	10/17 (0.60)	4[g]

DR = dietary restriction, AL = ad libitum, DR/AL = ratio of DR value to that of the control.
[a]DR/AL ratio is in parentheses; several of these values were not reported but were calculated.
[b]1 = Harrison et al. 1984: female C57BL/6J, 1 month old at diet initiation, 6–12 months old when studied. 2 = Feuers et al. 1991: male B6C3F1, 3 months old at diet initiation, 12 months old when studied. 3 = Koizumi et al. 1992: female (SHN × C3H)F1, 21 days old at diet initiation, 13 months old when studied. 4 = Duffy et al. 1990: female B6C3F1, 3 months old at diet initiation, 28 months old when studied.
[c]Blood glucose levels and body temperature vary over the day.
[d]Glucose data were obtained ~2 hours after feeding.
[e]Values were the lowest of the day.
[f]Body temperatures are the lowest observed during the day (see Figure 2 for complete daily profiles).
[g]Food intakes and body weights derived from Witt et al. (1991).

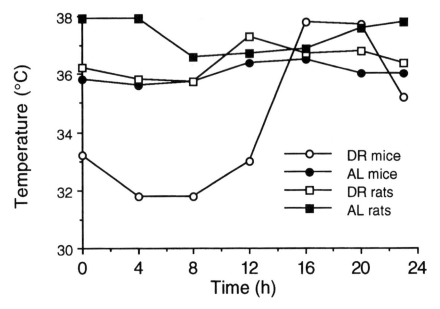

Figure 2. Mean body temperature in female Fischer 344 rats and B6C3F1 mice subjected to dietary restriction from 14 weeks of age. The rats were fed 60% of the ad libitum (AL) intake; mice were fed 61–67% of the AL intake. Each data point is the mean body temperature of the same 10 animals followed over a 24-hour period. For clarity, standard errors are not plotted but were quite small (0.2–0.4°C). Data are adapted from Duffy et al. (1990) (rats) and Duffy et al. (1991) (mice).

that, relative to AL animals, DR rats show only a very mild temperature reduction of ~2°C, whereas DR mice show daily reductions of ~6°C. Mice on more severe DR regimens (50% of AL) have been reported to enter daily torpor and cool to 23–24°C (Koizumi et al. 1992). Importantly, certain physiologic outcomes of DR, such as depressed cell proliferation rates (Koizumi et al. 1992) and lymphopenia (Koizumi et al. 1993), may depend on reduced body temperature, because these responses are attenuated by housing DR mice at 30°C, which prevents the mice from entering torpor.

The more robust temperature-lowering actions of DR in mice compared to rats suggests that DR in many rat studies is physiologically less severe than that occurring in mouse studies. Still, it must be emphasized that, even though body temperature is only mildly reduced in DR rats, survival is still improved. For example, female F344 rats from the National Center for Toxicological Research (NCTR) colony studied for body temperature had an average life span of 26 months for the AL group versus 31 months for the DR group (Duffy et al. 1990); 10th-decile survival was 34 months for the AL group versus 39 months for the DR group (Witt et al. 1991). These 10th-decile survival times are much less than the 50 months or more reported for both mice from long-lived F1 hybrids on DR (Weindruch et al. 1986, Harrison and Archer 1987) and F344 × Brown Norway F1 rats studied at the NCTR (R.J. Feuers, personal communication). It would be of great interest to investigate body temperatures in this latter model, which is not only very long-lived but offers a leaner presence at autopsy than do either of its parental strains (R.J. Feuers, personal communication).

In collaboration with Peter Duffy and other investigators at the NCTR, we are initiating studies to measure body temperature and other physiologic parameters around-the-clock in control and food-restricted rhesus monkeys. These data should improve understanding of the physiologic responses to long-term DR in primates.

Improved Glucoregulation

It has been suggested that DR could exert beneficial effects by lowering prevailing glucose levels, thereby reducing nonenzymatic glycosylation of proteins (Cerami 1985, Monnier 1990) and their oxidized derivatives (Dunn et al. 1989, Kristal and Yu 1992). Indeed, lower levels of fasting blood glucose have been reported for mice (Koizumi et al. 1989, Feuers et al. 1991), rats (Masoro et al. 1992), monkeys (Kemnitz et al. 1994) (Figure 3), and humans (Walford et al. 1992) subjected to chronic DR. In addition, plasma glucose levels in DR rats are lower around the clock. This is consistent with lower glycosylated hemoglobin concentrations, an indicator of prevailing glucose concentrations over a period of several weeks, in both rats (Masoro et al. 1989) and monkeys (Kemnitz et al. 1994). Lower glycosylated hemoglobin concentrations are associated with reduced cardiovascular disease in older women (Singer 1992).

Insulin concentrations are also lowered by DR in rats (Masoro et al. 1992) and monkeys (Kemnitz et al. 1994) (Figure 3) in the fasted state as well as following meals or intravenous carbohydrate challenge. Insulin sensitivity, the ability of endogenous insulin to promote glucose uptake and to inhibit gluconeogenesis, can be quantified by the Minimal Model Method, which uses mathematical models to account for the dynamic relationships between glucose and insulin during frequently sampled intravenous glucose tolerance tests (Bergman 1989). Insulin sensitivity is enhanced by DR in rhesus monkeys (Figure 3). Although the mechanism for this effect is not yet clear, changes in glucose and insulin concentrations and in insulin sensitivity were positively correlated with indicators of body fat content, suggesting a linkage between body composition and glucoregulatory parameters within individual animals (Kemnitz et al. 1994).

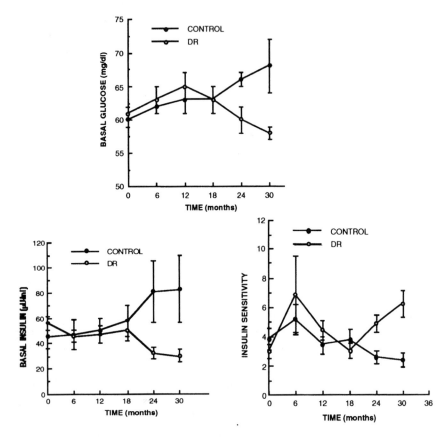

Figure 3. Influence of dietary restriction (DR) and age on fasting levels of glucose (top panel), insulin (left panel), and insulin sensitivity (right panel) in male rhesus monkeys. The controls were provided free access to food for 6–8 hours daily. Insulin sensitivity is derived from the Modified Minimal Model. It reflects the ability of insulin to promote glucose uptake and inhibit hepatic glucose production. The units for insulin sensitivity are (x10^4·min^{-1}·µU^{-1}mL^{-1}). Adapted from Kemnitz et al. (1994).

There is abundant evidence linking hyperinsulinemia/insulin resistance with the development of diabetes mellitus, hypertension, and atherosclerotic vascular disease (Reaven 1988, Haffner et al. 1992). Increase insulin secretion and hyperinsulinemia may lead to beta cell exhaustion, increased sympathetic tone, sodium retention, endothelial hyperplasia, and dyslipidemia. The present results support the possibility that DR may aid in disease prevention by improving insulin sensitivity and reducing circulating insulin levels.

To conclude, several physiologic measures are markedly influenced by DR. Further, the effects of DR appear to be more overt in mouse experiments than in studies using rats. An initial analysis exploring this possibility is presented in Tables 1 and 2. Although body composition data for rodents on long-term DR are surprisingly limited, one can see that Sprague-Dawley rats are quite fat, whereas C57BL/6 mice are rather lean. The result is that rats from this commonly used strain subjected to DR show similar body fat levels as do AL C57BL/6 mice. These data support the view that a 40% reduction of food intake of an AL and somewhat fat rat would be less of a physiological stress than that resulting from a 40% DR in a nonobese AL mouse. The data for blood glucose and body temperature reductions support this possibility.

Antipathologic Outcomes

It is well known that diverse spontaneous and induced diseases occur less often and/or later in life in mice and rats subjected to DR. Interspecies and even intraspecies comparisons of spontaneous diseases can be quite difficult, imprecise, and minimally informative because of differences among animal models in the susceptibility for developing a given disease. From the standpoint of safety testing, it seems important to consider pathologic outcomes that are of most concern and to select and maintain the animal model in a manner that maximizes the sensitivity and minimizes the variance of the bioassay. For example, if an agent is suspected to be nephrotoxic, it would make little sense to use AL-fed male F344 rats, because unrestricted feeding stimulates the development of nephropathy. This result occurs when either a casein-containing purified diet (Yu et al. 1982) or a nonpurified (NIH-31) diet (Witt et al. 1991) is fed; however, AL feeding of a purified diet containing soy protein produces far less nephropathy in this rat model (Iwasaki et al. 1988).

The argument can be reasonably advanced that the influence of DR on neoplasia in most of the mouse strains studied is stronger than that exerted in rats. Note that this difference only concerns neoplasia and not nephropathy, which is markedly opposed by DR in rats (Yu et al. 1982, Witt 1991). In contrast to DR's strong ability to oppose the development of neoplasia in mice, the appreciation of an inhibitory influence of 40% DR on neoplasia in F344 rats has required use of fairly complex statistical approaches to reveal clear dietary influences (Shimokawa et al. 1991, 1993). Although no attempt is made here to provide the background to support this contention (but see Weindruch and Walford 1988, Chapter 3), this

Table 3. Incidence of major age-associated pathologic changes in rhesus monkeys[a]

Pathology (%)	Age (years)/n			
	15–19/110	20–25/107	26–29/34	30+/27
Coronary sclerosis	10	36	62	89
Myocardial fibrosis	6	7	15	41
Emphysema	2	42	50	70
Osteoporosis	0	11	18	44
Neoplasm				
malignant	4	20	18	48
benign	3	22	9	37
Wasting disease				
(idiopathic)	2	12	12	17

[a]Data are from 278 autopsies conducted at the Wisconsin Regional Primate Research Center from 1980–1993. The population was ~80% females/20% males and was fed Purina Monkey Chow ad libitum.

proposed difference could be an expression of the more robust physiological consequences of DR in mice.

Variations in disease expression can be addressed by considering autopsy data obtained since 1980 on 262 rhesus monkeys at the Wisconsin Regional Primate Research Center that died at ages 15 to 37 years (Table 3). These data include sacrificed monkeys as well as others that died "naturally." These data will serve as a background for the future evaluation of DR's effects on age-related pathologies in this species.

Summary

Despite similar reductions below the AL caloric intake in rat and mouse studies, it appears that in most DR studies the physiological and anti-pathological outcomes are somewhat more pronounced in mice than in rats. In addition to species differences (i.e., a rat is not a mouse), there are other reasonable explanations for this observed differential response to DR. One possibility is a decreased energy expenditure for rats due to cage size-induced limitations in the ability of a rat to move around. In contrast, an individually housed mouse has ample room for active movement. A second and related factor is the possibility of greater obesity in conventionally fed rats from several of the commonly used strains (Sprague-Dawley, Wistar, F344) than in mice from nonobese strains. A reduced physiological severity of DR in most rat studies would also accord with the observation that the very longest-lived DR rodents are mice (not rats), despite the fact that the survival of AL mice and rats from most of the long-lived strains used in aging research do not differ overtly (Weindruch and Walford 1988). A glaring exception to this generalization would be

F344 × Brown Norway F1 rats, which are quite long-lived and leaner than either of the parent strains.

Current data support the view that the physiologic responses of rhesus monkeys to long-term DR resemble those of mice and rats subjected to similar dietary regimens. This observation suggests that rodents and primates adapt in similar ways to a major dietary alteration and bolsters confidence in the suitability of rodent models for human aging and toxicological studies.

It is our view that use of DR increases the relevance of animal studies for human health risk assessment. The logic supporting this conclusion is based on the fact that most surveys find that ~35% of the U.S. population is obese, whereas nearly all AL Sprague-Dawley rats are obese. Likewise, very high incidences of obesity can occur in many of the other commonly studied rodent models. Therefore, the rodent on mild DR provides a better model for the 65% or so of U.S. citizens who are not obese. Also of enormous importance to risk assessment is the fact that DR provides a powerful way to minimize the variance of the bioassay.

Acknowledgment

Our research is supported by the NIH (R01 AG10536, R01 AG07831, P01 AG11915, and P51 RR00167).

References

Bergman RN (1989) Towards a more physiological understanding of glucose tolerance: minimal model approach. Diabetes 38:1512

Cerami A (1985) Glucose as a mediator of aging. J Am Geriatr Soc 33:626

Duffy PH, Feuers RJ, Leakey JEA, Hart RW (1991) Chronic caloric restriction in old female mice: changes in the circadian rhythms of physiological and behavioral variables. In Fishbein L (ed), Biological effects of dietary restriction. Springer-Verlag, Berlin, pp 245–263

Duffy PH, Feuers R, Nakamura KD, et al. (1990) Effect of chronic caloric restriction on the synchronization of various physiological measures in old female Fischer 344 rats. Chronobiol Int 7:113

Dunn JA, Patrick JS, Thorpe SR, Baynes JW (1989) Oxidation of glycated proteins: age-dependent accumulation of N-(carboxymethyl)-lysine in lens protein. Biochemistry 28:9464

Feuers RJ, Casciano DA, Shaddock JG, et al. (1991) Modifications in regulation of intermediary metabolism by caloric restriction in rodents. In Fishbein L (ed), Biological effects of dietary restriction. Springer-Verlag, Berlin, pp 198–206

Feuers RJ, Weindruch R, Hart RW (1993) Caloric restriction, aging, and antioxidant activities. Mutat Res 295:191

Gonzales-Pacheco DM, Buss WC, Alpert SS (1993a) Reply to the letter of McCarter (1993b). J Nutr 123:1936

Gonzales-Pacheco DM, Buss WC, Koehler KM, et al. (1993b) Energy restriction reduces metabolic rate in adult male Fischer-344 rats. J Nutr 123:90

Haffner SM, Valdez RA, Hazuda HP, et al. (1992) Prospective analysis of the insulin-resistance syndrome (syndrome X). Diabetes 41:715

Harman D (1981) The aging process. Proc Natl Acad Sci U S A 78:7124

Harrison DE, Archer JR (1987) Genetic differences in the effects of food restriction on aging in mice. J Nutr 117:376

Harrison DE, Archer JR, Astle CM (1984) Effects of food restriction on aging: separation of food intake and obesity. Proc Natl Acad Sci U S A 81:1835

Iwasaki K, Gleiser CA, Masoro EJ, et al. (1988) The influence of dietary protein source on longevity and age related disease processes of Fischer rats. J Gerontol Biol Sci 43:B5

Kemnitz JW, Roecker EB, Weindruch R, et al. (1994) Dietary restriction increases insulin sensitivity and lowers blood glucose in rhesus monkeys. Am J Physiol 266:E540

Kemnitz JW, Weindruch R, Roecker EB, et al. (1993) Dietary restriction of adult male rhesus monkeys: design, methodology and preliminary findings from first year of study. J Gerontol Biol Sci 48:B17

Koizumi A, Roy NS, Tsukada M, Wada Y (1993) Increase in housing temperature can alleviate decreases in white blood cell counts after energy restriction in C57BL/6 female mice. Mech Ageing Dev 71:97

Koizumi A, Tsukada M, Wada Y, et al. (1992) Mitotic activity in mice is supressed by energy restriction-induced torpor. J Nutr 122:1446

Koizumi A, Wada Y, Tsukada M, Hasegawa J (1989) Low blood glucose levels and small islets of Langerhans in the pancreas of calorie-restricted mice. Age 12:93

Kristal BS, Yu BP (1992) An emerging hypothesis: synergistic induction of aging by free radicals and Maillard reactions. J Gerontol Biol Sci 47:B107

Lynn WS (1993) Reply to the letter of McCarter (1993). J Nutr 123:1938

Lynn WS, Wallwork JC (1992) Does food restriction retard aging by reducing metabolic rate? J Nutr 122:1917

Masoro EJ (1992) Retardation of aging processes by food restriction: an experimental tool. Am J Clin Nutr 55:1250S

Masoro EJ, Katz MS, McMahan CA (1989) Evidence for the glycation theory of aging from the food-restricted rodent model. J Gerontol Biol Sci 44:B20

Masoro EJ, McCarter RJM (1991) Aging as a consequence of fuel utilization. Aging 3:117

Masoro EJ, McCarter RJM, Katz MS, McMahan CA (1992) Dietary restriction alters characteristics of glucose fuel use. J Gerontol Biol Sci 47:B202

McCarter RJM (1993) Comment on the papers by Gonzales-Pacheco et al. (1993) and Lynn and Wallwork (1992): energy restriction and metabolic rate. J Nutr 123:1934

McCarter RJM, Masoro EJ, Yu BP (1985) Does food restriction retard aging by reducing the metabolic rate? Am J Physiol 248:E488

McCarter RJM, McGee J (1989) Transient reduction of metabolic rate by food restriction. Am J Physiol 257:E175

Monnier VM (1990) Minireview: nonenzymatic glycosylation, the Maillard reaction and the aging process. J Gerontol Biol Sci 45:B105

Nolen GA (1972) Effect of various restricted dietary regimens on the growth, health and longevity of albino rats. J Nutr 102:1477

Reaven GM (1988) Role of insulin resistance in human disease. Diabetes 37:1595

Rikimura T, Ichikawa M, Oozeki T, et al. (1988) Long-term effect of energy restriction at different protein levels on several parameters of nutritional assessment. J Nutr Sci Vitaminol 34:469

Sacher GA (1977) Life table modification and life prolongation. In Finch CE, Hayflick L (eds), Handbook of the biology of aging. Van Nostrand Reinhold, New York, pp 582–638

Shimokawa I, Yu BP, Masoro EJ (1991) Influence of diet on fatal neoplastic disease in male Fischer 344 rats. J Gerontol Biol Sci 46:B228

Shimokawa I, Yu BP, Masoro EJ (1993) Dietary restriction retards onset but not progression of leukemia in male F344 rats. J Gerontol Biol Sci 48:B68

Singer DE, Nathan DM, Anderson KM, et al. (1992) Association of HbA$_{1c}$ with prevalent cardiovascular disease in the original cohort of the Framingham heart study. Diabetes 41:202

Walford RL, Harris SB, Gunion MW (1992) The calorically restricted low-fat nutrient-dense diet in Biosphere 2 significantly lowers blood glucose, total leukocyte count, cholesterol, and blood pressure in humans. Proc Natl Acad Sci U S A 89:11,533

Weindruch R, Walford RL (1982) Dietary restriction in mice beginning at 1 year of age: effects on lifespan and spontaneous cancer incidence. Science 215:1415

Weindruch R, Walford RL (1988) The retardation of aging and disease by dietary restriction. Charles C. Thomas, Springfield, IL

Weindruch R, Walford RL, Fligiel S, Guthrie SD (1986) The retardation of aging by dietary restriction in mice: longevity, cancer, immunity and lifetime energy intake. J Nutr 116:641

Weindruch R, Warner HR, Starke-Reed PE (1993) Future directions of free radical research in aging. In Yu BP (ed), Free radicals and aging. CRC Press, Boca Raton, FL, pp 269–295

Witt WM, Sheldon WG, Thurman JD (1991) Pathological endpoints in dietary restricted rodents. In Fishbein L (ed), Biological effects of dietary restriction. Springer-Verlag, Berlin, pp 73–86

Yu BP (1990) Food restriction research: past and present status. In Rothstein M (ed), Review of biological research in aging, vol. 4. Wiley-Liss, New York, pp 349–371

Yu BP, Masoro EJ, McMahan CA (1985) Nutritional influences on aging of Fischer 344 rats: physical, metabolic and longevity characteristics. J Gerontol 40:657

Yu BP, Masoro EJ, Murata I, et al. (1982) Life span study of SPF Fischer 344 male rats fed ad libitum or restricted diets: longevity, growth, lean body mass and disease. J Gerontol 37:130

Dietary Restriction: Implications for the Design and Interpretation of Toxicity and Carcinogenicity Studies

A Panel Discussion

William T. Allaben
National Center for Toxicological Research

David A. Neumann
ILSI Risk Science Institute

As plans for this conference unfolded, it became apparent that the issue of restricting or controlling food intake by rodents in long-term toxicity and carcinogenicity studies was exceedingly complex. Much of this complexity was captured in the 10 questions (see Preface, this volume) that were circulated among the presenters as they prepared for the conference. Yet other questions and issues emerged in the course of the presentations and discussions. In anticipation of such developments and to place the information presented during the conference into perspective, the meeting concluded with a panel discussion that considered the implications of the findings reported during the conference with respect to safety assessment. The panel discussion was moderated by Dr. William T. Allaben (National Center for Toxicological Research, Food and Drug Administration); panel members included Drs. John Bucher (deputy director of the National Institute of Environmental Health Sciences' [NIEHS] Environmental Toxicology Program and chief of the Toxicology Branch in that program), David Hattan (deputy director, Division of Toxicology Review and Evaluation, Center for Food Safety and Applied Nutrition, FDA), Yuzo Hayashi (director, Biological Safety Research Center, National Institute for Health Sciences, Japan), David Kritchevsky (institute professor, The Wistar Institute), Margaret Miller (deputy director, New Animal Drug Division,

Center for Veterinary Medicines, FDA), Richard Robertson (executive director of toxicology and developmental biology, Merck Research Institute), and David Scales (director of toxicological sciences, Glaxo Group Research, Ltd., United Kingdom).

Dr. Allaben: As noted by Dr. Hart at the beginning of this conference (see Hart and Turturro, this volume), although scientists have made every effort to control the variables in the cancer bioassay, the one variable that has the greatest impact on the outcome of such studies, i.e., diet, remains uncontrolled. Data presented at this conference suggest that control of food intake prevents or slows age-associated declines in normal physiology and biochemistry. It is important to recognize that restricting food intake may modulate a biological response but that it is unlikely to result in the induction of unique processes that reduce spontaneous or induced pathologies. From this perspective, it may be time to take the initiative to control caloric intake and thus the growth of rodents that serve as surrogates for humans in many types of biomedical studies. A conference such as this affords the opportunity to develop an informed consensus on the application of dietary control when testing chemicals in toxicity and carcinogenicity bioassays.

Targeting body weight gain to approximate that observed in earlier bioassays could enhance the value of the bioassay. This approach would be consistent with the use of growth curves as described by Dr. Turturro (see Turturro et al., this volume) and could be achieved through modest and reasonable dietary control. This control would allow more of the test animals to be exposed to the chemical for a longer period of time, thus increasing the power of the bioassay to detect a true carcinogen. Such an approach would necessitate that all supporting mechanistic, pharmacokinetic, metabolic, and toxicologic data be collected under identical dietary conditions.

Dr. Bucher: Inequities in body weight between experimental and control groups can have a marked influence on the outcome of carcinogenesis studies. Studies at the NIEHS indicate that the sensitivity of the rodent bioassay to detect a carcinogenic response may be reduced when the study is performed using even a moderate diet restriction design (see Kari and Abdo, this volume). The loss of sensitivity is of significant concern (see Hayashi, this volume). The results of bioassays performed under standard National Toxicology Program (NTP) ad libitum feeding protocols can be grouped into three large categories: 1) approximately half of the studies were completely negative or gave equivocal results that likely represent "noise" in the assay, 2) between 5% and 10% of the studies were clearly positive, and 3) the remainder yielded evidence of neoplasia and were considered carcinogenic, but their potential hazard to humans was less immediate than that associated with chemicals that produced unequivocal outcomes. It is likely that some chemicals in this third category would be missed in bioassays performed under conditions of mild dietary restriction. Perhaps the relative potency of these chemicals as a function of diet should be factored into the way these types of chemicals are regulated.

The NTP also has studied a large number of pharmaceutical compounds that were mostly negative in the bioassay; most were nongenotoxic. Although their pharmacologic activity often precludes testing at high doses, some of the compounds are weakly carcinogenic at lower doses. Under conditions of dietary restriction the carcinogenic potential of such compounds may be missed. These observations make it difficult to fully endorse the use of only dietary restriction conditions in studies on these types of agents.

There may be merit in some of the other measures to control survival that were mentioned during the conference (see Rao, this volume; Haseman, this volume). These stop-gap measures may be more applicable to the Fischer rat than to the Sprague-Dawley rat, which may be beyond hope at this time. However, carefully designed and monitored and adequately controlled studies employing dietary restriction can provide useful information. Indeed, if resources permitted, the NTP might routinely conduct paired bioassays using both ad libitum and dietary restriction feeding protocols.

When interpreting data obtained under conditions of dietary restriction it would seem essential to consider the nature of the pharmaceutical or the chemical being studied in relation to the results obtained with chemicals of like biologic activity or similar pharmacological class under ad libitum feeding conditions. Examination of the strength of the carcinogenic response in tumor sites that are positive with this class of chemical should provide information about whether the bioassay would be adequate to detect a carcinogenic event in an organ that might be highly susceptible to the influence of a diet restriction regimen. If dietary restriction ultimately becomes the accepted practice for studies of this type, pharmacokinetic and other supporting studies should be performed on both ad libitum–fed and dietary restricted animals. It should be possible to design a bioassay that maximizes both the sensitivity and the specificity and that is predictive for human risk. That is the challenge. Control of caloric intake might be a good way to provide more consistent bioassay results, thus allowing for a more accurate identification of carcinogens and an easier comparison of their potency.

Dr. Hattan: This conference has facilitated a very useful discussion of how to interpret or compensate for changes in the incidence of disease and in the sensitivity of the carcinogenicity assessment process that are associated with an unstable animal model. In addition to agencywide concern about the sensitivity of carcinogenicity assessment, the FDA's Center for Food Safety and Applied Nutrition (CFSAN) is concerned about the influence that a change in nutritional status may have on study outcome. The results of studies employing dietary control or restriction protocols may be useful in the interpretation of studies of the new macroadditives used in foods.

The report that changes in conditions such as air flow rates within a laboratory or between laboratories can alter the caloric requirements of genetically identical test animals (see Wolf and Pendergrass, this volume) is cause for concern when considering test animal diet. Given the influence of such variables and the difficulty in recognizing and compensating or controlling for them, scientists from CFSAN and from the FDA's National Center for Toxicological Research (NCTR) are embarking on a study to examine the relationship between various levels of food restriction (15% and 25% of the ad libitum intake) and the rate of spontaneous tumor development. The findings of this study could provide benchmarks or calibration points for predicting the effects of lower levels of restriction.

In order to calibrate responses between laboratories or between studies it would be valuable to have an independent physiological measure for assessing the general nutritional status of the animals. A newly reported technique, total body electrical capacitance, may be the appropriate tool. This is a noninvasive procedure for assessing the relative size of the lean and fat compartments with respect to total body weight. However, the value of this procedure will become apparent only when a sufficient database has accrued.

Another concern is the significance for humans of these studies of the effects of dietary restriction in rodents. The report at this conference (see Weindruch et al., this volume) that the physiologic responses of nonhuman primates to restricted dietary intake are somewhat similar to those observed in rodents suggests a degree of commonality in response.

Dr. Kritchevsky: One of the important things in any scientific endeavor is to be able to compare results in a way that gives one a certain amount of confidence in the data. Because different strains and stocks of rats may give different results, there must be some way to compare data from different experiments. Perhaps there should be some kind of background information or standards available so that scientists who work with a certain type of animal can relate their findings to those of other investigators. Dr. Hattan's suggestion about using total body electrical capacitance as a standard is interesting; the method seems to perform satisfactorily in humans, but less so in animals.

Many of the studies described during the conference restricted food intake by 40% relative to that consumed by ad libitum–fed control animals. This may be excessive with respect to both rodents and humans. In studies at the Wistar Institute (see Kritchevsky, this volume), a 10% reduction did not affect chemically induced carcinogenicity but reduced weight gain by 10% and body fat by 16%. Such an approach may result in a healthier animal without altering the sensitivity of the test system by increasing longevity and hence exposure time. It is important that the general public and public officials understand that such studies do not give misleading information about cancer potential.

Consistency in the bioassay is the most important factor to be considered when discussing these types of studies. Assumptions about the uniformity of food consumption by gang-caged animals are unfounded, and single caging also may affect food consumption by rodents. Application of dietary restriction will likely produce a healthier animal that will live longer, and this can likely be achieved at modest levels, e.g., 10% or 15%, of restriction.

Dr. Miller: The FDA's Center for Veterinary Medicine (CVM) is responsible for regulating animal drug products, and its concern for carcinogenicity is based on the observation that when an animal is treated with a new drug there will be residues present in every tissue of that animal. It is the responsibility of the drug's sponsor to demonstrate the safety of those drug residues for consumers. CVM's safety evaluation is based on toxicology, residue and metabolism studies, and carcinogenicity testing. For noncarcinogens, a classic threshold approach is used and a NOEL [no-observed-effect level] is established with respect to the most sensitive effect in the most appropriate species as it relates to humans. The CVM uses mechanistic and human data when available.

When considering carcinogenic compounds, the Delaney anticancer clause comes into play, although CVM does have the DES proviso, which allows for the approval of carcinogenic compounds used in animals provided there is no residue present in edible tissues. "No residue" is defined as that dose that will result in residues that cause an incremental risk of cancer of less than one in a million. The CVM performs quantitative risk assessment on data from animal bioassays to establish such a risk. Thus, the CVM uses bioassay data in a somewhat different fashion than does CFSAN or the FDA's Center for Drug Evaluation and Research (CDER).

Because the CVM relies on the CFSAN document *Toxicological Principles for Safety Assessment of Direct Food Additives and Color Additives Used in Food* (FDA's Redbook), the center is interested in pursuing an agencywide approach to the use of dietary restriction in the performance of rodent bioassays. A guidance document or a "Points to Consider" document would be of value to CVM scientists. There is nothing in the CVM's current guidelines and regulations that would preclude the center from accepting a sponsor's data that had been obtained from studies employing dietary restriction. To date, the center has received only one request from a sponsor to perform a toxicity study using a caloric control design. Center scientists believe that the protocol design is appropriate for this specific animal drug and that by reducing background tumors and noise from aging-associated lesions, the interpretation of the results and, thus, regulatory decision making will be facilitated. However, it is important that drug sponsors discuss their plans to use dietary control/restriction with CVM scientists prior to doing such studies in order to gain consensus on the protocol.

The CVM is involved in international efforts to harmonize veterinary drug safety assessment. Currently, it is possible to do a carcinogenicity study that is acceptable in Europe but not in the United States, and vice versa. This often reflects country-to-country differences in test protocols, dosing regimes, and other factors. The issue of dietary control/restriction is sufficiently important to be introduced into the ongoing discussions about international harmonization. Because of the magnitude, length, and costs associated with carcinogenicity testing, the issue of diet may actually serve as a catalyst for advancing the international harmonization process.

Dr. Robertson: The term ad libitum is useless and meaningless in the absence of information about actual food intake. As noted by Dr. Keenan (see Keenan et al., this volume), the amount of food actually consumed varies widely under ad libitum feeding conditions. In Dr. Keenan's studies, rats restricted by 35% relative to the caloric intake of their ad libitum–fed counterparts had daily food intakes that were well within the reported ad libitum consumption range for rats in other laboratories.

Of the various mammalian species used for biomedical studies, only rodents are fed ad libitum; the caloric intake of dogs, minipigs, and nonhuman primates are all controlled. If beagles were allowed continuous access to food, the dogs would weigh 20 to 25 kilograms and have a 30% to 40% body fat content. It is unlikely that anyone in the audience would be comfortable evaluating data from such animals in the context of safety assessment. Yet this is how rodent carcinogenicity studies have been conducted for decades.

As described during this conference, controlling caloric intake is an effective means for reducing morbidity and mortality associated with long-term bioassays. The argument that the sensitivity of the bioassay is diminished under conditions of dietary restriction is reminiscent of the debate about the high rates of mycoplasma infections within rodent colonies nearly 30 years ago. Infected animals had an increased incidence of lung tumors. Some scientists advocated that nothing be done to treat the problem because it created a more sensitive animal that was very responsive to chemical insults. Wisdom seemed to prevail when the test protocol was amended to provide for the circulation of HEPA-filtered air within the animal facilities. This reduced or eliminated the incidence of mycoplasma infection, diminished background pathology in the lung, and increased survival.

Ad libitum–fed rats are not just obese, but also are endocrinologically bankrupt in the sense that they have a high incidence of pituitary tumors that secrete a variety of peptide hormones at nonphysiologic levels. The application of dietary control does not alter normal physiologic function; rather, it restores normal function to an animal that is potentially in a pathologic state.

Differences in the responses of animals fed calorically controlled diets with respect to those of obese, ad libitum–fed animals are to be expected. In some cases, e.g., the NCTR's acute toxicity data (see Duffy et al., this volume), sensitivity will increase while in others it will decrease. However, toxicology studies are conducted by dosing until a toxic effect is observed. Studies at Merck Research Laboratories suggest that while the dose required to see an effect may be different when dietary restriction is used, the nature of the effect, the target organ involved, and the resulting degenerative process do not change. That is a key point to remember when establishing doses for a chronic study. Thus, range-finding studies should be performed in animals that have the same caloric intake as those to be used in a carcinogenicity study, to allow for a valid comparison between studies.

Dr. Scales: The other panelists raised many important points. However, there are other questions that surround the issue of carcinogenicity testing that perhaps eclipse dietary restriction in terms of relevance or importance. Comparability between studies and laboratories may be enhanced by first agreeing on the strains and stocks of animals to be used for such studies. Next there should be a standardization of husbandry conditions followed by agreement on the composition of the diet. After these issues have been resolved, it would be appropriate to address the issue of dietary restriction.

Dr. Allaben: It is evident that while there are varying opinions regarding the utility of dietary control, there is general agreement that obesity, poor survival, and high background tumor incidence rates in control animals are significant problems. Although breeding practices that select for rapid growth may contribute to these problems, ad libitum feeding has provided the fuel supporting that energy demand for increased growth. In an effort to restore the body weight to historical norms, dietary control may be one of the most useful and effective means available.

Dr. Kevin Keenan (*Merck Research Laboratories*): Despite comments to the contrary, the Sprague-Dawley rat exhibits excellent survival when maintained according to restricted diet protocols and is not abused with calories.

It would seem that the regulatory agencies must be evaluating a diverse array of data that have been submitted in support of their respective registration responsibilities. There are a large number of chronic studies that are being performed that unintentionally inflict a late-onset dietary restriction on one or both sexes of Sprague-Dawley, Wistar, and Fischer rats. This de facto restriction, often resulting from the inability of the large animals to reach all of the food in their feeders, may contribute to improved survival or to an altered pattern of background tumor incidence. During the studies at Merck, the 35% restriction, surprisingly, was well within the range of published values for ad libitum feeding. Although tumor incidences were unchanged relative to the ad libitum–fed rats, survival was increased and degenerative disease was reduced among the restricted animals. Thus, the term ad libitum remains undefined and highly variable between laboratories. This confounds the interpretation and comparison of the results of different studies, including studies of the consequences of dietary restriction where restriction is expressed as a percentage of ad libitum intake.

Dr. Angelo Turturro (*National Center for Toxicological Research*): Dr. Wolf's observations (see Wolf and Pendergrass, this volume) clearly demonstrate how changes in behavior and/or husbandry can alter the animal's caloric requirements. Such observations support the suggestion that body weight changes or growth curves be used to guide animal nutrition. Thus, dietary modulation could be used to control the growth curve, which would take into account differences in activity, temperature, air flow, and other variables.

Dr. Bucher has suggested that potential toxicants may interact with normal background processes, giving rise to background tumors. If such processes are modified by differences in caloric intake or body weight, then it should be important to control these parameters; otherwise, achieving consistency between studies will remain elusive.

Dr. Bucher: The point is that when there is general agreement on an appropriate caloric control regimen, such an agreement will result in improved consistency. Performing quantitative risk assessments based on studies that are uncontrolled with respect to food intake is not a good idea.

Dr. Lewis Kinter (*SmithKline Beecham Pharmaceuticals*): Although there has been much discussion about controlling body weight through caloric restriction, body weight is an outcome of a diverse set of physiological processes. Body weight cannot be controlled directly by the investigator, but indirectly through control of the presentation of calories. This, rather than body weight, should be the focus of efforts to control variability in the bioassay.

There appears to be some ambiguity as to the purpose of the 2-year cancer bioassay. On the one hand, it has been ascribed the specific objective of assessing carcinogenic potential in a defined mammalian system and, on the other, as a procedure for predicting carcinogenic activity in humans. When protocols for conducting the 2-year bioassay were formulated, technology for controlling all of the variables as well as scientific understanding were insufficient to distinguish between these objectives. Data presented at this conference suggest that perhaps now such distinctions can be made and that if some recommendation or guidance is to emerge from this conference, it should include a redefinition of the objective of the 2-year carcinogenicity bioassay in rodents.

Dr. Allaben: Originally, the 2-year bioassay was intended to screen a large number of chemicals relatively quickly to determine if they posed potential hazards. Over time, through numerous modifications and refinements, the bioassay has become the basis for local, state, and federal regulatory authorities to perform risk assessments to predict potential human harm. This usage raises the question of the adequacy of the information provided by the bioassay, not to mention the uncertainties associated with extrapolating findings in animals to predict human risk. Although many of the variables, except food consumption, have been controlled for and other refinements introduced into the bioassay protocol, the current use of bioassay data for risk assessment and risk management decisions goes beyond the original intent of the assay.

Dr. Esther Rinde (*Office of Pesticide Programs, U.S. Environmental Protection Agency*): If the use of dietary restriction is more effective in reducing the incidence of spontaneous tumors than in reducing the incidence of chemically induced tumors, bioassays conducted under conditions of dietary restriction would be more sensitive than the bioassay as currently performed.

Dr. Kritchevsky: There is considerable evidence that dietary restriction can inhibit the growth of spontaneous tumors.

Dr. Robertson: Although reducing caloric intake is known to slow the progression of spontaneous tumors, the effect of dietary restriction on chemically induced tumors cannot be predicted. Restriction also reduces morbidity and increases survival, which is where the real power of using dietary control lies.

Dr. Bruce McCullough (*Schering-Plough Research Institute*): The issue seems to be one of definition, i.e., the quantities of food the rodents consume, in both quantitative and relative terms, and the time at which dietary restriction is imposed with respect to the ad libitum–fed controls. In addition to concern about the amount of food consumed, variation in the composition of the diet warrants careful consideration. It may be more appropriate to focus on the effects of variation in diet composition before addressing the effects of varying the amount of food consumed. As demonstrated by Dr. Keenan (see Keenan et al., this volume), the variation in the amount of food consumed under ad libitum feeding conditions is sufficiently great as to confound interpretation of the data. Such confusion could be rectified by incorporating information about dietary composition and food consumption in scientific and other publications.

Dr. Allaben: Information about dietary composition and food consumption is routinely recorded during many studies, but often are not reported. Dr. Duffy's report (see (Duffy et al., this volume) and others reinforce the need to monitor food and water consumption, particularly water consumption, during the first 3 months of the study.

Dr. McCullough: Part of the variability associated with ad libitum feeding may reflect the variety of feeding devices that are used.

Dr. Hattan: Defining dietary composition is done routinely in many laboratories, but requires considerable time and effort and is quite costly, particularly when evaluating trace components of the diet. Although such exercises assure the quality of the diet, the criteria for assessing the cost effectiveness of the analyses are elusive.

Dr. McCullough: Diets should be reasonably characterized to include reporting the number of calories consumed per day and information on composition that is provided by the manufacturers of certified diets. It also would be appropriate to report the age of the animals at the time at which dietary restriction was initiated.

Dr. Frank Kari (*National Institute of Environmental Health Sciences*)**:** Although this conference has focused on the use of dietary restriction or control to alter the body weight of test animals, it is clear that restricting food intake, as well as overfeeding, can alter a variety of factors and processes. The variable that most closely correlates with the experimental outcomes, i.e., survival and diminished incidence of spontaneous pathologies, is the ultimate body weight attained by the animal. There are many examples where the mass of food presented is independent of the eventual body weight. Many pharmaceutical agents that alter animal behavior or metabolism can uncouple the mass of food presented from the body weight. Similar findings have been reported for artificial sweeteners and other types of food supplements.

Because of differences in the caloric and nutrient densities of different feed formulations, proposals to implement dietary restriction protocols based on percentages of ad libitum intake appear premature. The apparent benefits of dietary restriction appear to be more closely linked to the eventual body weight than to the mass of food that is presented to the animals. There are probably 10 different diets that could be fed to animals to attain the same weight, yet the mass of each needed to achieve that weight varies. Inclusion of nonnutritive bulk material in the feed will decrease the caloric and nutrient density, thus necessitating consumption of large amounts of feed to attain a specific weight. This is independent of any effects of the test material or its route of administration, e.g., incorporation of the chemical in the food or drinking water or administration by gavage. Thus, controlling animal weight is critical to improving the rodent bioassay. Body weight can likely be altered by means other than dietary restriction; the challenge to the scientific community is to identify those alternatives.

Dr. Andrew Salmon (*Office of Environmental Health Hazard Assessments, California Environmental Protection Agency*)**:** The NTP approach to the rodent bioassay provides the best source of data available to risk assessors who do quantitative risk assessments. For this reason, the study protocol needs to be carefully defined and uncontrolled variables minimized or eliminated. Presentations at this conference clearly indicate that body weight is critical in determining the outcome of the studies, and this point is an important conclusion from this conference. Although dietary restriction is one approach to controlling body weight, other approaches such as changes in husbandry or genetic stock or some combination of approaches should be considered.

The issue of the sensitivity of the bioassay has been raised frequently during the conference and was addressed by Dr. Hayashi (see Hayashi, this volume). When there is severe dietary restriction that results in a final body weight below the historical range for the species or strain in question, the sensitivity of the bioassay with respect to observations of chemically induced tumors is substantially reduced. It would be irresponsible for any regulatory agency to endorse a modification of the study protocol that could result in such a loss of sensitivity. The message from this conference is that the body weight of the animals should be controlled to reflect the historical norm for each species and strain, possibly through the use of dietary restriction. This would allow the quantitative risk assessors to consider the results of new studies in much the same light by which the historical database is perceived.

Although there are problems and inconsistencies associated with the historical database, it remains the best available resource, and departing from it would be a mistake. The objective should be to design the best possible study and to carefully monitor body weight without seriously deviating from past practices. This would avoid the excesses of both overfeeding and underfeeding with their attendant perturbing effects. Although the issue of extrapolation to human risk remains controversial, by using such an approach the risk assessment community would at least have a stable database with which to work. A stable bioassay design is critical to performing accurate risk assessments.

It appears that ad libitum feeding has resulted in deviations from the standard design. The reasons for this are unknown, but clearly can result in extreme weight gain. Anything that can be done to restore the experimental design to the perceived norm should be encouraged.

Dr. Miller: Dr. Contrera of the FDA's Center for Drug Evaluation and Research (CDER) has also noted the apparent drift or deviations associated with the rodent bioassay that may be due in part to overeating. It may be appropriate to try to shift the test paradigm back to where it was 10 years ago. The CDER's interest in effecting such a shift has been reflected in discussions with scientists from the CVM about developing an agencywide guidance document that would limit this drift. It would be valuable to CVM scientists who perform quantitative risk assessments to have confidence that the parameters under consideration are consistent from study to study.

Dr. Salmon: It has become apparent that having a better understanding of body weight changes during a study would aid in understanding and resolving problems with individual studies where there are extreme body weight differences between groups.

Dr. Kritchevsky: Much of this type of data on weight gain and weight differences between groups is actually available and could be published if manuscripts were refereed with that in mind.

Dr. Keith Soper (*Merck Research Laboratories*): A basic precept in any scientific endeavor is that for an experiment to be valid, the results should be reproducible within the reporting laboratory as well as in other laboratories. The presentations during this conference indicate that the level of food consumption is an extremely important variable in the rodent bioassay. Food consumption varies greatly between laboratories and thus is out of control with respect to the bioassay. Instead of dietary restriction, perhaps the focus should be on regaining control of food consumption to enhance reproducibility and improve standardization of the bioassay.

Standardizing the bioassay on the basis of animal weight could be problematic in carcinogenicity studies where changes in weight occur among the treated animals. If body weight is to be standardized to specific growth curve values by restricting the amount of food given to control animals, it is unclear whether a 20% reduction in weight due to dietary restriction relative to ad libitum feeding is comparable to a 20% reduction in body weight due to the toxic effects of the test material.

In a 2-year bioassay where there is substantial mortality, body weights decline markedly in the weeks preceding death. This confounds the use of growth curves to standardize the bioassay. Similarly, the weights of female rats can be grossly influenced by mammary tumors that may weigh as much as 300–400 grams. Obviously, standardizing to body weight is fraught with problems and may be difficult to implement.

Dr. Erika Randerath (*Baylor College of Medicine*): If a guidance document is to be developed that will establish new parameters, it would be helpful to generate a databank encompassing study design and outcome information that would facilitate comparisons between studies and interpretation of their results.

Dr. Allaben: Dr. Sidney Siegel of the National Library of Medicine has been considering the development of such a computerized databank. The concept is still under discussion, but a resource like this could be very important to the scientific community. A potential problem is that information developed in support of product registration may not be available to the database because it is considered proprietary. Within the pharmaceutical industry, some companies readily share their data while others do not.

Dr. Robertson: Much of the information about the protocols and approaches used by Merck to modify the caloric intake of laboratory rodents has been, and will continue to be, made available to the scientific community. Studies demonstrating that altering caloric intake can impact drug disposition will continue to be published.

Dr. Ronald Hart (*National Center for Toxicological Research*): Because the definition of ad libitum feeding is arbitrary, it is meaningless to attempt to define dietary restriction. However, dietary control can be defined on the basis of measured food or caloric intake. The objective of dietary control is to maintain stability within the test system in order to reduce the variability and increase the reproducibility of toxicity studies. As described by Drs. Turturro and Rao (see Turturro et al., this volume; Rao, this volume) and others, body weight serves as a convenient measure of stability, although other measures may be used.

An important and often underappreciated concern was raised earlier in the conference by Dr. David Gaylor of the National Center for Toxicological Research. He observed that it would not be appropriate to control dietary intake to such an extent that spontaneous tumor incidence falls below some acceptable level. For example, Dr. Turturro (see Turturro et al., this volume) showed that if food intake by B6C3F1 mice is sufficiently reduced such that the body weight does not exceed 30 grams, the incidence of spontaneous liver tumors falls to zero. Such a finding would pose a real problem for risk assessors because the responsiveness of the animals is unknown as is the amount of chemical necessary to elicit a tumorigenic response.

Issues such as this reinforce the notion that a variety of factors need to be considered when attempting to set an appropriate level of dietary control. To do so will require the efforts of biometricians, pathologists, toxicologists, and other specialists who appreciate the implications of the various issues. If it is possible to determine an appropriate level of dietary control, it will then be important to develop guidelines and identify parameters that can be used to maintain consistency. To simply adopt dietary control measures without addressing these other concerns would be naive. Similarly, too narrow a focus on body weight without addressing these other issues would be ill advised.

Although other approaches to the current problems of spontaneous disease and survival in the 2-year bioassay have been proposed, it is clear from this conference that there is a considerable database available supporting the notion that these problems could be addressed in a timely fashion through the use of dietary control. Addressing these problems could lead to a more reasonable approach to risk assessment.

Dr. Norman Wolf (*University of Washington*): It is appropriate to discuss the use of dietary control to reduce certain negative features of the rodent bioassay as it is currently performed. However, the mechanisms responsible for these observed changes are only now being studied. Although the results obtained through the use of dietary control are readily apparent, the biological processes responsible for those results are less apparent. Indeed, different mechanisms may account for the different observed effects of dietary control. For example, dietary control seems to delay the onset of virtually all types of tumors, yet Dr. Duffy (see Duffy et al., this volume) demonstrated that ad libitum–fed animals respond at multiple dose levels, while animals on a restricted diet do not respond at any of the doses. Even when overwhelming doses are used, there is no evidence of long-lasting toxicity. Because findings such as this cannot be readily explained, it may be appropriate to study different classes of compounds in animals on restricted diets. Factors that contribute to the reduction in the number of tumors present at a certain point in time may be affected differently by different types of compounds. Understanding how dietary restriction affects the tumorigenic response to different classes of compounds may be a prerequisite to applying it broadly to all testing situations.

Dr. Patricia Lang (*Consultant*): Even by examining the effects of both dietary restriction and ad libitum feeding on the response of rodents to different classes of chemicals, it is still unclear which result is most valid for human health risk assessment.

The issue of the composition of the feed used in the laboratory remains a source of concern and there is evidence that the composition of certified diets is more variable than is generally assumed. For example, some suppliers specify that the protein content is not less than 20%, but that is not a particularly useful piece of information because not less than 20% can mean many things. It is also likely that the composition of the feed will vary with the source of the grain, growing conditions, and other factors.

It is also unclear how to implement a program for restricting food intake. Some laboratories apparently have feeding devices that can effectively reduce food intake. This information should be used in the implementation of restriction protocols rather than starting from scratch to develop a new feeding device.

Dr. Allaben: When the testing program began at the National Center for Toxicological Research during the 1970s, batches of animal feed were rigorously analyzed to determine the concentrations of nutrients and micronutrients and to detect contaminants. Carload lots of materials were rejected if they failed to meet specific criteria. Subsequently, the NCTR began using certified chow, yet analysis revealed that the composition of these feeds often varied considerably from the certification. The NCTR has resumed testing to assure that the feeds meet the center's criteria.

Dr. Bucher: Technical reports on feeding studies conducted by the NIEHS provide food consumption information for all dose groups and controls. The reports also present the results of extensive analysis of the feed used during the study, including protein and fat content as calculated from the various ingredients.

Dr. Ghanta Rao (*National Institute of Environmental Health Sciences*): In many cases manufacturers provide actual analysis of the fat, protein, fiber, calcium, phosphorus, and some of the other major nutrients in the feed. Data on trace elements and trace nutrients are occasionally provided. Under the NTP program each batch of feed is analyzed for macronutrients and random samples are analyzed for trace elements. There is also an extensive program for analyzing for contaminants that provides for testing for contaminants of concern to the public. All of this information is included in the NTP technical reports. The cost of performing these analyses is equal to about 50% of the actual cost of the feed.

Dr. Turturro: The growth curve was proposed as the metric to guide body weight attainment until the animals were 12 months of age. Mammary and other tumors will clearly affect weight, as will the development of abscesses and other pathological conditions that precede death. Data from the NCTR and elsewhere, including mechanistic data, suggest that many of the processes contributing to extended survival and modulation of pathological changes occur early and that by the time the animals are 12 months old, many attributes are in place. Interestingly, substantial dietary restriction implemented even when the animals are 12 months old has some beneficial effect.

As noted by Dr. Hart and others, it is important that dietary control of weight gain be carefully regulated to provide for some level, perhaps up to 5%, of background tumor incidence. This represents the low end of the range currently observed in the NTP studies and approximates the rate observed in bioassays performed 15 years ago. Such an approach would address the concerns expressed on behalf of the CDER as well as relevant toxicologic criteria.

Dr. Richard Sprott (*National Institute on Aging*)**:** Many of the questions raised during this conference were first asked 20 or more years ago, and they have been asked with increasing frequency in the last decade. While the question of restriction versus control is important, either would offer a vast improvement over our current approach to feeding. The data presented here suggest that the consequences of ad libitum feeding are complex and only partially understood. Perhaps it's time to seriously address approaches to the feeding of laboratory rodents.

Part of this discussion has been about understanding current feeding practices and trying to bring precision to the feeding protocols. Resistance to changing feeding practices has been ascribed to economic issues and to the potential inconvenience to laboratory staff. It is easy to disagree about such points, to quibble over how much restriction or control is appropriate, and to debate when and how to implement control, but such arguments simply avoid the difficult task of improving how science is conducted. Scientists at the NIA and elsewhere are learning to manipulate the genome and using viruses to target particular genes within the genome to produce very precise effects. That is the cutting edge of science, and if the participants at this conference are unwilling to apply the very best available scientific knowledge to do the most precise research, this community may become the first to be deemed irrelevant by its scientific peers, the American public, and the Congress. That would be shameful.

Dr. Bucher: While maintaining the sensitivity of the bioassay is important, the NIEHS is involved in the development of short-term assays that can predict carcinogenic effects in a much shorter time frame and in a specific manner. Work to date with the p53 hemizygous mice is reminiscent of the development of the strain A mouse lung tumor assay and the Syncar mouse. It may be appropriate to follow Dr. Hayashi's suggestion (see Hayashi, this volume) of performing the bioassays with healthy animals and then conducting short-term mechanistic studies with p53 transgenic mice or *ras*-mutated mice in order to begin to understand the positive results.

Dr. Robertson: It seems unreasonable to believe that 10 years from now the safety assessment community will be performing the same number of bioassays and in the same way as they are currently performed. As long as the bioassay continues to be part of the safety evaluation process, animal husbandry practices should be modified so that these rodent studies are performed in a manner consistent with that used for other species.

Dr. Kritchevsky: One way of addressing the relationship between dietary restriction and body weight is to consider feeding efficiency, because weight gain is a function of how much was eaten. This approach raises the issue of measuring food consumption, but this offers insight into utilization. Studies at the Wistar Institute employing various levels of restricted and ad libitum feeding indicated that feeding efficiency was the best predictor of the number of observed tumors.

Dr. Allaben: It is evident that the situation is out of control with respect to the variability associated with the feeding of rodents in the 2-year bioassay. There is a need to regain control of this variable, and this conference could develop a recommendation or procedure for addressing one of the conference's objectives, i.e., to consider the feasibility and practical implications of the routine use of dietary restriction in safety assessment. Is there interest, particularly among scientists from the regulatory community, in developing a document addressing the issue of controlling food consumption during chronic toxicity and carcinogenicity studies? Such an approach may be very timely given Dr. Sprott's observation that the Congress may become more involved in managing risk assessment.

Dr. Miller: There are several approaches available to the Food and Drug Administration to address diet-associated variability. The most appropriate is for scientists from the NCTR or one of the other centers to draft a "Points to Consider" document that would address agencywide concerns about diet-associated variability in the bioassay. After thorough review by the various FDA centers, the document would be published as a *Federal Register* notice with a solicitation of comments from the public. The ensuing discussion could lead to the development of final guidance on feeding protocols. An alternative is for a product sponsor to submit data from a study performed using dietary control, with that study protocol then serving as a model for other sponsors. The development of a "Points to Consider" document would offer a more valid scientific approach to the issue and would prevent the development of different guidance by each of the centers.

Dr. Hattan: Data presented during this conference and the experience of CFSAN scientists with bioassay data suggest that there is no alternative but to develop a position on or an approach to the problem of diet-associated variability in the bioassay. The process of developing a guidance document might include CFSAN scientists with expertise in nutrition. Although a process involving multiple disciplines can lead to interminable debate, it would seem prudent to approach this in a comprehensive manner to avoid overlooking important considerations.

Dr. Hayashi: Because many safety studies in Japan are performed with fish, Japanese scientists are not concerned about the issue of dietary restriction or dietary control in rodent bioassays. However, considering the importance ascribed to diet-related variability in the bioassay by scientists in the United States, it is likely to become an important research topic for Japanese scientists.

Dr. Allaben: Given the presentations during this conference and the ensuing discussion, it seems appropriate that scientists from the FDA review these issues on an agencywide basis and develop a "Points to Consider" document. Following review and comment by staff from each of the centers, an appropriate process for securing public comment could be determined.

Dr. Bucher: Those scientists involved in carcinogenesis testing need to maintain consistent approaches as much as possible. The development of protocols to promote consistency is an activity in which scientists from the NIEHS would be willing to participate.

Dr. Robertson: Scientists from Merck Research Laboratories would also be interested in participating in such an activity.

Index

In this index, page numbers followed by the letter *f* designate figures; page numbers followed by the letter *t* designate tables.